W9-CSF-256

ADSORPTION FROM SOLUTION

Proceedings of a Symposium
held at
The School of Chemistry, University of Bristol
8th – 10th September, 1982

Professor D.H. Everett, M.B.E., D.Phil., D.Sc., F.R.S.

Colloid and Interface Science Group
Faraday Division – Royal Society of Chemistry
and
Colloid and Surface Chemistry Group
Society of Chemical Industry

Symposium on

ADSORPTION FROM SOLUTION

in honour of
PROFESSOR D.H. EVERETT, F.R.S.

Edited by

R.H. OTTEWILL, F.R.S.
University of Bristol, Bristol, U.K.

C.H. ROCHESTER
University of Dundee, Dundee, U.K.

A.L. SMITH
Unilever Research Laboratories, Wirral, U.K.

1983

ACADEMIC PRESS
A Subsidiary of Harcourt Brace Jovanovich, Publishers
London New York
Paris San Diego San Francisco São Paulo
Sydney Tokyo Toronto

ACADEMIC PRESS INC. (LONDON) LTD.
24/28 Oval Road
London NW1

United States Edition published by
ACADEMIC PRESS INC.
111 Fifth Avenue
New York, New York 10003

British Library Cataloguing in Publication Data
Adsorption From Solution.
1. Adsorption—Congresses
2. Solution (Chemistry)—Congresses
I. Ottewill, R.H.
II. Rochester, C.H.
III. Smith, A.C.
541.3'453 QD547

ISBN 0-12-530980-5

Printed in Great Britain

LIST OF CONTRIBUTORS

BEE, H.E., *I.C.I. Australia, Melbourne.*

BERGMANN, M., *Institut für Technische Chemie der Technischen Universität München, Lehrstuhl für Makromolekulare Stoffe, Lichtenbergstraße 4, D 8046 Garching, BRD.*

CHANDER, S., *Department of Materials Science and Mineral Engineering, University of California, Berkeley, CA 94720, U.S.A.*

COSGROVE, T., *Department of Physical Chemistry, University of Bristol, Bristol BS8 1TS, U.K.*

CROSS, S.N.W., *Department of Chemistry, University of Nottingham NG7 2RD, U.K.*

CROWLEY, T.L., *Department of Physical Chemistry, University of Bristol, Bristol BS8 1TS, U.K.*

EVERETT, D.H., *Department of Physical Chemistry, University of Bristol, Bristol BS8 1TS, U.K.*

DENOYEL, R., *Centre de Thermodynamique et de Microcalorimétrie du CNRS, 26, rue du 141ème R.I.A., 13003 Marseille France.*

FINDENEGG, G.H., *Institute of Physical Chemistry, Ruhr-University Bochum, 4630 Bochum, West Germany.*

FLEER, G.J., *Laboratory for Physical and Colloid Chemistry, Agricultural University, De Dreijen 6, 6703 BC Wageningen The Netherlands.*

FUERSTENAU, D.W., *Department of Materials Science and Mineral Engineering, University of California, Berkeley, CA 94720, U.S.A.*

GRAHAM, J., *Department of Chemistry, University of Nottingham Nottingham NG7 2RD, U.K.*

HALL, D.G., *Unilever Research, Port Sunlight Laboratory, Quarry Road East, Bebington, Wirral, Merseyside L63 3JW, U.K.*

HARRIS, N.M., *Saint-Gobain Recherche, Aubervilliers, France.*

HIGUCHI, A., *Department of Polymer Science, Tokyo Institute of Technology, O-Okayama, Meguro-ku, Tokyo, Japan.*

HULL, M., *Unilever Research Laboratories, Port Sunlight, Wirral, Merseyside, U.K.*

JAYSON, G.G., *Department of Chemistry and Biochemistry, Liverpool Polytechnic, Byrom Street, Liverpool L3 3AF, U.K.*

KILLMAN, E., *Institut für Technische Chemie der Technischen Universität München, Lehrstuhl für Makromolekulare Stoffe Lichtenbergstraße 4, D 8046 Garching, BRD.*

KLEIN, J., *Cavendish Laboratory, Madingley Road, Cambridge CB3 OHE, U.K., also Polymer Department, Weizmann Instiute of Science, Rehovot, Israel.*

KOCH, C., *Institute of Physical Chemistry, Ruhr-University Bochum, 4630 Bochum, West Germany.*

KOOPAL, L.K., *Laboratory for Physical and Colloid Chemistry, Agricultural University, De Dreijen 6, 6703 BC Wageningen The Netherlands.*

KORN, M., *Institut für Technische Chemie der Technischen Universität München, Lehrstuhl für Makromolekulare Stoffe Lichtenbergstraße 4, D 8046 Garching, BRD.*

KRINGS, P., *HENKEL Laboratories, Düsseldorf, West Germany.*

LAGALY, G., *Institut für anorganische Chemie der Universität Olshausenstraße 40/60, 2300 Kiel, Germany.*

LANE, J.E., *Department of Physical Chemistry, University of Melbourne, Parkville, Victoria, Australia 3052.*

LIPHARD, M., *Institute of Physical Chemistry, Ruhr-University Bochum, 4630 Bochum, West Germany.*

LUCKHAM, P.F., *Cavendish Laboratory, Madingley Road, Cambridge CB3 OHE, U.K.*

LYKLEMA, J., *Laboratory for Physical and Colloid Chemistry, Agricultural University, De Dreijen 6, 6703 BC Wageningen The Netherlands.*

MARRA, J., *Laboratory for Physical and Colloid Chemistry, Agricultural University, De Dreijen 6, 6703 BC Wageningen The Netherlands.*

MICALE, F.J., *Center for Surface and Coatings Research, Sinclair Laboratory No.7, Lehigh University, Bethlehem, PA 18015, U.S.A.*

OTTEWILL, R.H., *School of Chemistry, University of Bristol, Bristol BS8 1TS, U.K.*

PAPENHUIJZEN, J., *Laboratory for Physical and Colloid Chemistry, Agricultural University, De Dreijen 6, 6703 BC Wageningen, The Netherlands.*

PENDLETON, P., *Center for Surface and Coatings Research, Sinclair Laboratory No.7, Lehigh University, Bethlehem, PA 18015, U.S.A.*

RANCE, D.G., *I.C.I. Petrochemicals and Plastics Division, Welwyn Garden City, Hertfordshire AL7 1HD, U.K.*

RICHARDSON, R.A., *School of Chemistry, University of Bristol, Bristol BS8 1TS, U.K.*

RIGBY, D., *Department of Materials Science and Metallurgical Engineering, University of Cincinnati, Cincinnati, Ohio 45221, U.S.A.*

ROCHESTER, C.H., *Department of Chemistry, University of Dundee, Dundee DD1 4HN, U.K.*

ROUQUEROL, F., *Centre de Thermodynamique et de Microcalorimétrie du CNRS, 26, rue du 141ème R.I.A., 13003 Marseille, France.*

ROUQUEROL, J., *Centre de Thermodynamique et de Microcalorimétrie du CNRS, 26, rue du 141ème R.I.A., 13003 Marseille, France.*

RUDHAM, R., *Department of Chemistry, University of Nottingham, Nottingham NG7 2RD, U.K.*

SANDER, H., *Institut für anorganische Chemie der Universität Olshausenstraße 40/60, 2300 Kiel, Germany.*

SCHWUGER, M.J., *HENKEL Laboratories, Düsseldorf, Germany.*

SHCHUKIN, E.D., *The Institute of Physical Chemistry, Academy of Sciences of the U.S.S.R., Moscow, U.S.S.R.*

SMITH, A.L., *Unilever Research Laboratories, Port Sunlight, Wirral, Merseyside, U.K.*

STEPTO, R.F.T., *Department of Polymer and Fibre Science, The University of Manchester Institute of Science and Technology, Manchester M60 1QD, U.K.*

STIGTER, D., *Department of Materials Science and Mineral Engineering, University of California, Berkeley, CA 94720, U.S.A.*

SUDER, B.J., *Chemistry Department, Virginia Polytechnic Institute and State University, Blacksburg, Virginia 24061, U.S.A.*

TAMAMUSHI, B., *Nezu Chemical Institute, Musashi University, Nerimaku, Tokyo 176, Japan.*

TERASHIMA, H., *Institute of Applied Physics, Tsukuba University, Japan.*

THOMPSON, G., *Department of Chemistry and Biochemistry, Liverpool Polytechnic, Byrom Street, Liverpool L3 3AF, U.K.*

TREBILCO, D.-A., *Department of Chemistry, University of Nottingham, Nottingham NG7 2RD, U.K.*

VAN DER SCHEE, H.A., *Laboratory for Physical and Colloid Chemistry, Agricultural University, De Dreijen 6, 6703 BC Wageningen, The Netherlands.*

VON RYBINSKI, W., *HENKEL Laboratories, Düsseldorf, Germany.*

VINCENT, B., *Department of Physical Chemistry, University of Bristol, Bristol BS8 1TS, U.K.*

WHITE, J.W., *Physical Chemistry Laboratory, University of Oxford, Oxford OX1 3QZ, U.K.*

WIGHTMAN, J.P., *Chemistry Department, Virginia Polytechnic Institute and State University, Blacksburg, Virginia 24061, U.S.A.*

WITTER, R., *Institut für anorganische Chemie der Universität Olshausenstraße 40/60, 2300 Kiel, Germany.*

YAMINSKY, V.V., *The Institute of Physical Chemistry, Academy of Sciences of the U.S.S.R., Moscow, U.S.S.R.*
YONG, G.H., *Department of Chemistry, University of Nottingham, Nottingham NG7 2RD, U.K.*
ZETTLEMOYER, A.C., *Center for Surface and Coatings Research, Sinclair Laboratory No.7, Lehigh University, Bethlehem, PA 18015, U.S.A.*

PREFACE

On the 31st July 1982, Douglas Hugh Everett retired from the position of Leverhulme Professor of Physical Chemistry in the University of Bristol, a post which he had held since 1954. This period of 28 years has seen marked changed in our University system and has also seen a great development of Colloid and Interface Science as a quantitative scientific discipline important both in industry and academe. A major role in this development has been played by Douglas Everett and this Symposium on Adsorption from Solution was held to honour Douglas Everett on the occasion of his formal retirement from the University.

Douglas Everett was born in Hampton, Middlesex, in December 1916, and after education at Hampton Grammar School, proceeded to the University of Reading to read chemistry. He graduated with first-class honours in 1938 and after a year as a research assistant at University College, Dundee, he went to Balliol College, Oxford, as a Ramsay Memorial Fellow, obtaining his D.Phil. in 1942. During this period he was much influenced by the late Lord Wynne-Jones and Professor R.P. Bell. After a period of wartime scientific service, for which he was awarded the M.B.E., he returned to Oxford as an I.C.I. Fellow in 1945 and then became a Fellow and Lecturer in Chemistry at Exeter College from 1947-1948. During this Oxford period the first series of fundamental papers on the thermodynamics of adsorption of gases on solids were written and later published in the Transactions of the Faraday Society. Following appointment to the position of Professor of Chemistry at University College, Dundee, in 1948, experimental work on this subject also began to develop. In 1954 Douglas Everett took up the position of Leverhulme Professor of Physical Chemistry at the University of Bristol in succession to Professor W.E. Garner who had held the chair since 1926; in fact Garner had succeeded J.W. McBain the first Leverhulme Professor who was appointed in 1919. Bristol provided the tradition and facilities for development of surface chemistry and there commenced in 1954 a long and fertile period of research into various aspects of interface and colloid chemistry. Amongst these topics came a succession of papers on Adsorption from Solution, firstly putting the thermodynamic analysis on to a sound footing and then developing experimental techniques for the precise determination of the extent of adsorption of small molecules on to various adsorbents. Later came an interest in the orientation of larger hydrocarbon molecules on to surfaces and then the development of operational methods for treating the adsorption of small polymer molecules and for considering the influence of adsorption on surface-surface interactions. A survey of this work is given in the Sir Eric Rideal Lecture of this Symposium. For his outstanding contributions to Colloid and Interface Science, Douglas Everett was appointed Tilden Lecturer of the Chemical Society in 1955, was the first recipient of the Chemical Society Award in Surface and Colloid Chemistry in 1971 and was elected a Fellow of the Royal Society in 1980. He was also President of the Faraday Division of the Royal Society of Chemistry from 1976-1978. Internationally his work has been recognised by Visiting Professorships in West Germany, in Canada, in the U.S.A. and Israel and many invitations to give Plenary Lectures. On the 1st August 1982, Douglas Everett became Emeritus Professor of Physical Chemistry in the University of Bristol.

This Symposium Volume is dedicated to Douglas Everett as a token of esteem from many friends, collaborators and scientific groups in many parts of the world. With it we express our thanks for the inspiration, clarification of thought, advice and scientific insight which we have received over the years. It is our hope that for Douglas Everett many more years of good health and research into the fascinating world of Colloid and Interface Science lie ahead.

R.H. Ottewill
Bristol
January, 1983

CONTENTS

THE SIR ERIC RIDEAL LECTURE

ADSORPTION FROM SOLUTION

Douglas H. Everett

Department of Physical Chemistry, School of Chemistry, University of Bristol, U.K.

INTRODUCTION

This lecture surveys some of the more significant advances made during the last twenty years, in the study of adsorption from solution by solids.

John Kipling's book (1), published in 1965, gives a very full account of progress up to the early 1960's, and it was about that time that, largely as a result of discussions with him, I became interested in liquid/solid systems. His book shows that, despite the recognised importance of adsorption phenomena, and the extensive measurements made during the first half of this century largely on systems of practical interest, rather limited progress had been made towards a fundamental understanding of adsorption from solution. On the experimental side this was in part because the importance of working with well-characterised surfaces was insufficiently appreciated, and partly because the experimental techniques were of inadequate precision to enable any detailed theories to be tested. It is, however, surprising that theoretical studies of the solid/liquid interface were neglected since the thermodynamics of a simple model of the vapour/solution interface had been formulated in 1932 by Butler (2) and developed by several authors (3). Extension to the liquid/solid interface was not made until much later (4) even though this involves no major new concepts other than consideration of the added contributions to the energy of the system from interactions between the solid and molecules in the liquid phase. The last two decades have seen major progress in both the range and significance of experimental work and in its theoretical understanding, derived from work by many research groups worldwide (5,6,7,8). In this lecture, however, I want to make particular reference to the contributions made by workers in Bristol, whose names are listed in the Appendix.

The main topics dealt with are the following. First, the role of simple models and of elementary thermodynamic and statistical mechanical arguments is discussed in relation to the concept of a 'surface phase'. Monolayer models of the adsorption region are often inadequate and thicker surfaces have to be assumed. A sharp boundary between surface and bulk phases is thermodynamically unacceptable and it is necessary to develop multilayer models and extend them to mixtures of molecules of different size. Although approaches of this kind form a useful basis for the interpretation of the behaviour of relatively simple systems, real systems exhibit more complex behaviour. It is not easy to perceive the modifications which need to be incorporated into the simpler models without guidance as to the source of the inadequacy of the theoretical models. It is therefore important to examine the usefulness of describing adsorption behaviour in purely thermodynamic terms without reference to any specific physical model. The results of such an analysis emphasize the importance of the interplay between enthalpic and entropic contributions to the free energy, and help to direct attention to the ways in which the simpler models need to be developed to provide more realistic theories of adsorption phenomena. Evidence is found for significant structural changes in the interfacial region, and a brief account is then given of some of the more striking examples. Finally, the role of adsorption from solution in controlling other phenomena such as wetting and liquid-liquid displacement, and the forces between colloidal particles is outlined.

SIMPLE MONOLAYER MODELS OF ADSORPTION

In developing a theoretical model for the solid/liquid interface it was natural initially to apply concepts and methods of approach analogous to those employed in discussing bulk phase equilibria. Thus the surface region was looked upon as a separate "surface phase". The continuous distribution relating the average local concentrations x of different species to distance from the surface z was replaced by a step function (Figure 1). The so-called 'surface phase' occupies a definite region of space within which the concentration is uniform and different from the bulk. It

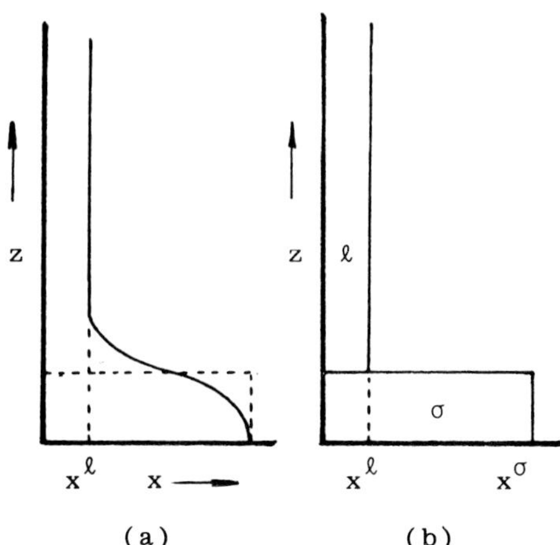

Fig. 1. Approximation of continuous variation of concentration near a surface by a step function defining a 'surface phase', σ.

was, however, appreciated (9) that the properties of such a phase are determined not only by forces between molecules in the surface, but also by interactions with the adjacent bulk phases: 'surface phases' are not truly autonomous.

The condition of equilibrium between bulk and surface regions is that of constancy of chemical potential of each species on passing from locations adjacent to the solid into the bulk liquid phase. To apply this condition to the surface phase model it is necessary to express the surface chemical potential in terms of the local concentration in the surface phase, and to relate this local concentration to the observable adsorption parameters. One conveniently measured quantity (5, 10, 11) is the surface excess of one component (say 2) defined by $n^o\Delta x_2^{\ell}$, where Δx_2^{ℓ} is the change in the mole fraction of component 2 in the bulk solution resulting from equilibration of a mass, m, of solid with an amount n^o of solution of initial mole fraction x_2^o. This may be expressed either as the specific surface excess $n^o\Delta x_2^{\ell}/m$, or, if the specific surface area, a_s, of the solid is known, as the areal surface excess:

$$\Gamma_2^{(n)} = n^o \Delta x_2^{\ell}/m\ a_s . \tag{1}$$

Since $\Gamma_1^{(n)} = -\Gamma_2^{(n)}$, this emphasizes the competitive nature of the adsorption. For a binary mixture, $\Gamma_2^{(n)}$ is related to the Gibbs relative adsorption of component 2 with respect to 1:

$$\Gamma_2^{(1)} = \Gamma_2^{(n)}/x_1^{\ell}. \tag{2}$$

If the surface phase occupies t molecular layers, then the mole fraction x_2^{σ} in the surface can be calculated from the equation:

$$x_2^{\sigma} = \frac{x_2 + \left(\frac{a_1^{o}}{t}\right)\Gamma_2^{(n)}}{1 - \frac{(a_2^{o}-a_1^{o})\Gamma_2^{(n)}}{t}}, \tag{3}$$

where a_1^{o} and a_2^{o} are the areas subtended on the surface by molecules of types 1 and 2 respectively (molecular or molar surface areas). In the particular case of monolayer adsorption of molecules of the same size

$$x_2^{\sigma} = x_2^{\ell} + a\,\Gamma_2^{(n)}, \tag{4}$$

where $a_1^{o} = a_2^{o} = a$.

Within these somewhat crude approximations one can envisage systems in which both bulk liquid and surface phases are ideal (11) in the sense that the energy of interaction between unlike molecules of equal size at their equilibrium separation (ε_{12}) is the arithmetic mean of those ($\varepsilon_{12}, \varepsilon_{22}$) between like molecules. The chemical potentials in the liquid (μ_2^{ℓ}) and surface (μ_2^{σ}) phases relative to the chemical potential of pure liquid as standard state ($\mu_2^{\ell\ominus}$) can then be written

$$\mu_2^{\ell} = \mu_2^{\ell,\ominus} + RT\ \ln x_2^{\ell} + (p - p_2^{*})v_2; \tag{5(a)}$$

$$\mu_2^{\sigma} = \mu_2^{\ell,\ominus} + RT\ \ln x_2^{\sigma} - (\sigma - \sigma_2^{*})a_2. \tag{5(b)}$$

In equation 5(a) p and p_2^{*} are the vapour pressures of solution and pure 2 respectively and v_2 is the molar volume of 2. The last term in this equation can usually be neglected, but the analogous term in 5(b) cannot. In this term σ is the 'interfacial tension' at the solid/solution interface defined by $(\partial G^{\sigma}/\partial A_s)_{T,n_i^{\sigma}}$ where G^{σ} is the Gibbs free energy of the surface and A_s its area; and σ_2^* refers to the solid/pure liquid interface. It is then assumed that the various interaction energies ε_{ij} are the same for molecules in the bulk and adsorbed phases and that the solid interacts appreciably only with molecules in the surface phase.

The resulting equilibrium conditions may be looked upon as defining an adsorption equilibrium constant, K_a, for the following exchange process taking place on a given area of solid (11):

$$(1)^{\sigma} + (2)^{\ell} \rightleftharpoons (1)^{\ell} + (2)^{\sigma}, \tag{6}$$

which has the simple form

$$K_a = \frac{x_2^{\sigma} x_1^{\ell}}{x_1^{\sigma} x_2^{\ell}}, \tag{7}$$

with $$\ln K_a = -(\sigma_2^* - \sigma_1^*)a/RT. \tag{8}$$

Taking equations (4) and (7) together leads to a relation between the observed specific surface excess and the bulk liquid composition:

$$\frac{x_1^{\ell} x_2^{\ell}}{n^o \Delta x_2^{\ell}/m} = \frac{m}{n^{\sigma}} \left[x_2^{\ell} + \frac{1}{(K_a - 1)}\right], \tag{9}$$

where $n^{\sigma}/m = a_s/a$ is the amount of either component required to cover the surface of unit mass of solid. Surface excess isotherms corresponding to this simple case have an inverted U-shape, the sharpness of the maximum increasing and its position moving to lower values of x_2^{ℓ} as K_a increases. It is often found that experimental data for simple organic mixtures adsorbed by graphitised carbon or silica gel conform to the linear relation (9), from which values of K_a and n^{σ}/m can be derived. If the size of the molecules is known, an estimate can be made of the specific surface area of the solid. In a surprisingly large number of examples the areas so obtained are in good agreement with those calculated from nitrogen gas adsorption by the BET method (12).

However, equation (9) is not a very sensitive test of the ideality of the system, and its use can be misleading. For example, there are cases in which linear relationships are observed experimentally even when the bulk solution is known to be non-ideal.

The model may be extended by introducing activity coefficients for both bulk and surface phases, (13), when the equilibrium constant adopts the more general form:

$$K_a = \frac{x_2^\sigma \gamma_2^\sigma}{x_2^\ell \gamma_2^\ell} \cdot \frac{x_1^\ell \gamma_1^\ell}{x_1^\sigma \gamma_1^\sigma} . \tag{10}$$

Conformity with equation (9) could then result from the cancellation of $\gamma_2^\sigma/\gamma_2^\ell$ with $\gamma_1^\sigma/\gamma_1^\ell$.

The relationship between activity coefficients in the bulk and surface regions has been the subject of considerable speculation. Thus several authors (14) have supposed that experimental data are consistent with the assumption that even when the bulk phase is non-ideal the surface phase is ideal. This contention is, however, not generally true, and it is of interest that in this Symposium Lane (15) investigates the consequences of supposing that the surface and bulk activity coefficients are related and that when $x_2^\sigma = x_2^\ell$,

$$\ln \gamma_2^\sigma(x_2^\sigma) = t \ln \gamma_2^\ell(x_2^\ell) . \tag{11}$$

Here t is a proportionality factor* such that when t = 0 the surface is ideal, while when t = 1 the surface and bulk activity coefficients are the same.

The non-ideality of the surface phase can be assessed by combining (10) and (4) to give (13,16)

$$K_a \frac{\gamma_1^\sigma}{\gamma_2^\sigma} = \frac{x_1^\ell \gamma_1^\ell}{x_2^\ell \gamma_2^\ell} \cdot \frac{x_2^\ell + n^o \Delta x_2^\ell / m\, n^\sigma}{x_1^\ell - n^o \Delta x_2^\ell / m\, n^\sigma} . \tag{12}$$

If n^σ is known (from a_s and a) then the right-hand side can be calculated from the adsorption isotherm and knowledge of the bulk solution activity coefficients. By plotting $\ln(K_a\gamma_1^\sigma/\gamma_2^\sigma)$ against x_2^σ and integrating from $x_2^\sigma = 0$ to 1, K_a can be found since $\int_0^1 \ln(\gamma_1^\sigma/\gamma_2^\sigma)\, dx_2^\sigma = 0$. Hence $\gamma_1^\sigma/\gamma_2^\sigma$ can

---

*This does not mean they cancel in (10) since at equilibrium $x_2^\sigma \neq x_2^\ell$. Note that t here is different from that in equ. (3).

be calculated as a function of x_2^σ and compared with theoretical predictions. Alternatively, individual activity coefficients can be computed from the following thermodynamically derived equation: (17)

$$\ln \gamma_2^\sigma = \ln \frac{x_2^\ell \gamma_2^\ell}{x_2^\sigma} + a \int_{x_2^\ell}^{x_2^\ell=1} \frac{\Gamma_2^{(n)}}{x_1^\ell x_2^\ell \gamma_2^\ell} d(x_2^\ell \gamma_2^\ell). \quad (13)$$

The theoretical calculation of surface activity coefficients by statistical mechanical methods in terms of a lattice model follows essentially the same lines as that for bulk mixtures (13,18). In a monolayer model, molecules in the surface layer interact not only with nearest neighbours in that layer, of whom there are ℓz (z is the lattice coordination number), but also with the solid surface on one side and with mz neighbours in the adjacent layer of bulk solution ($\ell + 2m = 1$). A simple treatment in terms of the Bragg-Williams approximation leads to surface excess isotherms whose shape now depends both on K_a and on the regular solution parameter, $\alpha = N_a z\left[\varepsilon_{12} - \frac{1}{2}(\varepsilon_{11} + \varepsilon_{22})\right]$, where N_a is Avogadro's constant. For certain combinations of these parameters reversals of the sign of adsorption of the kind often observed experimentally are predicted (13). The surface activity coefficients are given by equations of the form:

$$\ln \gamma_2^\sigma = \frac{\ell\alpha}{RT}\left(x_1^\sigma\right)^2 + \frac{m\alpha}{RT}\left(x_1^\ell\right)^2 , \quad (14)$$

which demonstrates the non-autonomy of the surface phase whose properties are in part determined by those of the liquid phase, while the ratio of activity coefficients is given by

$$\ln K_a \frac{\gamma_1^\sigma}{\gamma_2^\sigma} = \ln K_a - \frac{\alpha}{RT}\left[\ell(1 - 2x_1^\sigma) + m(1 - 2x_1^\ell)\right]. \quad (15)$$

The applicability of the theory is conveniently tested by plotting the left-hand-side of (15) calculated from equation (12) against the quantity in square brackets, choosing ℓ =1/2 ,m =1/4 corresponding to a close packed lattice. For the system [bromobenzene + chlorobenzene]/Graphon,

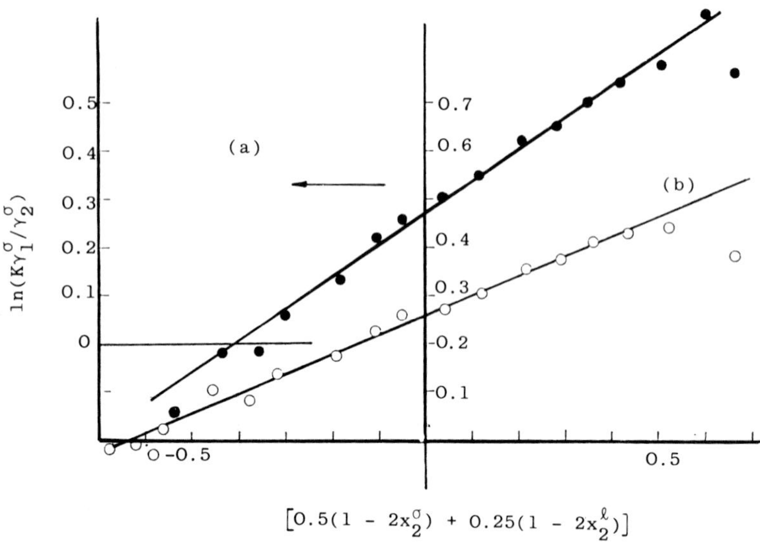

Fig. 2. Test of equation (15) for the system [1,2-dichloroethane + benzene]/ Graphon at (a) 273K, (b) 333 K.

which is nearly ideal in the bulk, the value of α derived from such a graph ($\alpha/R \approx 10$ K) is essentially the same as that characterising the bulk mixture (16). On the other hand, for the system [1,2 - dichloroethane + benzene]/ Graphon, which is also nearly ideal in the bulk ($\alpha/R \approx 10$ K) analysis in terms of equation (15) yields <u>negative</u> and temperature dependent values of α/R varying from -170 K at 273 K to -130 K at 333 K (16), figure 2. These values of α/R when inserted in equation (14) yield values of $\ln \gamma_2^\sigma$ in close agreement with those derived from equation (13). This illustrates that near-ideal behaviour in the bulk can be accompanied by strongly non-ideal behaviour in the surface. That the presence of a surface may modify the forces between adsorbed molecules is known from work on gas/solid systems (19), although no calculations have been made for liquid/ solid interfaces. In the particular example considered here it is probable that specific effects arising from differences between the preferred conformations of dichloroethane in the bulk and surface regions may be important. Any more detailed theory would have to take this into account and to consider the possibility that different values of α should, in equation (14), be associated with the surface-surface and surface-bulk interactions. It is also important to stress that the concept of surface activity coefficients is useful only within the framework of the surface phase model since their calculation involves knowledge of x_2^σ.

The monolayer model described above is a useful way of describing adsorption from relatively simple systems of molecules of roughly equal size. It becomes less easily applied to more complex systems. If the size difference is appreciable then the exchange process must be generalised (13,20):

$$r(1)^\sigma + (2)^\ell \rightleftharpoons r(1)^\ell + (2)^\sigma, \qquad (16)$$

where the adsorption of one molecule of (2) involves displacing r molecules of the smaller component. When restricted to the monolayer model this process only makes sense if component 2 is an r-mer of segments of the same size as the solvent, and lies wholly in the surface layer (parallel-layer model(20)). The adsorption equilibrium constant now has the form

$$K_a = \left(\frac{x_2^\sigma \gamma_2^\sigma}{x_2^\ell \gamma_2^\ell}\right)\left(\frac{x_1^\ell \gamma_1^\ell}{x_1^\sigma \gamma_1^\sigma}\right)^r . \qquad (17)$$

The activity coefficients in both the bulk and surface regions can be discussed in terms of a Flory-Huggins type theory, and adsorption isotherms computed. The data of Bown (21) on the system [n-hexane + n-hexadecane]/Graphon can be interpreted successfully (22) in terms of the resulting equations, the interesting feature being that the values of r which fit the adsorption data are closely similar to those which provide a satisfactory account of the bulk liquid activity coefficients (23).

However, there are a number of reasons for doubting the applicability of the monolayer model to more complex systems. In the first place, on general grounds one does not expect all adsorbed molecules to lie flat on the surface.

A second inconsistency of the monolayer concept arises from the necessary requirements (Rusanov (24)) that x_2^σ be less than unity and that $(\partial x_2^\sigma/\partial x_2^\ell)$ be positive, otherwise the surface will exhibit two-phase behaviour. It is frequently found that, using equation (4) with physically realistic values of a_s and a, these criteria are not satisfied. The only simple way of avoiding this difficulty is to suppose that the adsorbed phase occupies t layers, and to use equation (3) to find the smallest value of t which satisfies the criteria. For example, for several [alcohol + hydrocarbon]/Graphon systems values of t up to 4 may be required (25). Since only a minimum value of t can be estimated, and since it is usually taken as an integer, the physical significance of a surface phase model applied to such systems begins to fade, and the equations become little more than convenient ways of representing the data.

Further reasons for rejecting a surface phase model arise from the fact that the equations derived for regular solution monolayer adsorption are thermodynamically inconsistent with the Gibbs adsorption equation (26,27). This discrepancy is absent for ideal systems, but increases as deviations from ideality increase. It can be removed only by abandoning the concept of a surface phase and considering a multilayer model in which a discontinuous step in concentration is replaced by a continuous concentration profile (27). The physical reason for this is that according to the monolayer model, molecules in the adjacent liquid layer interact both with nearest neighbours in the monolayer where the composition is x_2^σ, and with neighbours in its own and in the next liquid layer at a concentration x_2^ℓ. Unless $\alpha = 0$ such a layer cannot be in equilibrium with one in the bulk liquid where all interactions are with molecules at the liquid composition x_2^ℓ. Thus the equilibrium concentration in the second layer must be perturbed from x_2^ℓ; similarly this perturbation must affect the third layer and so on. Thus independently of the range of the surface forces themselves, the local concentration change at the surface must be propagated out into the bulk solution. The range of the perturbation depends on K_a and on α/RT: for typical values it extends out for some 3 - 4 molecular layers (27,28,29). When α is negative implying that the components tend to form association complexes there may be an alternation of preferential adsorption in successive layers because enhancement of the concentration of component 2 in the first layer will tend to enhance that of component 1 in the second layer and so on.

MULTILAYER MODELS

Further progress requires the development of more realistic multilayer theories of adsorption. For molecules of equal size a regular solution model has been developed by Ono (27), Lane (28) and others (29) which satisfies the criteria of thermodynamic consistency. However, although such theories demonstrate the existence of concentration profiles extending out into the liquid, no simple procedures have been worked out for fitting experimental data to these models.

For molecules of different sizes, one of which is regarded as a flexible r-mer of the smaller component, Ash, Findenegg and I (30) developed exact lattice model equations taking account of the various configurations which a molecule can take up at the surface. For $r = 2,3$ and 4 it was possible to compute both the total surface excess isotherms and the frequency of occurence of the different configurations near the surface. For larger size ratios the computer time needed to solve the equations would have become prohibitive. One interesting outcome of this theory was to illustrate the way in which in certain cases the energetic advantage of adsorption in the flat configuration is more than offset by the entropic disadvantage of the decreased number of arrangements on the surface compared with those for a perpendicular orientation.

This work has been extended to polymers by Roe (31) and by Scheutjens and Fleer (32) by making slight approximations in the statistics. The latter authors, on the basis of one additional reasonable assumption, calculate the segment density distribution of adsorbed trains,loops and tails. Their results in fact agree with our results when $r = 2,3$ and 4, suggesting that their assumption is indeed justified. Once again, however, no detailed comparison with experimental data on polymer adsorption has so far been made. Indeed it is unlikely that data of sufficient precision and covering an adequate temperature range can be obtained to enable the necessary parameters to be derived from fitting procedures. Other types of measurement (for example n.m.r. and neutron scattering (33,34)) may well be more important in this context.

Alternative approaches to adsorption at the solid/liquid interface are being developed (35), based on Monte Carlo, molecular dynamics and other computational techniques. They have the important advantage of breaking away from the lattice model, but so far they have only been able to deal with relatively simple systems.

THERMODYNAMIC ANALYSIS

Faced with the failure of simple theories to give an adequate account of more complex systems one is attracted to the alternative possibility of characterising adsorption equilibria in thermodynamic terms independent of any specific molecular model. This might enable one to identify the factors which should be taken into account in more realistic theories.

The thermodynamic approach involves three key equations (7). First, from the Gibbs adsorption equation it follows that the 'surface tension' difference $(\sigma - \overset{*}{\sigma}_2)$, between a solid surface in contact with a solution of composition x_2^{ℓ}, and the same solid in contact with pure component 2, is given by

$$\sigma - \overset{*}{\sigma}_2 = RT \int_{x_2^{\ell}}^{x_2^{\ell}=1} \frac{\Gamma_2^{(n)}}{x_1^{\ell}\, x_2^{\ell}\, \gamma_2^{\ell}}\, d(x_2^{\ell}\gamma_2^{\ell}). \qquad (18)$$

When integrated across the whole concentration range this equation gives $(\overset{*}{\sigma}_1 - \overset{*}{\sigma}_2)$. From the temperature dependence of $(\sigma - \overset{*}{\sigma}_2)$ one can then obtain the difference between the enthalpy of immersion of the solid in a solution of composition $x_2^{\ell\dagger}$, and in pure component 2:

$$\left[\frac{\partial}{\partial(1/T)}\left(\frac{\sigma - \overset{*}{\sigma}_2}{T}\right)\right]_{x_2^{\ell}} = \Delta_w\hat{h} - \Delta_w\hat{h}_2^{*}, \qquad (19)$$

where the circumflex indicates the enthalpy associated with unit area of solid (areal enthalpy of immersion). The corresponding entropy difference can be evaluated either from

$$\Delta_w\hat{s} - \Delta_w\hat{s}_2^{*} = -\,\partial(\sigma - \overset{*}{\sigma}_2)/\partial T \qquad (20)$$

or

$$\Delta_w\hat{s} - \Delta_w\hat{s}_2^{*} = \frac{1}{T}\left[(\Delta_w\hat{h} - \Delta_w\hat{h}_2^{*}) - (\sigma - \overset{*}{\sigma}_2)\right]. \qquad (21)$$

To apply this analysis one needs, in addition to the measured surface excess isotherms over a range of

---

† It is assumed that immersion takes place into a volume of liquid sufficiently large that the concentration of the solution does not change. A more explicit notation would be to use $\Delta_w\hat{h}(\infty)$ instead of $\Delta_w\hat{h}$ (see ref. 7).

temperatures, an estimate of the specific surface area of the solid, and a knowledge of the activity coefficients in the bulk liquid phase.

Using our improved method of measuring adsorption isotherms (36) we have now assembled a substantial body of data suitable for thermodynamic analysis. Unfortunately, progress in the detailed analysis of this work is seriously held up by the lack, even for relatively simple systems, of reliable bulk activity coefficient data over a sufficiently wide temperature range. In many cases one is obliged to interpolate and extrapolate from measurements (often by different workers) at two or at best three temperatures. In the absence of direct measurements various correlation schemes can be employed but their usefulness is limited. Nevertheless, despite the uncertainties arising from activity coefficient data, a number of interesting features have already emerged.

In considering the results of a thermodynamic analysis it is useful to bear in mind the thermodynamic properties of the ideal monolayer model which for molecules of equal size are summarised in the following equations

$$\sigma - \sigma_2^* = x_1^\sigma(\sigma_1^* - \sigma_2^*) + \frac{RT}{a}\left[x_1^\sigma \ln \frac{x_1^\sigma}{x_1^\ell} + x_2^\sigma \ln \frac{x_2^\sigma}{x_2^\ell}\right], \tag{22}$$

$$\Delta_w\hat{h} - \Delta_w\hat{h}_2^* = x_1^\sigma(\Delta_w h_1^* - \Delta_w h_2^*), \tag{23}$$

$$T(\Delta_w\hat{s} - \Delta_w\hat{s}_2^*) = x_1^\sigma(\Delta_w\hat{s}_1^* - \Delta_w\hat{s}_2^*) - \frac{RT}{a}\left[x_1^\sigma \ln \frac{x_1^\sigma}{x_1^\ell} + x_2^\sigma \ln \frac{x_2^\sigma}{x_2^\ell}\right]. \tag{24}$$

The terms in square brackets are configurational terms arising from the difference between the entropy of mixing terms in the solution and surface phases. Typically they contribute up to about 10% to the total. The first terms on the right-hand side of these equations may be looked upon as the product of the difference between the properties of the two pure component/solid interfaces and a compositional factor related to the concentration change near the surface. In non-ideal systems we may still expect a similar sub-division, and that deviations from ideality may arise from both terms. Thus if the configurations

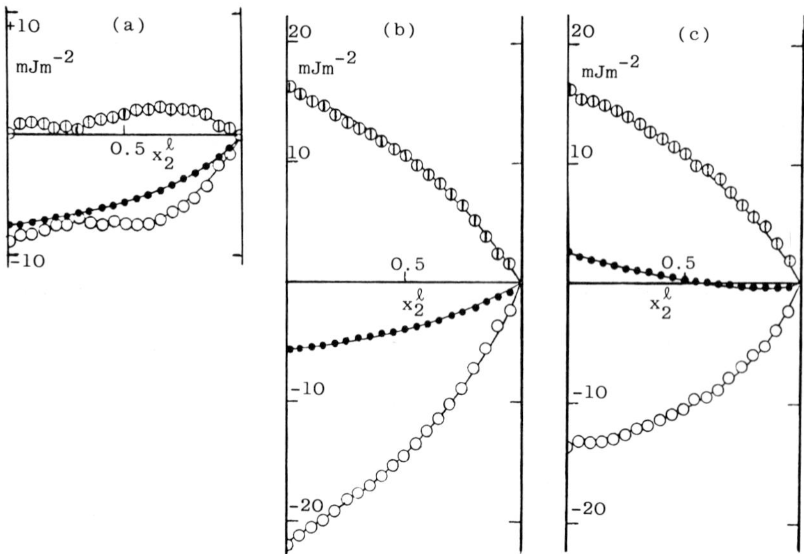

Fig. 3. $(\sigma - \sigma_2^*)$ ● ; $(\Delta_w\hat{h} - \Delta_w\hat{h}_2^*)$ ○ *and* $-T(\Delta_w\hat{s} - \Delta_w\hat{s}_2^*)$ ⦶ *as functions of* x_2^ℓ *for the systems (a) [benzene(1) + cyclohexane(2)]/Graphon, (b) [n-heptane(1) + cyclohexane(2)]/ Graphon, and (c) [n-heptane(1) + benzene(2)]/Graphon.*

available, say, to a molecule of type 1 near the surface depend on whether it is present as pure component or in a mixture with 2, then the configurational term will be changed. Similarly, if the energy of a molecule of 1 depends on the nature of its neighbours then an additional term will appear in equation (23) and be reflected in (22). Moreover, since in general x_2^σ has no physical meaning outside the surface phase model, the formulation of this factor needs careful consideration.

With these general observations in mind it is of interest to consider some typical examples (37).

The adsorption of benzene (1) from cyclohexane (2) solution by Graphon (Figure 3a) (38) is controlled almost entirely by enthalpy terms. The temperature dependence of $(\sigma - \sigma_2^*)$ is barely outside experimental error, and at most indicates values of $(\Delta_w\hat{s} - \Delta_w\hat{s}_2^*)$ of about the magnitude calculated for the configurational contribution. The preferentially adsorbed component is that (benzene) exhibiting the more exothermic enthalpy of immersion in the pure state. The same is true for the [n-heptane(1) + cyclohexane(2)]/Graphon system(Figure 3 b) (38) except that now the enthalpic term is opposed by a large entropic term arising from the difference of immersion entropies of Graphon in the two pure components. However, the enthalpic term dominates and $(\sigma - \sigma_2^*)$ remains negative with a positive value of $d(\sigma - \sigma_2^*)dx_2^\ell$, implying preferential adsorption of heptane. In the case of [n-heptane(1) + benzene(2)]/Graphon (Figure 3c) (38) the larger (more negative) enthalpy of immersion in heptane than in benzene would suggest preferential adsorption of heptane. However, the enthalpy difference is smaller than for heptane + cyclohexane, while the entropy term is almost identical in the two cases. Consequently over most of the concentration range the entropy term exceeds the enthalpy term: $(\sigma - \sigma_2^*)$ is positive, $d(\sigma - \sigma_2)/dx_2^\ell$ is negative and, except at high concentrations and low temperatures, benzene is preferentially adsorbed. Here the adsorption equilibrium is dominated by entropy effects. At higher temperatures $T(\Delta_w\hat{s} - \Delta_w\hat{s}_2^*)$ is larger than $(\Delta_w\hat{h} - \Delta_w\hat{h}_2^*)$ at all concentrations: $(\sigma - \sigma_2^*)$ is positive and benzene is adsorbed, over the whole concentration range. In these last two systems the more negative entropy of immersion in n-heptane compared with benzene presumably reflects the restriction imposed by the solid on the motions of an alkane chain in close proximity to it. If this is so one might expect the magnitude of the effect to depend on the chain length of the alkane. Recent measurements by Davis (39) on mixtures of benzene with n-pentane and iso-pentane and of n-butylbenzene with n-heptane illustrate this point (Figure 4). The curves of $-T(\Delta_w\hat{s} - \Delta_w\hat{s}_2^*)$ against x_2^ℓ show that the entropic contributions for n-pentane are less than for heptane, and that iso-pentane (with fewer conformational isomers) exhibits an even lower entropy curve. In the case of n-butyl benzene + n-heptane the entropy of wetting by n-butyl benzene includes an effect from restrictions on the motions of the butyl side chain which brings its value closer to that for heptane, leading to a smaller overall entropy difference. With a somewhat larger alkylbenzene the entropy effects might cancel exactly.

More complex behaviour is shown by [alcohol + hydrocarbon]/ Graphon systems. A careful reanalysis, using more detailed activity coefficient data, of the results of Brown, Everett and Morgan (25) for [ethanol(1) + benzene(2)] and [ethanol(1) + heptane(2)] has been made. It now appears that $(\sigma - \sigma_2^*)/T$ as a function of $1/T$ is curved: the enthalpies and entropies of immersion are temperature dependent. The difference between the heat capacity changes on immersion in a solution and in pure ethanol $(\Delta_w\hat{c}_p - \Delta_w\hat{c}_{p1}^*)$ is shown for the former system as a function

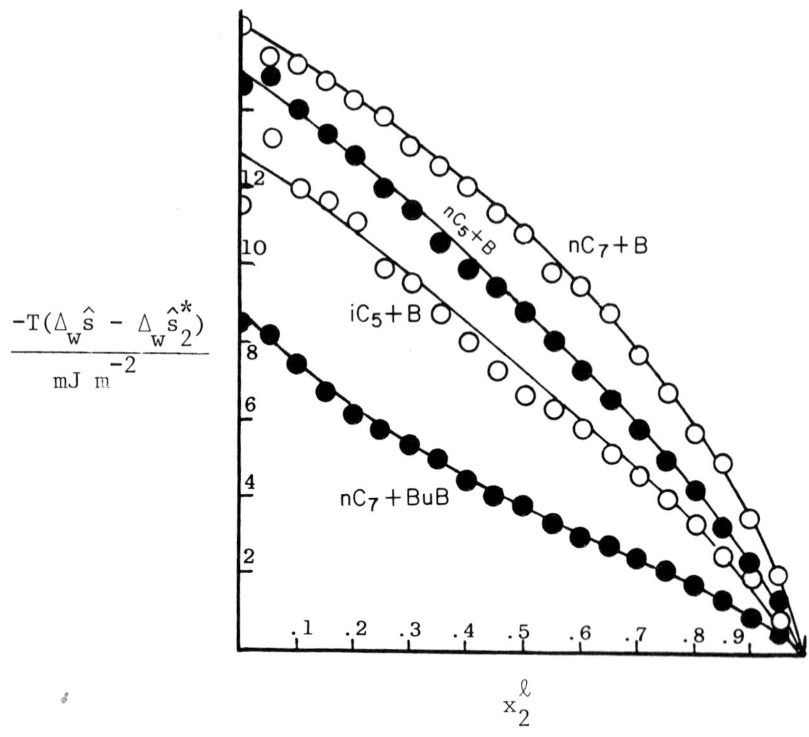

Fig. 4. $-T(\Delta_w\hat{s} - \Delta_w\hat{s}^*_2)$ *as function of* x^ℓ_2 *for [n-heptane (1), n-pentane(1), isopentane(1) + benzene(2)]/Graphon and [n-heptane(1) + n-butylbenzene(2)]/Graphon.*

of x^ℓ_1 in figure 5. The resulting variation in the enthalpy and entropy terms with temperature is such that the control of the adsorption passes from the entropy term at low temperatures to the enthalpy term at high temperatures. This is illustrated in figure 6 in which the data are now shown as a function of the activity, $\gamma^\ell_2 x^\ell_2$ of benzene in the bulk solution. At 10°C the entropy term dominates and makes the major contribution to the surface tension difference (figure 6(a)), while at 60°C the position is essentially reversed (figure 6(b)). At 10°C both enthalpy and entropy terms show a near-discontinuity near $\gamma^\ell_2 x^\ell_2 \simeq 0.90$ $(x^\ell_1 \sim 0.45)$, and this is associated with the sharp peak in the integrand of equation (18) at this point shown in the upper part of the diagram. At 60°C the changes in all three quantities are smoothed out. Figures 5 and 6 thus point to substantial structural changes in the interfacial layer at a critical

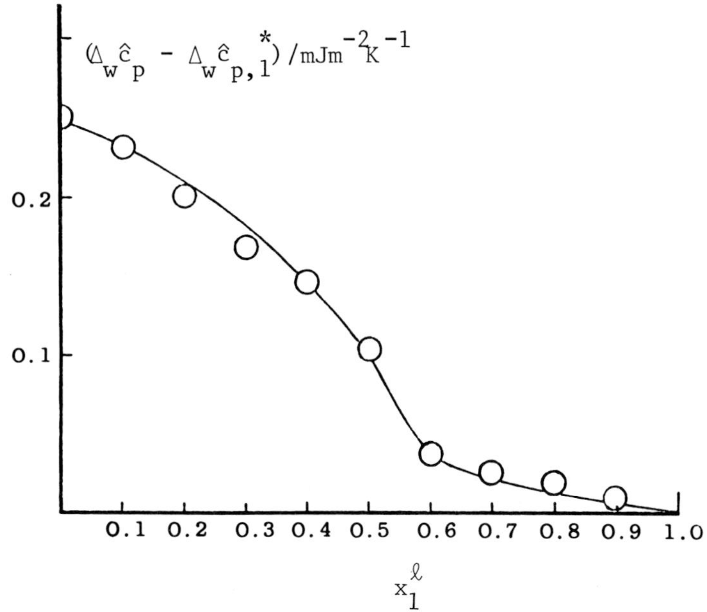

Fig. 5. $(\Delta_w \hat{c}_p - \Delta_w \hat{c}^*_{p1})$ *as function of* x_1^ℓ *for* [*ethanol(1) + benzene(2)*]/*Graphon.*

composition in the region of $x_1^\ell \approx 0.4 - 0.5$ or $\gamma_2^\ell x_2^\ell \simeq 0.90$, which at the lower temperatures has something of the character of a phase separation phenomenon in the surface region. The exact nature of these phenomena needs further investigation. Incidentally, the Rusanov criteria suggests that the interfacial layer increases in thickness up to 3-4 molecular layers as the temperature is decreased. Somewhat similar behaviour, which has not yet been fully analysed, is observed with the [methanol + benzene] (40) and [n-butanol + n-heptane]/Graphon (41) systems.

STRCTURAL CHANGES IN ADSORBED LAYERS[42]

The above analysis suggests that even in relatively simple systems there is evidence for structural changes in adsorbed layers. This evidence becomes much more striking in systems involving long chain alkanes, alcohols and alkanoic acids. Thus figure 7 shows data for [n-pentane + n-decane]/Graphon and [n-hexane + n-hexadecane]/Graphon derived from Bown's measurements (21). Both the enthalpy and entropy terms depend strongly on temperature, but in both cases at all temperatures the sign of $(\sigma_2^* - \sigma_1^*)$ is

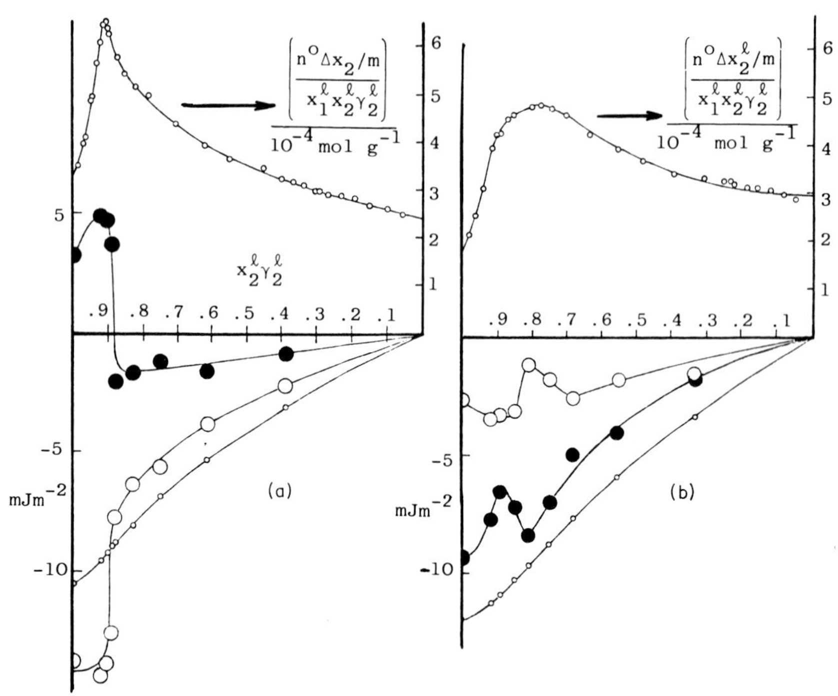

Fig.6. Thermodynamic functions for system[ethanol(1) + benzene(2)]/Graphon as function of benzene activity, $x_2^\ell \gamma_2^\ell$ at (a) 10°C, (b) 60°C ∘ $\sigma - \sigma_1^*$; ● $\Delta_w \hat{h} - \Delta_w \hat{h}_1^*$; ○ $T(\Delta_w \hat{s} - \Delta_w \hat{s}_1^*)$.

determined by the enthalpy term. However, the heat capacity term is large and temperature dependent, $(\Delta_w \hat{c}_{p2}^* - \Delta_w \hat{c}_{p1}^*)$/ mJ m^{-2}K^{-1} varying from 1.75 to 0.30 for the former system and from 3.1 to 0.85 for the latter over the ranges of temperature studied. The magnitudes and behaviour of the thermodynamic functions in comparison, for example, with those for the pentane and hexane + benzene systems all point to major structural changes in the adsorbed layers.

The earliest evidence for structural changes at the alkane/graphite interface came from direct calorimetric measurements of the enthalpies of immersion of graphitised carbons in pure alkanes (43). Much further work involving immersion calorimetry (44,45,46), and measurements of the adsorption of pure liquids at the graphite surface by Findenegg and his colleagues (47) has confirmed the general picture. For the higher alkanes the data are consistent with the formation at the surface of solid-like material

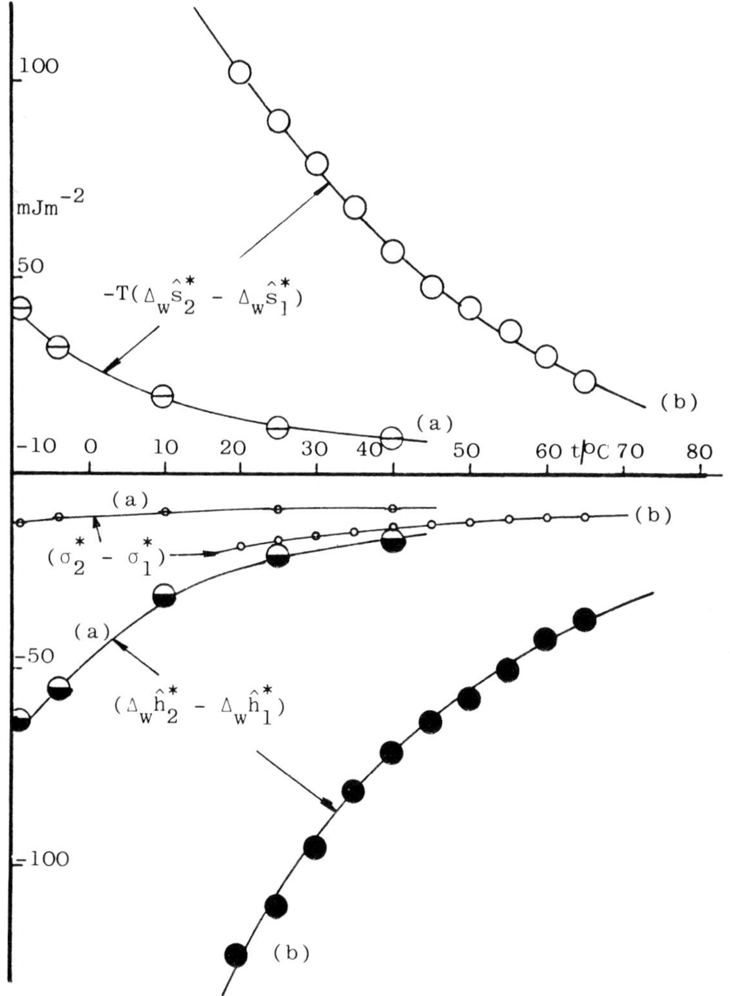

Fig.7. Thermodynamic functions for systems (a) [n-pentane (1) + n-decane(2)]/Graphon and (b) [n-hexane + n-hexadecane] /Graphon as function of temperature.

(possibly two-dimensional liquid crystals) at low temperatures, with the gradual breakdown of these structures as the temperature is raised. Evidence that similar structuring occurs in the adsorption of higher alkanes from solution comes from Groszek's measurements (48) of displacement enthalpies. Further support is provided by measurements of the enthalpies of immersion of Graphon in solutions

(44,45). More recently, using a Calvet-type calorimeter and a greatly improved bulb-breaking technique (49,50,51), a closer examination has been made of enthalpies of immersion in alkane mixtures which are generally consistent with the results of thermodynamic analysis of adsorption data, but in some instances provide evidence for more subtle effects (51).

Much more dramatic effects occur with higher alcohol + hydrocarbon mixtures. For the lower alcohols in mixtures with heptane, heptane is preferentially adsorbed, while from n-hexanol, upwards the alcohol is preferentially adsorbed (22,41,52) and the surface excess isotherms show a remarkable change in shape as the temperature is changed. For example, figure 8 shows the isotherms for the [n-heptane + n-hexanol]/Graphon system (41). The change in shape as the temperature is increased is brought out most clearly by considering the variation with temperature of the

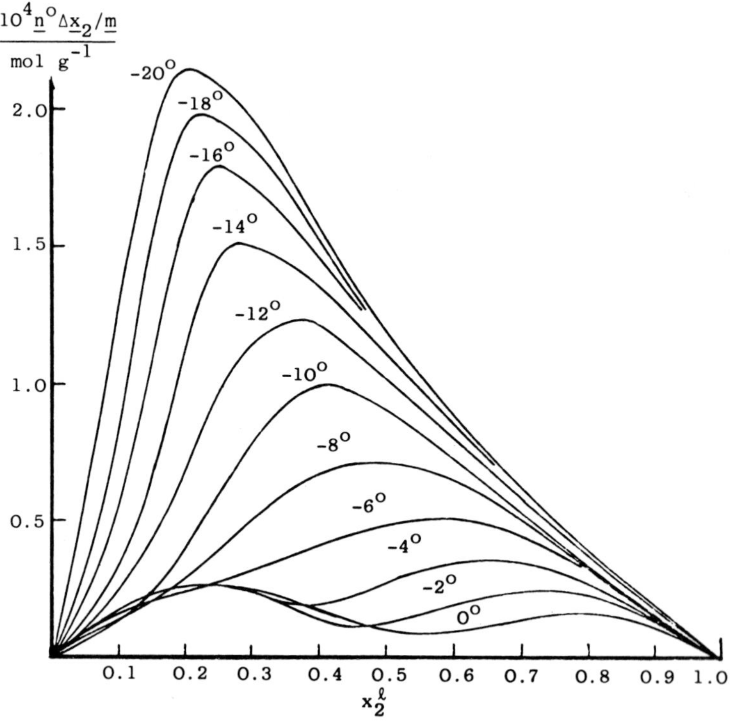

Fig. 8. Specific surface excess isotherms for [n-heptane(1) n-hexanol(2)]/Graphon at various temperatures.

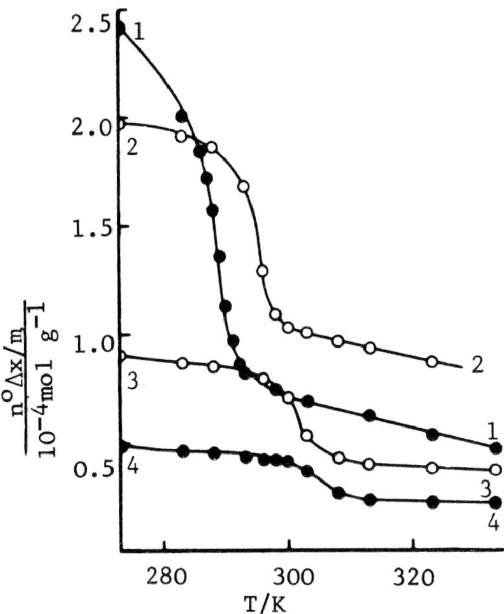

Fig. 9. Adsorption as a function of temperature at constant liquid mole fraction for system [n-heptane(1)+n-octanol(2)] Graphon (1) x_2^ℓ = 0.068,(2)0.232,(3)0.614,(4)0.860.

adsorption from a solution of constant composition (45,52) (figure 9). At each concentration the adsorption falls off sharply over a small temperature range, the mid-point of this range being a function of solution concentration, and tending, for the pure component, to a value roughly 40°C above the m.p. of the alkanol. A linear relationship is found between the logarithm of the activity of alcohol in the solution and the reciprocal of the absolute 'mid-transition' temperature. We interpreted this (45) as indicating that at low temperatures the alcohol is present as quasi-solid and that the transition is a smeared-out melting taking place much more sharply than in the case of n-alkanes. In the present Symposium, Findenegg *et al.*(53) interpret a similar relationship for the [heptane + dodecanol] /graphitised Vulcan system in terms of a phase transition into two pure adsorbed components below the transition temperature. Since their work refers to very dilute solutions, the difference between the alternative pictures is minimal. Evidence for structural changes, based on adsorption measurements, is also found for the systems [heptane + dodecanol]/Graphon (54) (agreeing broadly with Findenegg *et al.*); [heptane + n-hexanol]/Graphon (41); [benzene + stearic acid]/Graphon (55); [benzene + octanoic acid]/Graphon (55); [n-heptane + n-tetracosane]/Graphon (54). Calorimetric measurements of the enthalpies of immersion of graphite into higher alkanols and alkanoic acids also indicate the presence of structuring (49). The nature of these transitions is brought out very strikingly by the heat capacity measurements of Martin (56) on a slurry of n-octanol + Graphon (figure 10) which confirm the presence of a lambda-type surface transition. Further study of surface phase transitions should prove interesting and fruitful; it is unfortunate that facilities for heat capacity studies, which are particularly revealing, are available in so few laboratories.

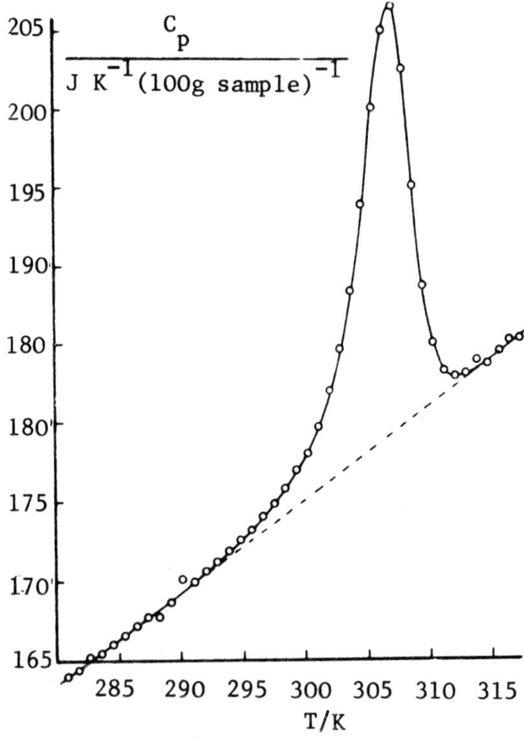

Fig. 10. Heat capacity of slurry of n-octane 61.1% + Graphon 38.9% by weight as function of temperature.

SOME APPLICATIONS OF ADSORPTION FROM SOLUTION

In addition to being of interest in its own right, the phenomenon of adsorption from solution has important implications for other phenomena. This arises from the relationship between adsorption and the free energy of surfaces. Three examples may be quoted.

It is well known that there are no simple mechanical methods of measuring the surface tension of a solid surface. The Gibbs adsorption equation, however, provides a thermodynamic route for the evaluation of changes in surface tension of solids brought about by adsorption. Thus vapour adsorption measurements carried out from vacuum to the saturation vapour pressure in principle enable the surface tension difference between clean solid (σ^{so}) and solid in equilibrium with saturated vapour (σ^{sv}) to be calculated:

$$\sigma^{sv} - \sigma^{so} = RT \int_0^{p^o} \Gamma \, d \ln p/p^{\dagger} , \qquad (25)$$

while the solid/saturated vapour and solid/liquid surface tensions are related through Young's equation:

$$\sigma^{sv} - \sigma^{s\ell} = \sigma^{\ell v} \cos\theta . \qquad (26)$$

Here $\sigma^{s\ell}$ should be the same as σ^* in equation (18). It should thus be possible to calculate, from vapour adsorption measurements, values of $(\sigma_1^* - \sigma_2^*)$ for comparison with those obtained from adsorption from solution measurements on solutions of 1 and 2. Although attempts have been made to confirm this relation (57) the situation is not entirely clear, mainly because of the need to obtain precise vapour adsorption data at very low coverages and the difficulty of extrapolating to saturation vapour pressure. Further study of this problem is needed (39).

Secondly, adsorption from solution measurements can be of value in discussing wettability and liquid-liquid displacement problems (58). Thus if of three substances A,B and C, A and B, and B and C are completely miscible whereas A and C are immiscible, then adsorption measurements on a given solid lead to values of $(\sigma_A^* - \sigma_B^*)$ and of $(\sigma_B^* - \sigma_C^*)$. The difference between these gives $(\sigma_A^* - \sigma_C^*)$ which is related to the contact angle θ at the A/C/solid line of contact by Young's equation:

$$\sigma_A^* - \sigma_C^* = \sigma_{AC}\cos\theta, \tag{27}$$

where σ_{AC} is the interfacial tension between A and C. Experiments to assess the validity of this procedure have recently been completed by Fletcher (40), who has studied the triads benzene, ethanol, water; n-heptane, ethanol, water and benzene, methanol, water adsorbed by Graphon. Contact angle measurements on freshly cleaved synthetic graphite were made by an interferometric technique in collaboration with Dr. I. Callaghan at B.P. Research Centre, Sunbury. For the heptane/water/graphite system the observed contact angle was $7^o \pm 2^o$, while that calculated from adsorption data was 8°. For benzene/water/graphite no contact angle was observed, while the adsorption data gave a value of '$\cos\theta$'> 1, indicating that heptane displaces water from a graphite surface. This work has suggested that one can now draw up a wettability table (Figure 11) in which the surface tensions of various pure liquids in contact with graphite are presented relative to benzene. Whether or not a given liquid X displaces water from graphite is immediately assessed by comparing $(\sigma_{H_2O}^* - \sigma_X^*)$ with $\sigma_{H_2O,X}$. The data in figure 11 show some interesting correlations with molecular structure which there is not space here to discuss in detail. Current work (59) is aimed at preparing a similar table for clay surfaces for which, however, macroscopic contact angle measurements are not practicable.

Finally, reference should be made to the influence of adsorption on the forces between colloidal particles. Here one is concerned with the way in which adsorption depends on the distance (h) between the particle surfaces, the key equation being (60)

$$f(h) - f_o(h) = 2\int_{-\infty}^{\mu_2} (\partial\, \Gamma_2^{(1)}/\partial h)\,d\mu_2, \tag{28}$$

where f is the force at separation h in the solution and f_o that in the presence of pure liquid 1. A positive value of f represents repulsion between the particles. This shows that if moving the particles together (dh-ve) leads to an increase in adsorption, then adsorption effects make an attractive contribution to the force, while if $(\partial\Gamma_2^{(1)}/\partial h$ is positive then a repulsive contribution results.

Experiments to confirm and illustrate these considerations are difficult to devise. Recent work on the forces between macroscopic objects immersed in solutions are

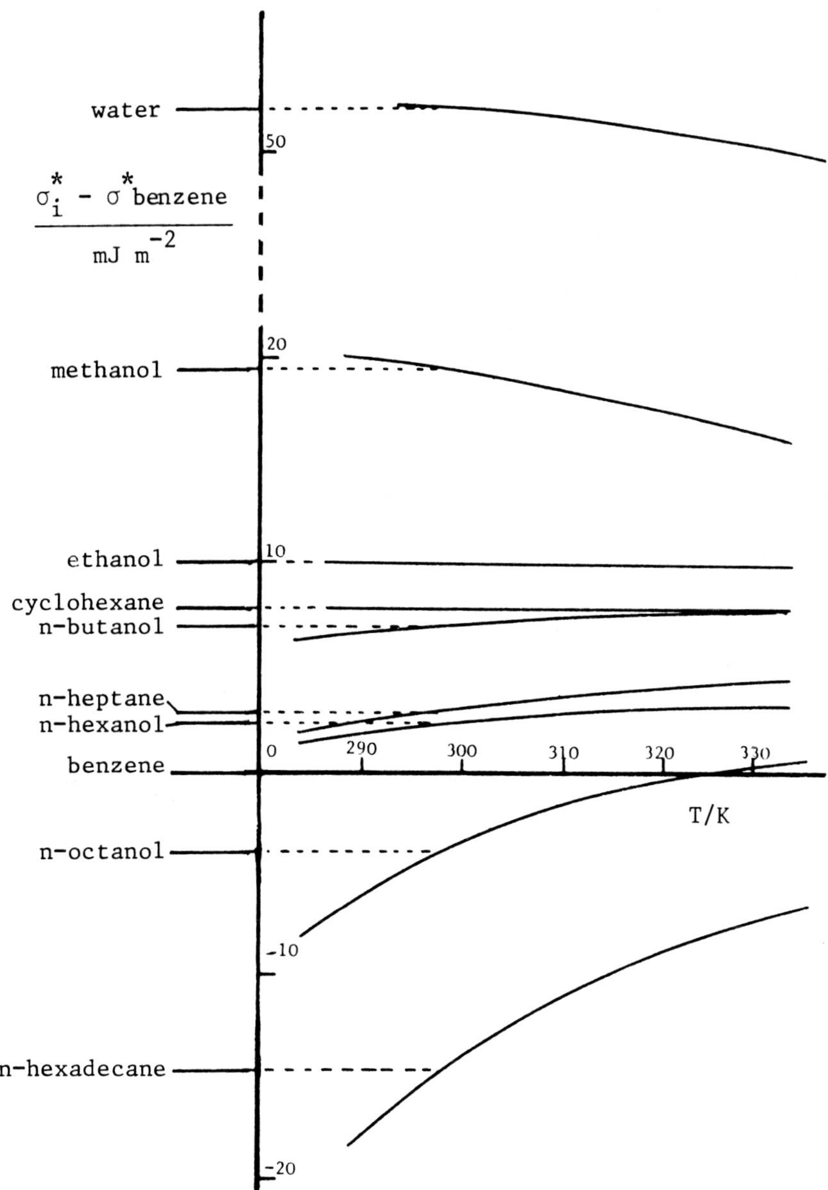

Fig.11. Surface tensions of liquid i/graphite interface relative to that of the benzene/graphite interface. Left hand: at 298K; right hand: as function of temperature. Data from refs. 16,25,36,38,39,40,41,52.

already providing interesting data on the forces involved (61), but the problem of measuring adsorption as a function of separation has not been solved.

ACKNOWLEDGEMENTS

The work which I have summarised in this lecture has depended to a major extent on my collaborators listed in the appendix; I am most grateful to them for their enthusiasm and thoroughness with which they have worked. I also express my thanks to the S.R.C. (now SERC) for their support with equipment, postdoctoral funds and research studentships. B.P. Research have also provided financial aid while much of the work on the development of immersion calorimetry has been supported by generous grants from Procter and Gamble.

REFERENCES

1. Kipling, J.J. (1965). "Adsorption From Solutions of Non-Electrolytes", Academic Press, London and New York.
2. Butler, J.A.V. (1932). *Proc.Roy.Soc.*, A135, 348.
3. Schuchowitzky, A. (1944). *Acta Physicochim. U.R.S.S.*, 19, 176; Belton J.W. and Evans M.G.(1945).*Trans.Faraday Soc.*, 41,1; Guggenheim, E.A. (1945). *Trans.Faraday Soc.*, 41,150.
4. Schuchowitzky, A. (1938). *Acta Physicochim. U.R.S.S.*, 8, 531; Schay, G. (1956). *Acta Chim.Acad.Sci.Hung.*, 10,281; Siskova, M. and Erdös, E (1960). *Coll.Czech. Chem.Comm.*, 25, 1729,3086; Everett, D.H. (1964). *Trans. Faraday Soc.*, 60, 1803.
5. Everett, D.H. (1973). *In* "Colloid Science" (ed. D.H. Everett). Vol.1, Chap. 2, Specialist Periodical Reports, The Chemical Society, London.
6. Everett, D.H. and Brown, C.E. (1975). *In* "Colloid Science" (ed. D.H. Everett). Vol.2, Chap. 2, Specialist Periodical Reports, The Chemical Society, London.
7. Everett, D.H. and Podoll, R.T. (1979). *In* "Colloid Science" (ed. D.H. Everett). Vol.3, Chap. 2, Specialist Periodical Reports, The Chemical Society, London.
8. Everett, D.H. and Davis, J. (1982). *In* "Colloid Science" (ed. D.H. Everett). Vol. 4, Chap. 3, Specialist Periodical Reports, The Royal Society of Chemistry, London.
9. Defay, R.(1934) *Etude thermodynamique de la tension superficielle*, Paris; Defay, R. and Prigogine, I. (1950) *Trans.Faraday Soc.*, 46, 199.

10. IUPAC "Manual of Symbols and Terminology for Physico chemical Quantities and Units, Appendix II, Definitions, Terminology and Symbols in Colloid and Surface Chemistry", Part I, (1973). *Pure Appl.Chem.*, 31, 579.
11. Everett, D.H. (1964). *Trans. Faraday Soc.*, 60, 1803.
12. e.g. Schay, G. (1970). *In* "Proceedings of the International Symposium on Surface Area Determination", (Eds D.H. Everett and R.H. Ottewill), p.272, Butterworths, London.
13. Everett, D.H. (1965). *Trans.Faraday Soc.*, 61, 2478.
14. e.g. Kiselev, A.V. and Khopina, V.V. (1969). *Trans. Faraday Soc.*, 65, 1936 and earlier papers; Nagy, L.G. and Schay, G, (1963). *Acta Chim.Acad.Sci.Hung.*, 39, 365; Myers, A.L. and Sircar, S. (1971). *Amer.Inst.Chem. Engineers J.*, 17, 186.
15. Lane, J.E., This Volume, p.51.
16. Everett, D.H. and Podoll, R.T. (1981). *J.Coll.Interface Sci.*, 82, 14.
17. Schay, G, Nagy, L.G. and Szekrenyesy,T.(1962). *Periodica Polytechnica*, 6, 91.
18. Defay, R., Prigogine, I., Bellemans, A. and Everett, D.H. (1965). "Surface Tension and Adsorption", chap.XII, Longmans, London.
19. e.g. Sinanoglu, O and Pitzer, K.S. (1960). *J.Chem. Physics*, 32, 1279; cf. Everett, D.H. (1965). *Discuss. Faraday Soc.*, 40, 177.
20. Ash, S.G., Everett, D.H. and Findenegg, G.H. (1968), *Trans.Faraday Soc.*, 64, 2639.
21. Bown, R, (1973). Ph.D. Thesis, Bristol.
22. Everett, D.H., (1978). *Progr.Coll.Polymer Sci.*, 65, 103.
23. Everett, D.H. and Munn, R.J. (1964). *Trans.Faraday Soc.*, 60, 1951.
24. Rusanov, A.I. (1967). "Phase Equilibrium and Surface Phenomena", Chapter VI, Chimia, Leningrad.
25. Brown, C.E., Everett, D.H. and Morgan C.J. (1975). *J.C.S. Faraday I*, 71, 883.
26. Defay, R. and Prigogine, I. (1950). *Trans.Faraday Soc.*, 46, 199.
27. Ono, S. (1947). *Mem.Fac.Eng.,Kyushu Univ.*, 10, 195; Murakami, T, Ono, S., Tamura, M, and Kurata, M. (1951). *J.Phys.Soc.Japan*, 6, 309; Ono, S. and Kondo S. (1960). *In* "Encyclopaedia of Physics" (ed. Flügge), Vol. X, p.264, Springer, Berlin.
28. Lane, J.E. (1968). *Austral.J.Chem.*, 21, 827.
29. Altenberger, A.R. and Stecki, J. (1970). *Chem.Phys. Letters*, 5, 29.

30. Ash, S.G., Everett, D.H. and Findenegg, G.H. (1968). *Trans.Faraday Soc.*, 64, 2645; *ibid.* (1970). 66, 708.
31. Roe, R-J. (1974). *J.Chem.Phys.*, 60, 4192; (1975). *J.Colloid Interface Sci.*, 50, 64.
32. Scheutjens, J.M.H.M., and Fleer, G.J. (1979). *J. Phys. Chem.*, 83, 1619; (1980). 84, 178.
33. Harris, N.M., Ottewill, R.H. and White, J.W. This volume p. 139.
34. Cosgrove, T., Crowley, T.L. and Vincent, B. This volume, p. 287.
35. see e.g. Faraday Symposium No. 16, (1981). "Structure of the Interfacial Region", Royal Society of Chemistry, London.
36. Ash, S.G., Bown, R. and Everett, D.H. (1973). *J.Chem. Thermodynamics*, 5, 239.
37. Everett, D.H. (1981). *J.Phys.Chem.*, 85, 3263.
38. Ash, S.G., Bown, R. and Everett, D.H. (1975). *J.C.S. Faraday I*, 71, 123.
39. Davis, J. (1983). Ph.D. Thesis, Bristol, in preparation.
40. Fletcher, A.J.P. (1982). Ph.D. Thesis, Bristol.
41. Smith, R.W. (1979). Ph.D. Thesis, Bristol.
42. Everett, D.H. (1975). *Israel J.Chem.*, 14, 267.
43. Robert, L. (1963). *Compt.Rend.*, 256, 655; (1967). *Bull. Soc.Chim.Fr.*, 2039; Clint, J.H., Clunie, J.S., Goodman J.S. and Tate, J.R. (1969). *Nature*, 223, 51; Everett, D.H. and Findenegg, G.H. (1969). *ibid.*, 223, 52; Everett, D.H., Findenegg, G.H. and Cram, P.J. (1969). *J.Chem.Thermodynamics*, 1, 573.
44. Thorne, P.E. (1974).Ph.D. Thesis, Bristol.
45. Brown, C.E., Everett, D.H., Powell, A.V. and Thorne, P.E. (1975). *Faraday Disc.Chem.Soc.*, 59, 97.
46. Parfitt, G.D. and Tideswell, M.W. (1981). *J.Colloid Interface Sci.*, 79, 518.
47. Ash, S.G. and Findenegg, G.H.(1970). *Spec.Disc.Faraday Soc.*, 1, 105; Findenegg, G.H. (1971). *J.Colloid Interface Sci.*, 35, 249; (1972). *J.C.S. Faraday I*, 68, 1799; (1973). 69, 1069.
48. Groszek, A. (1970). *Proc.Roy.Soc.*, A314, 473.
49. Wightman, J.P., unpublished.
50. Langdon, A.G., unpublished.
51. Maher, P. (1983). Ph.D. Thesis, Bristol, in preparation.
52. Brown, C.E., unpublished.
53. Findenegg, G.H., Koch, C. and Liphard, M. This volume, p.87.
54. Greenleaf, D.J. (1982). Ph.D. Thesis, Bristol.
55. Podoll, R.T., unpublished.
56. Martin, J.F., unpublished; see ref. 45.

57. See summary of work by Larionov, *et al.*, ref. 7, pp.83-91.
58. Everett, D.H. (1981). *Pure Appl.Chem.*, 53, 2181.
59. Nunn, C. and Foster, C., work in progress.
60. Everett, D.H. (1971). Proc.Int.Conf.Thermochem., Marseilles Colloqu.Int. CNRS, No. 201, p.45 CNRS(Paris) 1972; Ash, S.G., Everett, D.H. and Radke, C.J. (1973). *J.C.S. Faraday II*, 69, 1256; Hall, D.G. (1972), *J.C.S. Faraday II*, 68, 2169; Everett, D.H. and Radke, C.J. (1975). *In* "Adsorption at Interfaces", (ed. K.L. Mittal) Amer.Chem.Soc.Symp.Ser. Vol 8, p.1 Amer.Chem.Soc., New York; Everett, D.H. (1976). *Pure Appl.Chem.*, 48, 419.
61. Israelachvili, J.N. (1982). *Adv.Coll.Interface Sci.*, 16, 31; Klein, J. (1982). *Adv.Coll.Interface Sci.*, 16, 101.

APPENDIX

Contributors to work on Adsorption from Solution at Bristol 1962-1982

J.E. Lane	E.E.H. Wright	D.F. Billett	S.G. Ash
J.P. Wightman	A.V. Powell	R. Bown	C.J. Morgan
A.G. Langdon	G.H. Findenegg	P.E. Thorne	P.J. Szanto
	C.E. Brown	R.W. Smith	J. Davis
	R.T. Podoll	D.J. Greenleaf	A.J.P. Fletcher
	C. Nunn	P. Maher	C. Foster

THERMODYNAMICS OF ADSORPTION FROM MIXED SOLVENTS

Denver G. Hall

Unilever Research Port Sunlight Laboratory
Quarry Road East, Bebington,
Wirral, Merseyside L63 3JW

ABSTRACT

A general thermodynamic treatment of adsorption from mixed solvents is provided. It is designed to apply in situations where adsorbed species may be regarded as distinct from their bulk counterparts. Clausius-Clapeyron type expressions are derived which relate differential and integral enthalpies of adsorption to the temperature dependence of the adsorption equilibria concerned. The differential enthalpies refer to the transfer of material from bulk to adsorbed states under the condition that non-specifically adsorbed material is allowed to re-equilibrate. The integral enthalpies refer to the enthalpy of formation of the interface from the bulk solution at constant interfacial and bulk intensive properties. They correspond closely to heats obtainable from calorimetric measurements. The treatment is applied to the adsorption of a nonionic surfactant from a binary mixture and to the adsorption of anionic surfactant from an electrolyte solution with a common ion.

INTRODUCTION

Formal thermodynamic treatments of adsorption have essentially two main aims. One is to derive rigorous thermodynamic expressions which relate the results of different kinds of experiment such as calorimetric measurements and the temperature dependence of adsorption equilibria. The second is to estimate quantities of fundamental interest at a molecular level from experimental data. Although the pioneering work of Gibbs was published over 100 years ago (1) it was not until about 1950 that the general framework currently used to achieve these aims was first forwarded. At about this time Everett (2) and Hill (3), separately, stated precisely the distinction between the various adsorption heats and their relationship to other thermodynamic quantities. They also discussed in some detail the important case of gas adsorption. Their results for this case are readily adaptable to adsorption from dilute solutions in which the solvent is a single pure component (4). Further extensions include concentrated liquid mixtures (5), multi component monolayers (6) and electrochemical systems (7). However some aspects of adsorption from mixed solvents do not appear to have been fully covered. The present paper is an attempt to fill this gap.

Previous treatments of adsorption thermodynamics have been based on the analysis of an excess surface free energy G^σ given by

$$G^\sigma = \sum_i \mu_i \Gamma_i + \sigma \qquad (1)$$

where the summation extends over all independent components, σ denotes surface tension, μ denotes chemical potential and the Γ_i denote surface excesses defined according to a suitable Gibbs convention (e.g. $\Gamma_{solvent} = 0$). For mixed solvents the summation in equn 1 includes the minor solvent components whose Γ_i thus appear in the variables describing interfacial composition. A consequence of this is that the various partial molar quantities of adsorption as usually defined may not correspond to processes of practical interest and as such have only formal significance. In particular when charged species are present some Γ_i may be negative. In this paper we adopt the view that the surface phase should be considered to consist only of components which without ambiguity can be said to exist in an adsorbed state. In addition we regard adsorbed and bulk forms of the same component as distinct and base our treatment on an analysis of partial equilibrium states in which the chemical potentials of these distinct forms may differ. We recognise that such a viewpoint is not in itself original but believe that some aspects of the exposition which follows are novel.

BASIC EQUATIONS

Let the system of interest consist of two fluid phases. We suppose for simplicity that one of these phases contains but a single component, o, which is inert in the sense used by Guggenheim (8) and as such is not present in the second phase. Let this second phase consist of a principal solvent component, 1, a series of other minor solvent components, α, β, etc. which are not adsorbed in the sense outlined above and a series of solute species i,j,k etc. which are present both in an adsorbed state, which we denote by superscript σ, and in bulk solution, which we denote by superscript b. The partial equilibrium states of interest are determined by the variables T,p, the amounts of the above species and the interfacial area A. Hence we write the fundamental equation of our entire system as

$$d(^G/T) = -\frac{H}{T^2}\,dT + \frac{V}{T}\,dp + \frac{\mu_o}{T}\,dN_o + \frac{\mu_1}{T}\,dN_1 + \sum_\alpha \frac{\mu_\alpha}{T}\,dN_\alpha + \sum_i \frac{\mu_i^\sigma}{T}\,dN_i^\sigma + \sum_i \frac{\mu_i^b}{T}\,dN_i^b + \frac{\sigma}{T}\,dA \qquad (2)$$

where G denotes Gibbs free energy, H denotes enthalpy, V denotes volume and N denotes amount. At constant T and p,G, H and V are linear homogeneous functions of the various Ns and A. In particular at constant T and p

$$dH = H_o dN_o + h_1 dN_1 + \sum_\alpha h_\alpha dN_\alpha + \sum_i h_i^\sigma dN_i^\sigma + \sum_i h_i^b dN_i^b + \left(\frac{\partial H}{\partial A}\right)_{T,p,N} dA \qquad (3)$$

where the constant N in the derivative denotes that all amounts are held constant. From equns 2 and 3 it is apparent that

$$h_i^\sigma = \left(\frac{\partial H}{\partial N_i^\sigma}\right)_{T,p,N\neq N_i^\sigma,A} = -T^2\left(\frac{\partial \mu_i^\sigma/T}{\partial T}\right)_{p,N,A} = -T^2\left(\frac{\partial \mu_i^\sigma/T}{\partial T}\right)_{p,\Gamma_i^\sigma,m_i,m_\alpha}$$

$$h_i^b = \left(\frac{\partial H}{\partial N_i^b}\right)_{T,p,N\neq N_i^b,A} = -T^2\left(\frac{\partial \mu_i^b/T}{\partial T}\right)_{p,N,A} = -T^2\left(\frac{\partial \mu_i^b/T}{\partial T}\right)_{p,\Gamma_i^\sigma,m_i,m_\alpha}$$

$$h_\alpha = \left(\frac{\partial H}{\partial N_\alpha}\right)_{T,p,N\neq N_\alpha,A} = -T^2\left(\frac{\partial \mu_\alpha/T}{\partial T}\right)_{p,N,A} = -T^2\left(\frac{\partial \mu_\alpha/T}{\partial T}\right)_{p,\Gamma_i^\sigma,m_i,m_\alpha}$$

$$\left(\frac{\partial H}{\partial A}\right)_{T,p,N} = -T^2\left(\frac{\partial \sigma/T}{\partial T}\right)_{p,N,A} = -T^2\left(\frac{\partial \sigma/T}{\partial T}\right)_{p,\Gamma_i^\sigma,m_i,m} \qquad (4a\text{-}d)$$

where the m_i and m_α denote the bulk concentrations of the species concerned expressed as mole ratios with respect to component 1*.

---

* To show that h_α is indeed the bulk partial molar enthalpy of species α we note that

$$\Gamma_\alpha A = N_\alpha - m_\alpha N_1$$

and that

$$\left(\frac{\partial \mu_\alpha/T}{\partial T}\right)_{p,N,A} = \left(\frac{\partial \mu_\alpha/T}{\partial T}\right)_{p,m_i,m_\alpha} + \frac{1}{T}\sum_i\left(\frac{\partial \mu_\alpha}{\partial m_i}\right)_{T,p,m_j m_\alpha}\left(\frac{\partial m_i}{\partial T}\right)_{p,N,A}$$

$$+ \frac{1}{T}\sum_\beta\left(\frac{\partial \mu_\alpha}{\partial m_\beta}\right)_{T,p,m_i,m_\alpha}\left(\frac{\partial m_\beta}{\partial T}\right)_{p,N,A}$$

The first term on the RHS of this expression is $-h_\alpha/T^2$. Also

$$\left(\frac{\partial m_\beta}{\partial T}\right)_{p,N,A} = -\frac{A}{N_1}\left(\frac{\partial \Gamma_\beta}{\partial T}\right)_{p,N,A}$$

and approaches zero when A/N_1 becomes very large, which is usually the case in studies of isolated interfaces. Similar arguments apply to bulk i and to the quantity $(\partial H/\partial A)_{T,p.N}$.

$(h_k^\sigma - h_k^b)$ is the change in enthalpy when one molecule or ion of type k is transferred from bulk solution to the adsorbed state at constant T,p,A and all other N_i^σ when the interface is allowed to equilibrate in all other respects.

The enthalpy of forming unit area of interface at constant intensive properties but in a closed system is given by H_s where

$$H_s = \left(\frac{\partial H}{\partial A}\right)_{T,p,\Gamma_i^\sigma,N_i,N_\alpha} = \sum_i (h_i^\sigma - h_i^\beta)\Gamma_i^\sigma - T^2\left(\frac{\partial \sigma/T}{\partial T}\right)_{p,\Gamma_i^\sigma,m_i,m_\alpha} \quad (5)$$

and where $N_i = N_i^b + N_i^\sigma$.

It is also apparent that

$$AH_s = H - N_o h_o - N_1 h_1 - \sum_i N_i h_i^b - \sum_\alpha N_\alpha h_\alpha \quad (6)$$

The Gibbs adsorption equation may be derived from equn 2 in the usual way. The convention we adopt is that used by Parsons (7) and by Hansen (9) whereby the Gibbs-Duhem equations for the two bulk phases are used to eliminate $d\mu_o$ and $d\mu_1$ from the Gibbs-Duhem equation for the entire system. The result is

$$d(\sigma/T) = -\frac{H^\sigma}{T^2}\,dT + \frac{V^\sigma}{T}\,dp - \sum_i \Gamma_i^\sigma d(\mu_i^\sigma/T)$$

$$- \sum_i \Gamma_i^b\, d\left(\frac{\mu_i^b}{T}\right) - \sum_\alpha \Gamma_\alpha\, d(\mu_\alpha/T) \quad (7)$$

where H^σ, V^σ and the Γs are excess quantities in accordance with the above convention. It is clear however that $\Gamma_i^\sigma = N_i^\sigma/A$ is independent of any choice of Gibbs type convention.

Clausius-Clapeyron equations for differential quantities of adsorption

We note that μ_i^b and μ_α depend only on T, p and bulk solution composition. We write after Guggenheim (8)

$$\left.\begin{aligned} d\left(\frac{\mu_k^b}{T}\right) &= -\frac{h_k^b}{T^2}\,dT + \frac{v_k^b}{T}\,dp + \frac{1}{T}\,D\mu_k^b \\ d\left(\frac{\mu_\beta}{T}\right) &= -\frac{h_\beta}{T^2}\,dT + \frac{v_\beta}{T}\,dp + \frac{1}{T}\,D\mu_\beta \end{aligned}\right\} \qquad (8a,b)$$

where

$$D\mu_k^b = \sum_i \left(\frac{\partial \mu_k^b}{\partial m_i}\right)_{T,p,m_i,m_\alpha} dm_i + \sum_\alpha \left(\frac{\partial \mu_k^b}{\partial m_\alpha}\right)_{T,p,m_i,m_\beta} dm_\alpha$$

$$D\mu_\beta = \sum_i \left(\frac{\partial \mu_\beta}{\partial m_i}\right)_{T,p,m_i,m_\alpha} dm_i + \sum_\alpha \left(\frac{\partial \mu_\beta}{\partial m_\alpha}\right)_{T,p,m_i,m_\beta} dm_\alpha \qquad (9a,b)$$

On the other hand we regard μ_k^σ/T as a function of T, p, Γ_i^σ, μ_i^b/T and μ_α/T and write

$$d\left(\frac{\mu_k^\sigma}{T}\right) = \left(\frac{\partial \mu_2^\sigma/T}{\partial T}\right)_{p,\Gamma_i^\sigma,\mu_i^b/T,\ \mu_\alpha/T} dT + \frac{1}{T}\left(\frac{\partial \mu_k^\sigma}{\partial p}\right)_{T,\Gamma_i^\sigma,\mu_i^b,\mu_\alpha} dp$$

$$+ \frac{1}{T}\sum_i \left(\frac{\partial \mu_k^\sigma}{\partial \Gamma_i^\sigma}\right)_{T,p,\Gamma_i^\sigma,\mu_i^b,\mu_\alpha} d\Gamma_i^\sigma + \sum_i \left(\frac{\partial \mu_k^\sigma}{\partial \mu_i^b}\right)_{T,p,\Gamma_i^\sigma,\mu_j^b,\mu_\alpha} d(\mu_i^b/T)$$

$$+ \sum_\alpha \left(\frac{\partial \mu_k^\sigma}{\partial \mu_\alpha}\right)_{T,p,\Gamma_i^\sigma,\mu_i^b,\mu_\beta} d(\mu_\alpha/T) \qquad (10)$$

However since

$$\left(\frac{\partial \mu_k^\sigma/T}{\partial T}\right)_{p,\Gamma_i^\sigma,m_i,m_\alpha} = \left(\frac{\partial \mu_k^\sigma/T}{\partial T}\right)_{p,\Gamma_i^\sigma\ \frac{\mu_i^b}{T},\ \frac{\mu_\alpha}{T}}$$

$$+ \sum_i \left(\frac{\partial \mu_k^\sigma}{\partial \mu_i^b}\right)_{T,p,\Gamma_i^\sigma,\mu_j^b,\mu_\alpha} \left(\frac{\partial \mu_i^b/T}{\partial T}\right)_{p,\Gamma_1^\sigma,m_i,m_\alpha}$$

$$+ \sum_\alpha \left(\frac{\partial \mu_k^\sigma}{\partial \mu_\alpha}\right)_{T,p,\Gamma_i^\sigma,\mu_i^b,\mu_\beta} \left(\frac{\partial \mu_\alpha/T}{T}\right)_{p,\Gamma_1^\sigma,m_i,m_\alpha}$$

and from equn 7 it is apparent that

$$\left(\frac{\partial \mu_k^\sigma}{\partial \mu_i^b}\right)_{T,p,\Gamma_i^\sigma \mu_j^b \mu_\alpha} = -\left(\frac{\partial \Gamma_i^b}{\partial \Gamma_k^\sigma}\right)_{T,p,\Gamma_j^\sigma \mu_i^b \mu_\alpha}$$

and

$$\left(\frac{\partial \mu_k^\sigma}{\partial \mu_\alpha}\right)_{T,p,\Gamma_i^\sigma \mu_1^b,\mu_\beta} = -\left(\frac{\partial \Gamma_\alpha}{\partial \Gamma_k^\sigma}\right)_{T,p,\Gamma_j^\sigma,m_i,m_\alpha}$$

it follows from equns 4a-c and 8 a,b, that

$$d\left(\frac{\mu_k^\sigma}{T}\right) = -\frac{h_k^\sigma}{T^2}\, dT + \frac{v_k^\sigma}{T}\, dp + \frac{1}{T} \sum_i \left(\frac{\partial \mu_k^\sigma}{\partial \Gamma_i^\sigma}\right)_{T,p,\Gamma_j^\sigma,m_i,m_\alpha} d\Gamma_i^\sigma$$

$$-\frac{1}{T} \sum_i \left(\frac{\partial \Gamma_i^b}{\partial \Gamma_k^\sigma}\right)_{T,p,\Gamma_j^\sigma m_i,m_\alpha} D\mu_i^b - \frac{1}{T} \sum_\alpha \left(\frac{\partial \Gamma_\alpha}{\partial \Gamma_k^\sigma}\right)_{T,p,\Gamma_j^\sigma,m_i,m_\alpha} D\mu_\alpha \qquad (11)$$

At equilibrium $\mu_k^\sigma = \mu_k^b$ and we may equate the RHS of equn 8a with that of equn 11 to obtain

$$0 = \frac{-(h_k^\sigma - h_k^b)}{T^2}\, dT + \frac{(v_k^\sigma - v_k^b)}{T}\, dp + \sum_i \left(\frac{\partial \mu_k^\sigma}{\partial \Gamma_i^\sigma}\right)_{T,p,\Gamma_j^\sigma,m_i,m_\alpha} d\Gamma_i^\sigma$$

$$-\frac{1}{T}D\mu_k^b - \frac{1}{T}\sum_i \left(\frac{\partial \Gamma_i^b}{\partial \Gamma_k^\sigma}\right)_{T,p,\Gamma_j^\sigma,m_i,m_\alpha} D\mu_i^b - \frac{1}{T}\sum_\alpha \left(\frac{\partial \Gamma_\alpha}{\partial \Gamma_k^\sigma}\right)_{T,p,\Gamma_j^\sigma,m_i,m_\alpha} D\mu_\alpha \tag{12}$$

Equation 12 leads to Clausius-Clapeyron type expressions which relate differential enthalpies of adsorption to the temperature dependence of adsorption equilibria. The differential enthalpies concerned have a clear physical significance which the corresponding quantities obtained from more traditional treatments may lack.

Integral quantities of adsorption

We take as our starting point the relation between H^σ and H_s. From equation 7 we obtain

$$T^2 \left(\frac{\partial \sigma/T}{\partial T}\right)_{p,\Gamma_i^\sigma,m_i,m_\alpha} = - H^\sigma + \sum_i \Gamma_i^\sigma h_i^\sigma + \sum_i \Gamma_i^b h_i^b + \sum_\alpha \Gamma_\alpha h_\alpha \tag{13}$$

which together with equn 5 gives

$$H_s = H^\sigma - \sum_i (\Gamma_i^\sigma + \Gamma_i^b)\, h_i^b - \sum_\alpha \Gamma_\alpha h_\alpha \tag{14}$$

hence equn 7 becomes

$$d\,(^\sigma/T) = -\frac{H_s}{T^2}\,dT + \frac{V_s}{T}\,dp - \sum_i \Gamma_i^\sigma\, d\left(\frac{\mu_i^\sigma - \mu_i^b}{T}\right)$$

$$-\frac{1}{T}\sum_i (\Gamma_i^\sigma + \Gamma_i^b)\, D\mu_i^b - \frac{1}{T}\sum_\alpha \Gamma_\alpha D\mu_\alpha \tag{15}$$

Let superscript o denote the properties of a system with the same m_α as that of interest but in which all Γ_i^σ and m_i are zero. For this system we have by analogy with equn 15

$$d\left(\frac{\sigma^o}{T}\right) = -\frac{H_s^o}{T^2}\,dT + \frac{V_s^o}{T}\,dp - \frac{1}{T}\sum_\alpha \Gamma_\alpha^o D\mu_\alpha^o \tag{16}$$

We note that in general the μ_α^o will differ from the μ_α. Subtracting equn 16 from equn 15 we obtain

$$d\left(\frac{\sigma-\sigma^o}{T}\right) = -\left(\frac{H_s-H_s^o}{T^2}\right)dT + \left(\frac{V_s-V_s^o}{T}\right)dp - \sum_i \Gamma_i^\sigma \, d\left(\frac{\mu_i^\sigma-\mu_i^b}{T}\right)$$

$$-\frac{1}{T}\sum_i (\Gamma_i^\sigma + \Gamma_i^b)\, D\mu_i^b - \frac{1}{T}\sum_\alpha \Gamma_\alpha D\mu_\alpha \qquad (17)$$

equn 17 may be rewritten as

$$d\left(\frac{\sigma-\sigma^o}{T}\right) = -\left(\frac{H_s-H_s^o}{T^2}\right)dT + \left(\frac{V_s-V_s^o}{T}\right)dp - \sum_i \Gamma_i^\sigma \, d\left(\frac{\mu_i^\sigma-\mu_i^b}{T}\right)$$

$$-\frac{1}{T}\sum_i X_i dm_i - \frac{1}{T}\sum_\alpha X_\alpha dm_\alpha \qquad (18)$$

where

$$X_k = \sum_i (\Gamma_i^\sigma+\Gamma_i^b)\left(\frac{\partial\mu_i^b}{\partial m_k}\right)_{T,p,m_j,m_\alpha} + \sum_\alpha \Gamma_\alpha \left(\frac{\partial\mu_\alpha}{\partial m_k}\right)_{T,p,m_j,m_\alpha}$$

and

$$X_\beta = \sum_i (\Gamma_i^\sigma+\Gamma_i^b)\left(\frac{\partial\mu_i^b}{\partial m_\beta}\right)_{T,p,m_i,m_\alpha} + \sum_\alpha \Gamma_\alpha \left(\frac{\partial\mu_\alpha}{\partial m_\beta}\right)_{T,p,m_i,m_\alpha}$$

$$-\sum_\alpha \Gamma_\alpha^o \left(\frac{\partial\mu_\alpha^o}{\partial m_\beta}\right)_{T,p,m_\gamma} \qquad (19a,b)$$

Equations 17 and 18 apply both in partial equilibrium situations and at equilibrium. In the latter situation $\mu_i^\sigma = \mu_i^b$ and it is apparent from equn 15 that H_s, the enthalpy of forming unit area of surface <u>at equilibrium</u> is given by

$$-T^2\left(\frac{\partial\sigma/T}{\partial T}\right)_{p,m_i,m_\alpha}$$

It can be obtained from the change in surface tension with temperature at constant bulk composition. This latter result for multi component systems appears to have been first stated by Guggenheim (8).

The quantity $(H_s-H_s^o)$ in equns 17 and 18 is readily obtainable from calorimetric measurements. The expression which shows how this can be done is similar to one given by Hansen (9) but has the important feature that all terms other than $(H_s-H_s^o)$ appear as measurable enthalpy differences. Multiplying equn 17 by A, adding $d\,\Sigma\, N_i^\sigma\,(\mu_i^\sigma-\mu_i^b)/T$ to both sides and rearranging we obtain

$$d\,\Sigma\, N_i^\sigma\left(\frac{\mu_i^\sigma-\mu_i^b}{T}\right)$$

$$= -\frac{A(H_s-H_s^o)}{T^2}\,dT + \frac{A(V_s-V_s^o)}{T}\,dp + \sum_i \frac{(\mu_i^\sigma-\mu_i^b)}{T}\,dN_i^\sigma$$

$$-\frac{A}{T}\sum_\alpha \Gamma_\alpha D\mu_\alpha - \frac{A}{T}\sum_i (\Gamma_i^\sigma+\Gamma_i^b)\,D\mu_i^b + \frac{A}{T}\sum_\alpha \Gamma_\alpha^o D\mu_\alpha^o - Ad\,\frac{(\sigma-\sigma^o)}{T} \quad (20)$$

Equation 20 enables us to define generalised integral quantities of adsorption such as $(h_i^*-h_i^b)$ where

$$(h_i^*-h_i^b) = \left(\frac{\partial(H_s-H_s^o)A}{\partial N_i^\sigma}\right)_{T,p,N_j^\sigma,m_i,m_\alpha,\frac{(\sigma-\sigma^o)}{T}}$$

$$= -T^2\left(\frac{\partial\left(\frac{\mu_i^\sigma-\mu_i^b}{T}\right)}{dT}\right)_{p,N_i^\sigma,m_i,m_\alpha,\frac{(\sigma-\sigma_o)}{T}}$$

$$= -T^2\left(\frac{\partial\left(\frac{\mu_i^\sigma-\mu_i^b}{T}\right)}{dT}\right)_{p,x_i^\sigma,m_i,m_\alpha,\frac{(\sigma-\sigma^o)}{T}} \quad (21)$$

where $x_i^\sigma = \Gamma_i^\sigma/\sum_k \Gamma_i^\sigma$.

From either equn 17 or equn 20 we may conclude that

$$(H_s - H_s^o) = \sum_i \Gamma_i^\sigma (h_i^* - h_i^b) \tag{22}$$

In the particular case that there is but a single adsorbed species it is apparent that equation 22 justifies the use of the term integral quantity in this context. It is also apparent from equns 18, 8 and 11 that

$$\left(\frac{\partial(H_s - H_s^o)}{\partial\Gamma_i^\sigma}\right)_{T,p,\Gamma_j^\sigma,m_i,m_\alpha} = -T^2 \left(\frac{\dfrac{\partial(\mu_i^\sigma - \mu_i^b)}{T}}{dT}\right)_{p,\Gamma_i^\sigma,m_i,m_\alpha} = (h_i^\sigma - h_i^b) \tag{23}$$

which shows clearly the relationship between $(H_s - H_s^o)$ and the differential enthalpies defined in the previous section. However for a single adsorbed component it does not follow that

$$(H_s - H_s^o) = \int_o^{\Gamma^\sigma} (h^\sigma - h^b)\, d\Gamma^\sigma \tag{24}$$

where the integration is performed at T,p,m_i and m_α. The reason is that the lower limit of integration does not correspond to the mixed solvent of interest. For equation 24 to hold it is required that

$$(\partial H_s / \partial m_i)_{T,p,m_j,m_\alpha,\Gamma^\sigma=0}$$

be zero for all m_i in the range of interest. It is apparent from equn 18 that this condition is equivalent to

$$\left(\frac{\partial X_i / T}{\partial T}\right)_{p,m_i,m_\alpha,\Gamma^\sigma=0}$$

being zero. This is unlikely to be generally true. Choosing the reference solution o as that with the same m_i and m_α as that of interest but with all $\Gamma_i^\sigma=0$ avoids this problem but poses the more serious difficulty that σ^o in this reference state is inaccessible.

Equation 20 shows that we may regard $(\mu_i^\sigma-\mu_i^b)/T$ as a function of T, p, x_i^σ, $\left(\frac{(\sigma-\sigma^o)}{T}\right)$ and bulk solution composition. By making use of equations 19 a and b we find that

$$
\begin{aligned}
d\left(\frac{\mu_k^\sigma-\mu_k^b}{T}\right)
&= \frac{-(h_k^*-h_k^b)}{T^2}\,dT + \frac{v_k^*-v_k^b}{T}\,dp + \frac{1}{T}\sum_i \left(\frac{\partial \mu_k^\sigma}{\partial x_i^\sigma}\right)_{T,p,x_j^\sigma,m_i,m_\alpha,\frac{(\sigma-\sigma^o)}{T}} dx_i^\sigma \\
&\quad - \frac{1}{T}\sum_i \left(\frac{\partial(\Gamma_i^\sigma+\Gamma_i^b)A}{\partial N_k^\sigma}\right)_{T,p,N_j^\sigma,m_i,m_\alpha,\frac{(\sigma-\sigma^o)}{T}} D\mu_i^b \\
&\quad - \frac{1}{T}\sum_\alpha \left(\frac{\partial \Gamma_\alpha A}{\partial N_k^\sigma}\right)_{T,p,N_j^\sigma,m_i,m_\alpha,\frac{(\sigma-\sigma^o)}{T}} D\mu_\alpha \\
&\quad + \frac{1}{T}\sum_\alpha \Gamma_\alpha^o \left(\frac{\partial A}{\partial N_k^\sigma}\right)_{T,p,N_j^\sigma,m_i,m_\alpha,\frac{(\sigma-\sigma^o)}{T}} D\mu_\alpha^o \\
&\quad - \frac{1}{T}\left(\frac{\partial A}{\partial N_k^\sigma}\right)_{T,p,N_j^\sigma,m_i,m_\alpha,\frac{(\sigma-\sigma^o)}{T}} d\frac{(\sigma-\sigma^o)}{T}
\end{aligned}
\qquad (25)
$$

For a single adsorbed species equn 25 reduces to equn 17 when the latter is divided by Γ^σ. At equilibrium equn 25 leads to Clausius-Clapeyron type expressions for integral quantities of adsorption akin to those which result from equn 12 for the differential quantities.

To conclude this general treatment we note the following points. (1) The above discussion has been concerned primarily with adsorption enthalpies. Changing the variable p to p/T enables these to be replaced by the corresponding internal energies. Also other extensive quantities (e.g. volumes) can be handled by similar methods. (2) When dealing with systems containing ions it is especially convenient to work with the quantities θ_i defined by

$$\theta_i = \mu_i - \frac{\nu_i \mu_c}{\nu_c} \qquad (26)$$

where ν denotes ionic valence and includes the sign and where species c is chosen for convenience but is otherwise arbitrary. The θ_i play the same role as the chemical potentials of uncharged species and when they are included in the above formalism where appropriate no further modifications are required. (3) Some of the derivatives in equations 12 and 25 require more variables to be held constant than is possible at equilibrium. To use these expressions when only equilibrium data are available it may be necessary to estimate these derivatives by arguing that some are negligible and that the others can be approximated by quantities which are obtainable solely from equilibrium data. Unfortunately restrictions on the length of the paper forbid further elaboration on this point at present.

Applications

1. Nonionic surfactant in a binary solvent

Let k denote the surfactant and let α denote the minor solvent component. At constant p and m_α equn 12 takes the form

$$o = \frac{-(h_k^\sigma - h_k^b)}{T^2}\, dT + \frac{1}{T}\left(\frac{\partial \mu_k^\sigma}{\partial \Gamma_k^\sigma}\right)_{T,p,m_k,m_\alpha} d\Gamma_k^\sigma$$

$$-\frac{1}{T}\left[\left(\frac{\partial \mu_k^b}{m_k}\right)_{T,p,m_\alpha} + \left(\frac{\partial \Gamma_\alpha}{\partial \Gamma_k^\sigma}\right)_{T,p,m_k,m_\alpha} \left(\frac{\partial \mu_\alpha}{\partial m_k}\right)_{T,p,m_\alpha}\right] dm_k \qquad (27)$$

according to the Kirkwood-Buff theory of solutions (10) we have at constant T and p

$$kT\, d\ln m_k = (1+N_{kk}^+)\, d\mu_k + N_{k\alpha}^+\, d\mu_\alpha \qquad (28)$$

where

$$N^+_{k\alpha} = N^*_{k\alpha} + m_\alpha (N^*_{k1} + N^*_{\alpha 1} - (1+N^*_{11})) \tag{29}$$

$$N^*_{k\alpha} = n_\alpha \int_0^\infty (g_{k\alpha}(r) - 1)\, 4\pi\, r^2 dr \tag{30}$$

n_α denotes number density and where α and k may be any two species including species 1 or, for that matter, the same species.

For solutions dilute in k it is apparent that $N^+_{kk} \simeq 0$ provided that k does not self associate, also $N^+_{k\alpha}$ may be regarded as the average amount of α associated with a bulk k molecule. Under these circumstances it is clear that

$$\left.\begin{aligned} \left(\frac{\partial \mu_k}{\partial m_k}\right)_{m_\alpha} &= \frac{kT}{m_k} - N^+_{k_\alpha} \left(\frac{\partial \mu_\alpha}{\partial m_k}\right)_{m_\alpha} \\ \left(\frac{\partial \mu_\alpha}{\partial m_k}\right)_{m_\alpha} &= \left(\frac{\partial \mu_k}{\partial m_\alpha}\right)_{m_k} = -N^+_{k\alpha} \left(\frac{\partial \mu_\alpha}{\partial m_\alpha}\right)_{m_k} \end{aligned}\right\} \tag{31a,b}$$

From these results it follows that the coefficient of dm_k in equation 27 is

$$\frac{1}{T}\left[\frac{kT}{m_k} + N^+_{k\alpha}\left(N^+_{k\alpha} - \left(\frac{\partial \Gamma_\alpha}{\partial \Gamma^\sigma_k}\right)_{T,p,m_k,m_\alpha}\right)\left(\frac{\partial \mu_\alpha}{\partial m_\alpha}\right)_{T,p,m_k}\right]$$

When m_k is small one expects the leading term in the above brackets to dominate the others in which case we have approximately

$$(h^\sigma_k - h^b_k) \simeq -kT^2 \left(\frac{\partial \ln m_k}{\partial T}\right)_{p,m_\alpha,\Gamma^\sigma_k} \tag{32}$$

which is exactly analogous to the result one gets for a dilute solution in a single component solvent. The neglected terms may however become significant when $N^+_{k\alpha}$ is fairly large and the solution is fairly dilute in α.

To obtain an expression for the integral enthalpy we divide equn 15 by Γ^σ_k at equilibrium and at constant p and m_α we obtain

$$0 = -\left(\frac{h_k^*-h_k^b}{T^2}\right)dT - \left[\left(1+\frac{\Gamma_k^b}{\Gamma_k^\sigma}\right)\left(\frac{\partial \mu_k^b}{\partial m_k}\right)_{T,p,m_\alpha} + \frac{\Gamma_\alpha}{\Gamma_k^\sigma}\left(\frac{\partial \mu_\alpha}{\partial m_k}\right)_{T,p,m_\alpha}\right] dm_k$$

$$- \frac{1}{\Gamma_k^\sigma}\, d\left(\frac{\sigma-\sigma_o}{T}\right) \qquad (33)$$

Substituting for the derivatives of μ_k and μ_α as given by equns 31a,b we find that in this case the coefficient of dm_k is given by

$$-\left[\frac{kT}{m_k}\left(1+\left(\frac{\Gamma_k^b}{\Gamma_k^\sigma}\right)\right) + \left[N_{k\alpha}^+\left(1+\frac{\Gamma_k^b}{\Gamma_k^\sigma}\right) - \frac{\Gamma_\alpha}{\Gamma_k^\sigma}\right] N_{k\alpha}^+\left(\frac{\partial \mu_\alpha}{\partial m_\alpha}\right)_{T,p,m_k}\right]$$

hence

$$\frac{-(h_k^*-h_k^b)}{kT^2} = \qquad (34)$$

$$\left[1+\frac{\Gamma_k^b}{\Gamma_k^\sigma} - \left(\frac{\Gamma_\alpha}{\Gamma_k^\sigma} - N_{k\alpha}^+\left(1+\frac{\Gamma_k^b}{\Gamma_k^\sigma}\right)\right) N_{k\alpha}^+ m_k\left(\frac{\partial \mu_\alpha}{\partial m_\alpha}\right)_{T,p,m_k}\right]\left(\frac{\partial \ln m_k}{dT}\right)_{pm_\alpha \frac{(\sigma-\sigma^o)}{T}}$$

although $\frac{\Gamma_\alpha}{\Gamma_k^\sigma}$ will usually become indefinitely large as $\Gamma_k^\sigma \to 0$ this does not cause any problems because Γ_k^σ/m_k remains finite in this limit. For dilute solutions of nonionic surfactants one expects that $\Gamma_k^b/\Gamma_k^\sigma$ will usually be very small and that when species k is very surface active equn 34 will become equivalent to the expression one obtains for a single component solvent. However when α itself is fairly surface active the correction terms may well become significant and are likely to be larger than the corresponding terms in the full expression for the differential enthalpy of adsorption.

Ionic surfactant in the presence of electrolyte with a common ion

In practice it is found that adsorption from solutions of inorganic salts alone at air water and oil water interfaces is small (11). Thus for the sake of simplicity we will suppose that in this case Γ_α^o in equn 17 is zero. Also when surfactant ions, k, and co-ions, α, bear the same charge it is probably reasonable to write (12)

$$\frac{\Gamma_k^b}{\Gamma_\alpha} = \frac{m_k}{m_\alpha} \tag{35}$$

and to suppose that the mean ionic activity coefficients of surfactant, γ_k, and electrolyte, γ_α, are equal when m_2 given by $m_2 = (m_k + m_\alpha)$ lies in the range $0 < m_2 < 0.1$ molar. When these conditions hold we find that

$$\left.\begin{aligned} \frac{1}{T} D\theta_k &= k \, d\ln m_k + k \, d\ln m_2 + 2k \, d\ln\gamma \\ \frac{1}{T} D\theta_\alpha &= k \, d\ln m_\alpha + k \, d\ln m_2 + 2k \, d\ln\gamma \end{aligned}\right\} \tag{36a,b}$$

where $\gamma = \gamma_\alpha = \gamma_k$ and where counterions, species 2, are taken as species c in equn 26.

Let β^σ and α be defined by

$$\beta^\sigma = \alpha\Gamma^\sigma = -2(\Gamma_k^b + \Gamma_\alpha)$$

at constant T and p equn 17 gives on dividing by Γ_k^σ followed by appropriate use of equns 35 and 36

$$0 = \frac{-(h_k^* - h_k^b)}{T^2} dT - k \, d\ln m_k - \frac{1}{\Gamma_k^\sigma} d\left(\frac{\sigma - \sigma^o}{T}\right)$$
$$- \left[(1-\alpha) + (2-\alpha)\left(\frac{\partial \ln\gamma}{\partial \ln m_2}\right)_{T,p}\right] k \, d\ln m_2 \tag{38}$$

from which we deduce that

$$(h_k^* - h_k^b) = -kT^2 \left(\frac{\partial \ln m_k}{\partial T}\right)_{p,\left(\frac{\sigma-\sigma^o}{T}\right)} - kT^2 \; J^* \left(\frac{\partial \ln m_2}{\partial T}\right)_{p,\left(\frac{\sigma-\sigma^o}{T}\right)} \qquad (39)$$

where

$$J^* = -\left(\frac{\partial \ln m_k}{\partial \ln m_2}\right)_{T,p,(\sigma-\sigma^o)} = \left[(1-\alpha) + (2-\alpha)\left(\frac{\partial \ln \gamma}{d \ln m_2}\right)_{T,p}\right] \qquad (40)$$

Equn 39 shows how the integral enthalpy of adsorption of an ionic surfactant from an electrolyte solution can be obtained from the dependence of the surface tension on m_k, m_2 and temperature. It is very simple to use because there is no need to calculate activity coefficients or to estimate amounts adsorbed.

At constant T,p and Γ_k^σ equn 12 gives

$$\frac{h_k^\sigma - h_k^b}{T^2} dT = -\frac{1}{T}\left(1 + \left(\frac{\partial \Gamma_k^b}{\partial \Gamma_k^\sigma}\right)_{p,m_k,m_\alpha}\right) D\theta_k^b - \frac{1}{T}\left(\frac{\partial \Gamma_\alpha}{\partial \Gamma_k^\sigma}\right)_{p,m_k,m_\alpha} D\theta_\alpha \qquad (41)$$

To simplify this expression we note that equn 38 enables us to deduce that $\beta^\sigma = \beta^\sigma(T,p,\Gamma_k^\sigma,m_2)$ and that

$$\left.\begin{aligned} \left(\frac{\partial \Gamma_k^b}{\partial \Gamma_k^\sigma}\right)_{T,p,m_k,m_\alpha} &= -\frac{m_k}{2m_2}\left(\frac{\partial \beta^\sigma}{\partial \Gamma_k^\sigma}\right)_{T,p,m_2} \\ \left(\frac{\partial \Gamma_\alpha}{\partial \Gamma_k^\sigma}\right)_{T,p,m_k,m_\alpha} &= -\frac{m_\alpha}{2m_2}\left(\frac{\partial \beta^\sigma}{\partial \Gamma_k^\sigma}\right)_{T,p,m_2} \end{aligned}\right\} \qquad (42)$$

Together with equns 35 and 36 equns 40 and 42 lead to

$$0 = \frac{-(h_k^\sigma - h_k^b)}{T^2} dT - k\, d\ln m_k - \left((1-\alpha_d) + (2-\alpha_d)\left(\frac{\partial \ln \gamma}{\partial \ln m_2}\right)_{T,p}\right) k\, d\ln m_2 \qquad (43)$$

where

$$\alpha_d = (\partial \beta^\sigma / \partial \Gamma_k^\sigma)_{T,p,m_2}$$

Hence

$$(h_k^\sigma - h_k^b) = -kT^2 \left(\frac{\partial \ln m_k}{dT}\right)_{p,\Gamma_k^\sigma} - J^\sigma kT^2 \left(\frac{\partial \ln m_2}{\partial T}\right)_{p,\Gamma_k^\sigma} \tag{44}$$

where

$$J^\sigma = -\left(\frac{\partial \ln m_k}{\partial \ln m_2}\right)_{p,\Gamma_k^\sigma} = \left((1-\alpha_d)-(2-\alpha_d)\left(\frac{\partial \ln\gamma}{\partial \ln m_2}\right)_{T,p}\right) \tag{45}$$

Equation 44 provides an expression which enables the differential enthalpy of adsorption of an ionic surfactant from an electrolyte solution to be obtained from the dependence of $(\sigma-\sigma^o)$ on T, m_k and m_2. In this case it is necessary to estimate Γ_k^σ. However as equn 38 shows this can be obtained from the expression

$$\left(\frac{\partial(\sigma-\sigma^o)}{\partial \ln m_k}\right)_{T,p,m_2} = -\Gamma_k^\sigma kT \tag{46}$$

By use of equns 38 and 43 it is straightforward to show that

$$(h_k^* - h_k^b) + \Gamma_k^\sigma \left(\frac{\partial(h_k^* - h_k^b)}{\partial \Gamma_k^\sigma}\right)_{T,p,m_2} = -kT^2 \left(\frac{\partial \ln m_k}{dT}\right)_{p,\Gamma_k^\sigma,m_2} = h_k^\sigma - h_k^b \tag{47}$$

Moreover we have assumed above that $\Gamma_\alpha^o = 0$ and that equn 35 holds. This is equivalent to assuming that H_s has the same value as for pure water when $\Gamma_k^\sigma = 0$. It follows that in this case

$$\Gamma_k^\sigma (h_k^* - h_k^b) = \int_o^{\Gamma_k^\sigma} (h_k^\sigma - h_k^b)\, d\Gamma_k^\sigma \tag{48}$$

where the integration is performed at constant T,p and m_2. Hence the differential and integral enthalpies of adsorption are related in the same way as on the corresponding quantities for gas adsorption.

Similar methods to the above are applicable to surfactant binding by macromolecules and to micellisation. In particular the equation which describes the latter (13) is identical in form to equn 38.

REFERENCES

1. Gibbs, J.W. (1961). Scientific papers Vol.1 Dover Inc., N.Y.
2. Everett, D.H. (1950). *Trans.Faraday Soc.*, 46, 453 942, 957.
3. Hill, T.L. (1949). *J.Chem.Phys.*, 17, 520. (1950). 18, 246.
4. Anderson, P.J. and Pethica, B.A. (1956). *Trans.Faraday Soc.*, 52, 1080; Betts, J.J. and Pethica, B.A. (1957). *Proc.2nd Intern.Congr.Surface Activity*, 1, 152.
5. Schay, G. (1973). *J.Colloid and Interface Sci.*, 42, 478.
6. Monomura, K. (1974). *J.Colloid and Interface Sci.*, 48, 307.
7. Parsons, R. (1959). *Canadian J.Chem.*, 37, 308.
8. Guggenheim, E.A. (1940). *Trans.Faraday Soc.*, 36, 397.
9. Hansen, R.S. (1962). *J.Phys.Chem.*, 66, 410.
10. Kirkwood, J.G. and Buff, F.P. (1951). *J.Chem.Phys.*, 19, 774; Hall, D.G. (1971). *Trans.Faraday Soc.*, 67, 2516.
11. Hey, M.J., Shield, D.W., Speight, J.M. and Will, M.C. (1981). *J.Chem.Soc., Faraday I*, 77, 123.
12. Hurwitz, H.D. (1965). *J.Electroanal.Chem.*, 10, 35; Hall, D.G., Pethica, B.A. and Shinoda, K. (1975). *Bull.Chem.Soc., Japan*, 48, 324.
13. Hall, D.G., (1972). *J.Chem.Soc., Faraday II*, 68, 1439.

SURFACE ACTIVITY COEFFICIENTS

J.E. Lane

Department of Physical Chemistry
University of Melbourne
Parkville, Victoria,
Australia, 3052.

1. INTRODUCTION

Everett and his co-workers have made a number of important contributions to the study of surface activity coefficients. They have concentrated on relating surface activity coefficients with adsorption at the solid/solution interface. It is the intent of this contribution to consolidate these achievements, and to complement them with an examination of the relationship between the surface activity coefficients and the surface tension.

2. MODEL

The system consists of two phases, α and β, phase α containing only components A and B. A Gibbs dividing surface is defined by,

$$\Gamma_A^{(n)} + \Gamma_B^{(n)} = 0 \tag{1}$$

where $\Gamma_i^{(n)}$ is the surface excess of component i. The surface phase is bounded by two parallel surfaces, one of which lies in phase α a distance τ^α from the Gibbs surface, the other in phase β at a distance τ^β from the Gibbs surface. Using equation (1), the amount Γ_i^s of component i in unit area of surface phase is given by,

$$\Gamma_i^s = \Gamma_i^{(n)} + \tau^\alpha c_i^\alpha + \tau^\beta c_i^\beta \tag{2}$$

with c_i^α and c_i^β being respectively the amount of i in unit volumes of phases α and β. It is expected that $\tau^\alpha \approx \tau^\beta$ and in the remainder of this article it is assumed that for components A and B, $c_i^{\alpha} >> c_i^{\beta}$, so that the last term of equation (2) may be neglected. It is unnecessary to consider phase β any further, so the superscript α is omitted for quantities associated with phase α. A superscript s identifies surface phase quantities.

The surface mole fraction x_A^s is given by

$$x_A^s = \Gamma_A^s/(\Gamma_A^s + \Gamma_B^s) = \Gamma_A^{(n)}/\tau c + x_A \tag{3}$$

where $\tau = \tau^{\alpha}$ and $c = c_A + c_B$. The area a_m^s per mole of mixture in the surface phase is given by,

$$a_m^s = 1/\tau c \tag{4}$$

Combining equations (3) and (4) gives the surface excess of component A as

$$\Gamma_A^{(n)} = (x_A^s - x_A)/a_m^s \tag{5}$$

and a similar relation is applicable to component B.

If a_A^s and a_B^s are respectively the areas per mole for pure components A and B, it is assumed that τ varies with composition so that

$$a_m^s = x_A^s a_A^s + x_B^s a_B^s \tag{6}$$

The bulk phase activity coefficients are defined by

$$\mu_i = \mu_i^o + RT \ln f_i x_i \tag{7}$$

and

$$f_i(x_i = 1) = 1 \tag{8}$$

where μ_i is the chemical potential of component i, and μ_i^o the chemical potential when phase α contains only i, R is the gas constant and T the temperature. Everett (1) defines the surface activity coefficients f_i^s by,

$$\mu_i = \mu_i^o + RT \ln f_i^s x_i^s + (\gamma_i^o - \gamma)a_i^s \tag{9}$$

and

$$f_i^s(x_i^s = 1) = 1 \tag{10}$$

where γ is the surface tension and γ_i^o its value when phase α contains only component i.

Equations (7) and (9) may be combined, and rearranged to give

$$\gamma = \gamma_i^o + (RT/a_i^s) \ln (f_i^s x_i^s / f_i x_i) \tag{11}$$

Equation (11) is separately applicable to components A and B. Combining both and eliminating γ gives on rearrangement,

$$(1/r) \ln (f_A^s x_A^s / f_A x_A) - \ln (f_B^s x_B^s / f_B x_B) - \ln K = 0 \tag{12}$$

where,

$$\ln K = (\gamma_B^o - \gamma_A^o) a_B^s / RT \tag{13}$$

and,

$$r = a_A^s / a_B^s \tag{14}$$

The component with the larger surface area is chosen as A so that $r \geqslant 1$. If γ_A^o and γ_B^o are not known, but the complete adsorption isotherm is available, then K is obtained by integrating, from $x_A = 0$ to $x_A = 1$, the Gibbs relation,

$$\partial\gamma/\partial x_A = - (\Gamma_A^{(n)} / x_B) \, \partial\mu_A / \partial x_A \tag{15}$$

If equation (12) can be solved, then the adsorption may be calculated through equation (5) and the surface tension through equation (11). In order to solve equation (12) it is necessary to use a model to provide a relationship between the f_i^s and x_i^s. Using the superscripts m and e to distinguish model and experimental quantities, then regardless of the model used, equations (5), (11) and (12) require,

$$\Gamma_A^{(n),m} = \Gamma_A^{(n),e} = 0, \quad x_A = 0, 1 \tag{16}$$

$$\gamma^m - \gamma^e = 0, \quad x_A = 0, 1 \tag{17}$$

Define a function $\phi(x_A)$ by

$$\phi(x_A) = \gamma^m - \gamma^e \tag{18}$$

with the derivative function $\phi'(x_A)$ given by equation (15) as,

$$\phi'(x_A) = -(\Gamma_A^{(n),m} - \Gamma_A^{(n),e})(1/x_B)\ \partial\mu_A/\partial x_A \qquad (19)$$

Noting that by equation (17) $\phi(x_A) = 0$ at $x_A = 0, 1$, then by Rolle's theorem (from the calculus), the derivative $\phi'(x_A)$ must vanish at some $x_A = \xi$ in the interval between 0 and 1. The term $1/x_B$ is always positive, while the partial derivative $\partial\mu_A/\partial x_A$ is subject to the thermodynamic condition of phase stability that it always be positive, so the right hand side of equation (19) can only vanish if,

$$\Gamma_A^{(n),m}\ (x_A = \xi) = \Gamma_A^{(n),e}\ (x_A = \xi);\quad 0 < \xi < 1 \qquad (20)$$

Restricting discussion to models for which there is a single value of ξ, the condition that $\phi'(x_A = \xi) = 0$ indicates a maximum or minimum in $\phi(x_A)$ at this point. Regardless of whether it is a maximum or minimum, it is evident that the solution concentration $x_A = \xi$ produces perfect agreement between the model and experimental adsorption and the worst agreement for surface tension. If $\phi(x_A)$ has a maximum, then the model surface tension always exceeds the experimental value and vice versa for a minimum. A maximum is indicated by a negative of the second derivative $\phi''(x_A = \xi)$, or the equivalent that $\Gamma_A^{(n),m} < \Gamma_A^{(n),e}$ for $x_A < \xi$.

The bulk activity coefficients must satisfy the thermodynamic requirement,

$$x_A d\ln f_A + x_B d\ln f_B = 0 \qquad (21)$$

Everett (1) has shown that the surface activity coefficients must satisfy the analogous expression

$$x_A^s d\ln f_A^s + x_B^s d\ln f_B^s = 0 \qquad (22)$$

A model is now proposed in which the surface activity coefficients at a surface concentration $x_i^s = \alpha$ are given by,

$$f_i^s(x_i^s = \alpha) = \{f_i(x_i = \alpha)\}^t \qquad (23)$$

where t is a constant determined by the choice of components and the temperature. As the f_i already satisfy equation (21), the f_i^s will automatically satisfy equation (22). Examination of equation (23) reveals that when $t = 1$, the non-ideality of the surface phase is the same as for the bulk phase, while $t = 0$ indicates an ideal surface phase.

An adsorption isotherm can be obtained for every choice of t. According to the preceding analysis, each one will intersect the experimental isotherm at some $x_A = \xi(t)$, ξ being a function of t. By repeating the application of Rolle's theorem to a pair of model adsorption isotherms (with different values for t), it is found that the definition (23) results in a situation where all possible model isotherms for a given system have a common point at $x_A = \xi^m$ and $x_A{}^s = \eta^m$. The value of ξ^m is found as the solution of,

$$\{f_A(x_A = \xi^m)\}^{1/r} = f_B(x_A = \xi^m) \tag{24}$$

and ξ^m is determined solely by r and the properties of the bulk solution phase.

At $x_A = \xi^m$, the model surface tension is given by equations (11) and (23) as

$$\gamma^m = C + (RT/a_A{}^s)\ln\{f_A(x_A = \eta^m)\}^t \tag{25}$$

where $C = \gamma_A{}^o - (RT/a_A{}^s)\ln[\{f_A(x_A = \xi^m)\}^t \xi^m/\eta^m]$ is a constant. This result indicates that, at the model adsorption crossover point, the surface tensions are ordered according to the parameter t, increasing with t if $f_A(x_A = \eta^m) > 1$ and decreasing otherwise. The converse of Rolle's theorem requires that, if there is only one solution to equation (24), no pair of model surface tension isotherms will intersect for $0 < x_A < 1$. The majority of bulk mixtures have only one solution of (24), and therefore have model surface tension isotherms that are ordered according to the surface tension at $x_A = \xi^m$. In particular, the model surface tension isotherm with t = 1 lies above all isotherms with t < 1, if $f_A(x_A = \eta^m) > 1$, and above those with t > 1 if $f_A(x_A = \eta^m) < 1$.

Finally, if at some $x_A = \theta$, the bulk activity coefficients satisfy

$$\{(f_A)^{1/r}/f_B\}^{t-1}_{x_A=\theta} = K \tag{26}$$

the model adsorption isotherm will show an azeotropic point ($x_A{}^s = \theta$). The model can only satisfy (26) if the bulk solution is non-ideal and $t \neq 1$.

3. RESULTS

In all of the following diagrams, the experimental data are represented by open circles. The broken curves indicate the model predictions with t = 1, the dotted curves t = 0, and the full curves the best fitting value of t. The best fit is obtained by treating t as a parameter and minimizing the sum of the squares of the differences between model and experimental data.

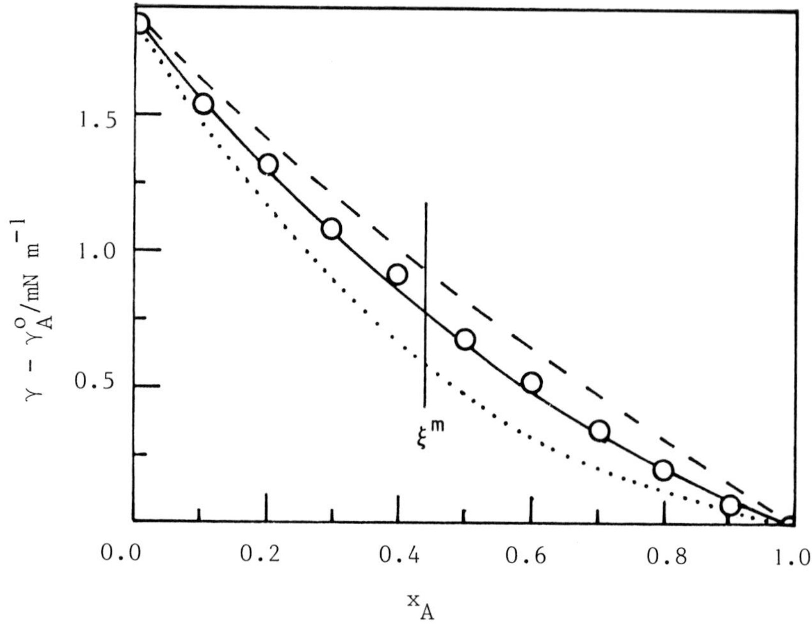

Fig. 1. Concentration dependence of surface tension for mixtures of cyclohexane (A) and carbon tetrachloride (B) at 298.15 K.

Figs. 1 and 2 respectively show the concentration dependence of surface tension and adsorption for mixtures of cyclohexane (A) and carbon tetrachloride (B) at a temperature of 298.15 K. The experimental surface tensions are those of Lane (2), while the "experimental" adsorption points were derived through the Gibbs adsorption isotherm, equation (15). The solution thermodynamic data is from Scatchard *et al.* (3). The molecules were assumed spherical

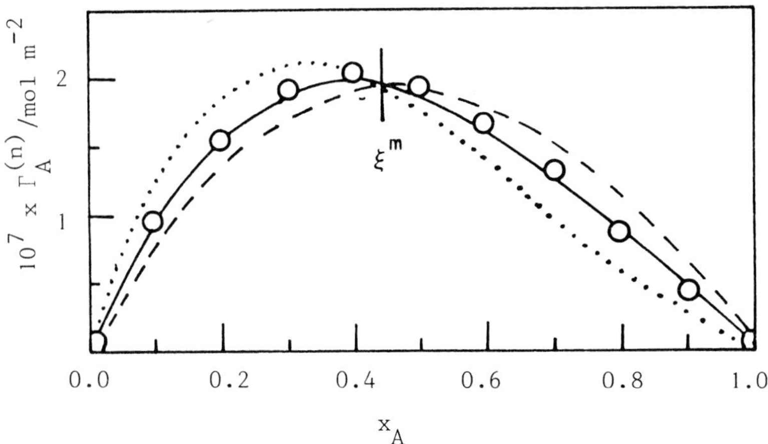

Fig. 2. Adsorption of cyclohexane (A) from mixtures with carbon tetrachloride (B) at 298.15 K.

so that $r = (V_A/V_B)^{2/3} = 1.08$, while the surface phase was assumed to be monomolecular in thickness giving $a_B{}^s = 1.95 \times 10^5$ m^2/mol. The value K = 1.15 was obtained from equation (13) using the difference in surface tension between the two pure components. The best fit between model and experiment is obtained with t = 0.58.

The experimental surface tension of mixtures of cyclohexane (A) and benzene (B) at the temperature of 298.15 K, as reported by Lam and Benson (4), are shown in Fig. 3. The solution thermodynamic data of Scatchard *et al.* (5) was used to compute the model surface tension. Again assuming spherical molecules, the molar volumes of the two components were used to obtain r = 1.14 and the monomolecular surface phase has $a_B{}^s = 1.84 \times 10^5$ m^2/mol, while K = 1.32. The best fit between model and experiment is obtained with t = 0.57. Molecular models suggest that benzene is far from spherical, and if the molecules lie flat in the surface phase, then they will occupy an area comparable with the cyclohexane molecules. Following this interpretation, and with r = 1, $a_B{}^s = 2.10 \times 10^5$ m^2/mol and K = 1.37, the best fit (not shown) between model and

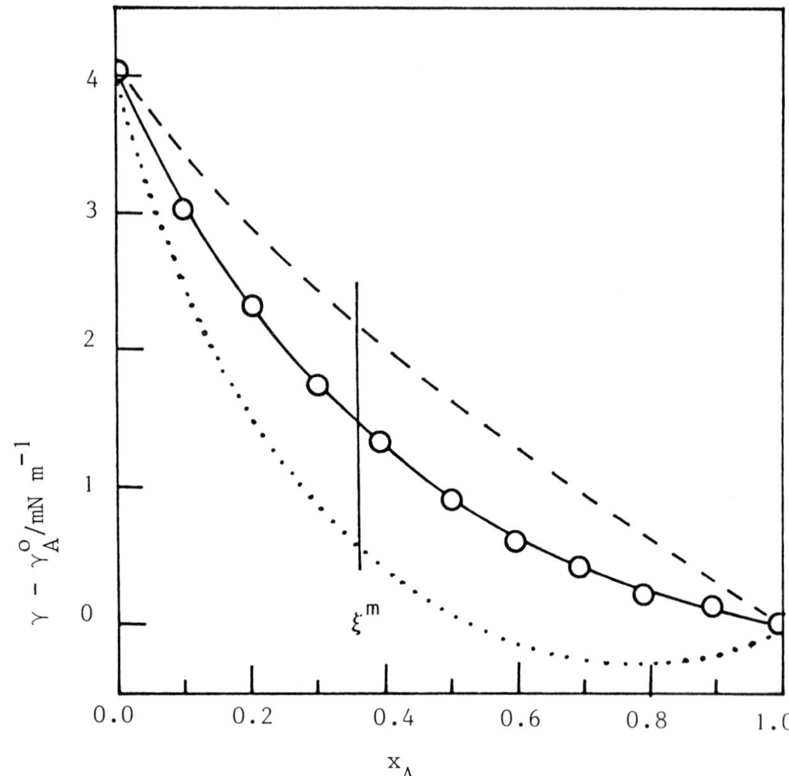

Fig. 3. Surface tension dependence on composition for mixtures of cyclohexane (A) with benzene (B) at 298.15 K.

experiment is given with t = 0.46. There is no essential difference between the goodness of fit between the two assumptions of molecular shape for benzene. Although the results are not shown here, the same procedure was followed for the surface tensions at a temperature of 303.15 K. The optimum values t = 0.54 were found for "spherical" benzene and t = 0.45 for "non-spherical" benzene in the flat orientation.

Jain *et al.* (6) have reported the surface tension for mixtures of n-heptane (A) and cyclohexane (B) at a temperature of 298.15 K, and these are shown in Fig. 4. In evaluating the model surface tensions, it was assumed that r = 1.42, which is the ratio obtained by vapour phase adsorption on graphon (7), and compensates for the non-spherical shape of the n-heptane. A monomolecular surface phase was assumed, with $a_B{}^s = 2.1 \times 10^5$ m^2/mol, and by using equation (13) K = 1.48. The thermodynamic data of

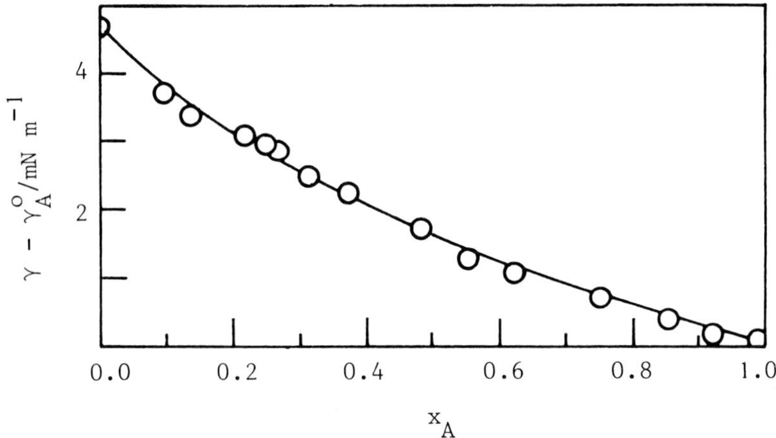

Fig. 4. Surface tension as a function of composition for mixtures of n-heptane (A) with cyclohexane (B) at 298.15 K.

Crutzen *et al.* (8) indicate that the solutions exhibit small, negative deviations from ideality at this temperature. Only the optimum model curve, with t = 4.3, is shown in the figure. The curves with t = 0 and t = 1 are very close to, and above, the best fitting curve, the curve with t = 0 being the highest.

The remaining figures all show adsorption at a solid/ solution interface. All experimental adsorption data are from Ash *et al.* (7, 9). The solid is graphon, with a surface area of 86 m^2/g. The values of the a_i^s (and hence r) are assumed to be the same as for monolayer adsorption at the graphon/vapour interface. The adsorption is reported as the experimental quantity $n^o \Delta x_i/m$; Δx_i being the change of mole fraction on adding n^o mol of mixture to m g of solid. With this particular adsorbent, $n^o \Delta x_i/m = 86\ \Gamma_i^{(n)}$. All values of K are obtained by combining the experimental adsorption data with the solution thermodynamic data and integrating equation (15).

The adsorption of benzene (B) from solutions with n-heptane (A) at a temperature of 283.15 K is shown in Fig. 5. The solution thermodynamic data of Brown and Ewald (10) was combined with r = 1.42, $a_B^s = 2.41 \times 10^5$ m^2/mol and K = 0.82 to produce the model adsorption curves. The best agreement between model and experiment was

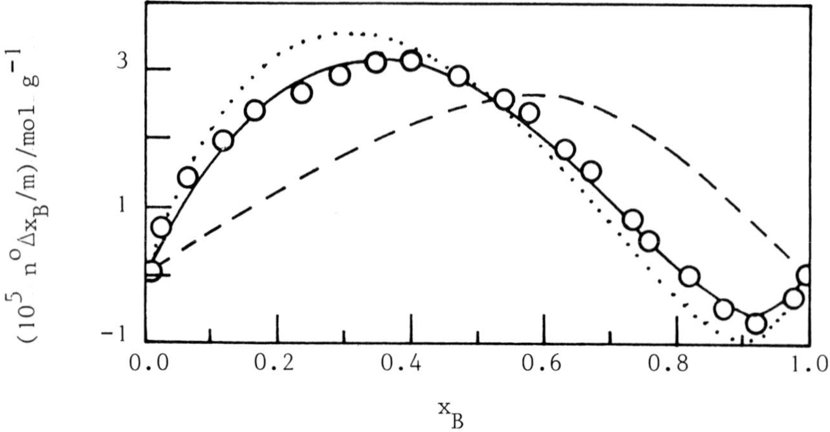

Fig. 5. Adsorption of benzene (B) for the system n-heptane (A)-benzene (B)/graphon at 283.15 K.

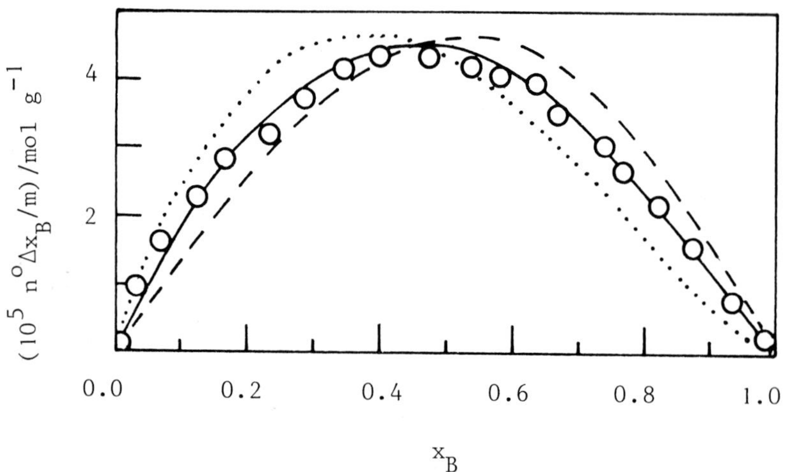

Fig. 6. Adsorption of benzene (B) for the system n-heptane (A)-benzene (B)/graphon at 343.15 K.

obtained with t = 0.29. Fig. 6 shows the same system, but at a temperature of 343.15 K. At this temperature, K = 0.66 and the best fitting curve is given by t = 0.53.

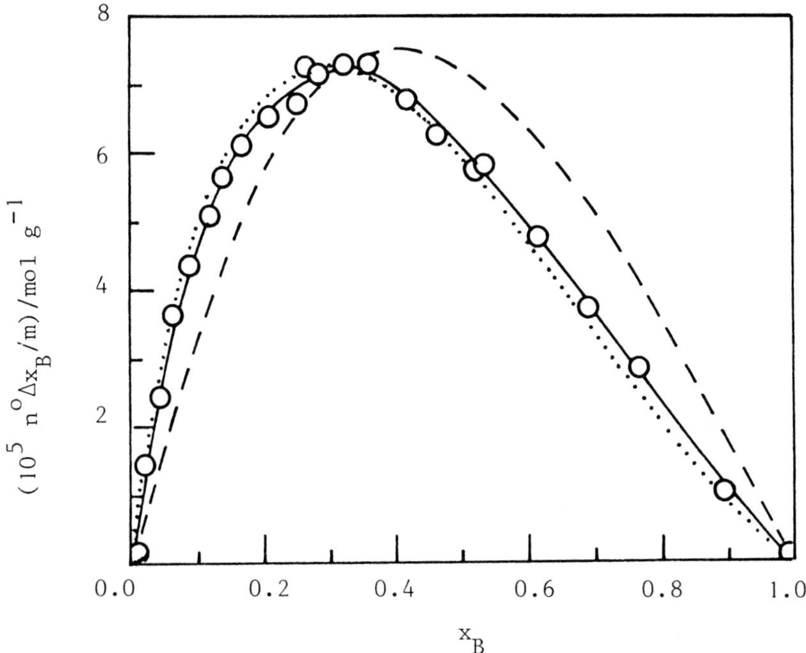

Fig. 7. Adsorption of benzene (B) from the system cyclohexane (A)-benzene (B)/graphon at 313.15 K.

Fig. 7 shows the adsorption of benzene (B) from mixtures with cyclohexane (A) at a temperature of 313.15 K. The model curves were produced by using r = 1, $a_B{}^s = 2.41 \times 10^5$ m^2/mol, K = 0.51 and the solution thermodynamic data (5). A value of t = 0.19 produced the best fitting curve.

The adsorption of n-heptane (A) from mixtures with cyclohexane (B) at temperatures of 283.15 K and 343.15 K is shown in Fig. 8. At both temperatures, r = 1.42 and $a_B{}^s = 2.41 \times 10^5$ m^2/mol. At the lower temperature K = 1.93, and as the solution thermodynamic data (8) indicates a pseudo-ideal solution ($G_m{}^E = 0$, $H_m{}^E \neq 0$), the model curve is independent of t. At the higher temperature K = 1.31, and although the solution is now non-ideal, the deviations from ideality are small. This has a tendency to crowd the

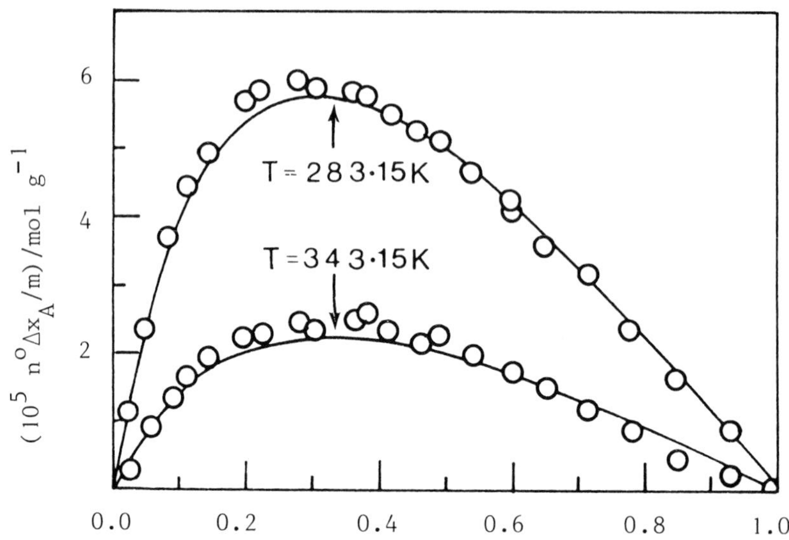

Fig. 8. Adsorption of n-heptane (A) from the system n-heptane (A)-cyclohexane (B)/graphon at temperatures of 283.15 K and 343.15 K.

model curves corresponding to different values of t, so only the best fitting (t = 2.2) curve is shown.

4. DISCUSSION

The results of the previous section show that, by choosing a suitable value of the parameter t, the model can reproduce the surface properties for a number of systems. Setting t = 1 gives the surface phase the same mixture properties as the bulk phase, whereas setting t = 0 results in an ideal surface phase. It is evident that neither choice of t is appropriate for any of the systems. Figs. 1-3 and 5-7 refer to solutions with positive deviations from ideality, and the optimum values of t range between 0 and 1. On this basis, it would be tempting, but erroneous, to speculate that surface phases are more ideal than their corresponding solution phases. In Figs. 4 and 8, the solution phases have negative deviations from ideality, and as the optimum t > 1, the surface phase has an even greater negative deviation from ideality. Nevertheless, in all systems, the optimum t produces surface activity coefficients that are less than the equivalent bulk activity coefficients. Noting the relationship between surface tension and surface activity coefficients derived in section 2, the following statement can be made.

"For all systems examined in this article, the experimental surface tension is less than that of the model surface tension with surface activity coefficients having the same concentration dependence as the bulk activity coefficients ($t = 1$)."

This result should not be a surprise, when related to the surface tension minimum requirement at thermodynamic equilibrium (11). The asymmetric molecular environment of the surface phase allows a higher probability of energetically favourable molecular orientations (and hence a lower surface tension) than the symmetric (average) molecular environment of the bulk phase.

It was shown in section 2 that the experimental and model adsorption isotherms must cross, irrespective of the value of t, but the same is not true for the surface tension isotherms. Furthermore, the concentration at the point at which the adsorption isotherms cross, is the concentration of maximum disagreement between the experimental and model surface tensions. These results are clearly illustrated in Figs. 1 and 2. Therefore, comparing model with experiment through the adsorption isotherm can give an illusion of better agreement than through the surface tension isotherm.

The arbitrary nature of the surface phase thickness is a weakness of this model. In Fig. 1, the optimum agreement between model and experiment is given by $t = 0.58$, when the surface phase is monomolecular. If the thickness is increased by a factor of 5.3, the optimum agreement is achieved with $t = 1$. There is little independent evidence to support such a thick surface phase.

REFERENCES

1. Everett, D.H. (1965). *Trans. Faraday Soc.* 61, 2478.
2. Lane, J.E. (1975). *Faraday Discuss. Chem. Soc.* 59, 55.
3. Scatchard, G., Wood, S.E. and Mochel, J.M. (1939). *J. Phys. Chem.* 61, 3206.
4. Lam, V.T. and Benson, G.C. (1970). *Can. J. Chem.* 48, 3773.
5. Scatchard, G., Wood, S.E. and Mochel, J.M. (1939). *J. Phys. Chem.* 43, 119.

6. Jain, D.V.S., Singh, S. and Wadi, R.K. (1974). *J. Chem. Soc., Faraday Trans. I.* 70, 961.
7. Ash, S.G., Bown, R. and Everett, D.H. (1975). *J. Chem. Soc., Faraday Trans. I.* 71, 123.
8. Crutzen, J.L., Haase, R. and Sieg, L. (1950). *Naturforsch* 52, 600.
9. Ash, S.G., Bown, R. and Everett, D.H. (1973). *J. Chem. Thermodyn.* 5, 239.
10. Brown, I. and Ewald, A.H. (1951). *Aust. J. Sci. Res.* A4, 198.
11. Gibbs, J.W. (1961). "Collected Works", Dover, New York, 1, 237-240.

WATER ON HYDROPHOBIC SURFACES

G. Lagaly, R. Witter, and H. Sander

Institut für anorganische Chemie der Universität
Olshausenstraße 40/60
2300 Kiel, Germany

ABSTRACT

Highly polar liquids (water, ethanol, FA, DMFA, DMSO) on silicate surfaces covered with alkylammonium ions are aggregated in clusters. This tendency increases with the chain packing density. Salts sensitively influence the cluster formations. In water and ethanol the effects follow the Hofmeister series in the opposite direction: The amount of adsorbed water increases, the amount of adsorbed ethanol decreases with increasing structure breaking. The adsorption of DMSO is sharply reduced at very low concentrations of breaking salts.

A model is proposed relating the hydrophobic/hydrophilic character of the surface to the clustering of the water molecules between the alkyl chains.

1. INTRODUCTION

Surfaces can be made hydrophobic by covering with amphiphilic long chain compounds. The critical alkyl chain length increases with the average chain distance. Long chain alkylammonium ions $C_nH_{2n+1}NH_3^+$ make the surface of mica-type layer silicates (vermiculites, beidellite, montmorillonites) hydrophobic (1) when the number n of carbon atoms in the alkyl chain is

$$n \geqslant 0.08 \cdot (2Ae).$$

The surface remains hydrophilic, when

$$n \leqslant 0.08\,(2Ae) - 2.$$

The value of 2Ae specifies the area on the surface (external basal surface) which binds one alkylammonium ion.

The outstanding character of the alkylammonium layer silicates is the intracrystalline swelling. Water and various kinds of organic liquids penetrate between the layers and increase the layer distance (basal spacing). From the studies reported one expects the interlayer water sorption to be impeded when the silicate becomes hydrophobic. However, quite contrarily, hydrophilic montmorillonites with short alkylammonium ions do not intercalate larger amounts of water, and hydrophobic montmorillonites with long alkylammonium ions do adsorb interlayer water. A first step in solving this challenging problem is studying the organization of the water molecules on the internal surfaces and between the alkyl chains.

2. MATERIALS and EXPERIMENTAL METHODS

Most of the studies were performed with vermiculite from South Africa, beidellite from Germany (B 17) and montmorillonites from USA (Wyoming, "Greenbond", M 40) and Germany (Niederschönbuch, M 39). Vermiculite is a high-charged mica-type layer silicate (average charge $\bar{\xi}$ = 0.75 eq/(Si,Al)$_4$O$_{10}$, surface charge density σ = 24.3 μCcm^{-2}). Beidellite has $\bar{\xi}$ = 0.42eq/(Si,Al)$_4$O$_{10}$ and σ = 14.5 μCcm^{-2}; the montmorillonites have $\bar{\xi}$ = 0.31 eq/(Si,Al)$_4$O$_{10}$, σ = 10.6 μCcm^{-2}.) The area per alkylammonium ion in the interlayer space (equivalent area) is A_e = 33 Å^2 (vermiculite), Ae = 57 Å^2 (beidellite) and A_e = 75 Å^2 (montmorillonites). On the external basal planes the area per alkylammonium ions is 2A_e (2,3).

The alkylammonium derivatives were prepared by ion exchange and then placed in contact with water and other highly polar liquids: formamide (FA), dimethyl formamide (DMFA), dimethyl sulphoxide (DMSO), ethanol (EtOH). The discussion of the interlayer organization is mainly based on basal spacing measurements and competitive adsorption studies (4).

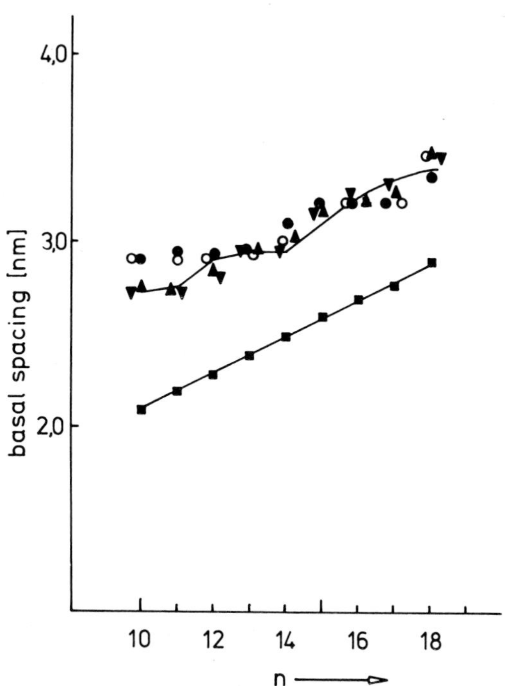

Fig. 1. Basal spacings of vermiculite with alkylammonium ions ($C_nH_{2n+1}NH_3^+$) on the surface (■) and under water (+), ethanol (x), FA (●), DMFA (▲), DMSO (▼) and 1:1-DMSO/H_2O (○).

3. RESULTS AND DISCUSSION

Liquids on the Internal Surfaces

Water expands the layer separation of alkylammonium vermiculites by 0.5 - 0.6 nm (Fig. 1). The changes were interpreted (Fig. 2) by reorientation of the alkyl chains from a 56°-orientation to the perpendicular followed by an additional expansion at shorter chains (n < 14).

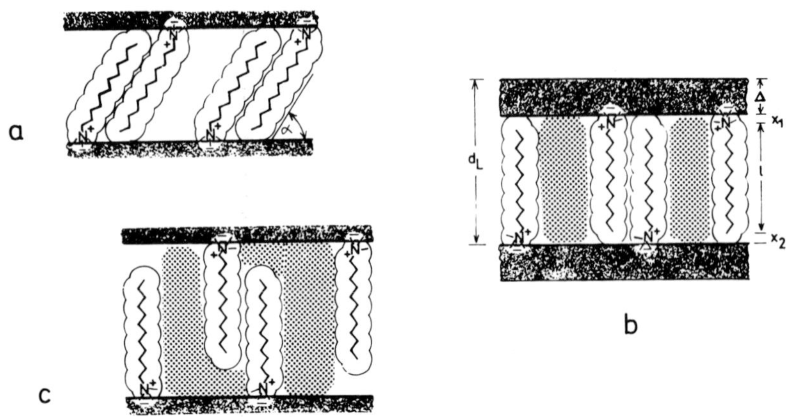

Fig. 2. Alkyl chains on the internal surfaces of vermiculite. a) V1(56°) structure, b) V1(90°) structure, after contact with liquids, n > 14, c) V1(90°)ex structure, after contact with liquids, n < 14.

The reason for the lattice expansion is seen in the formation of distinct water clusters surrounded by the alkyl chains. The volume of the clusters and, therefore, the number of water molecules per cluster is nearly independent of the alkyl chain length. Three unit cells contain two large cavities between the chains which can be filled up by water molecules (cf. Fig. 7 in (4)). A cluster then contains about 20 water molecules.

Compared with organic liquids water by no means reveals an exceptional behavior (Fig. 1). The basal spacing changes by DMFA, DMSO, and EtoH are comparable, only FA gives a distinctly larger expansion at short chains. The individual properties of the liquids are brought about at distinct chain lengths, for example at n = 10, 14, 18. At other chain lengths (n = 13, 16) the alkyl chains conceal the individual organization of the liquid clusters.

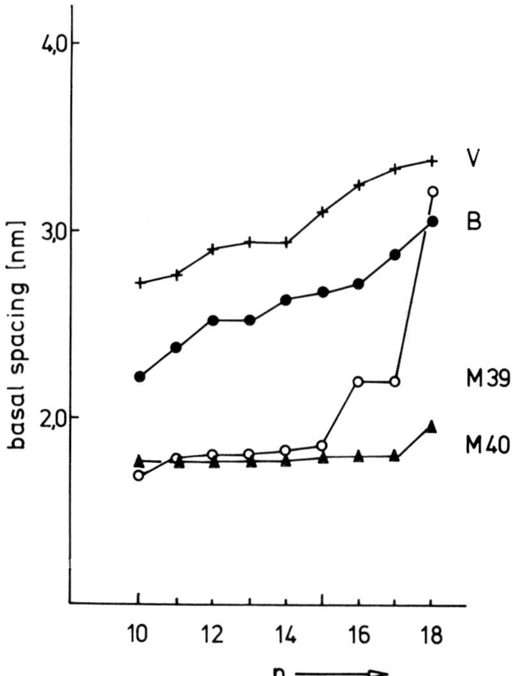

Fig. 3. Basal spacings under water. + vermiculite, ● beidellite, ○ montmorillonite (Germany), ▲ montmorillonite (Wyoming).

Increasing distance between the chains (vermiculite → beidellite → montmorillonites) shifts the basal spacings under water away from the organic liquids (Fig. 3) and augments the differences between the organic liquids. The montmorillonites only adsorb very small amounts of water.

In vermiculites most of the clusters are completely surrounded by alkyl chains. At lower surface charge densities (vermiculites → montmorillonites) some of the alkyl chains are removed and the clusters "flow" together. The liquid phase between the chains becomes more continuous and the prèsence of FA, DMFA and DMSO the structure V1(90°) is expanded. With water the interlayer separation decreases (beidellite) and finally collapses (montmorillonites). This leads to the conclusion that the height of the water clusters decreases with increasing chain distance. If the chains are located too distantly from each other, the water adsorption is strongly reduced.

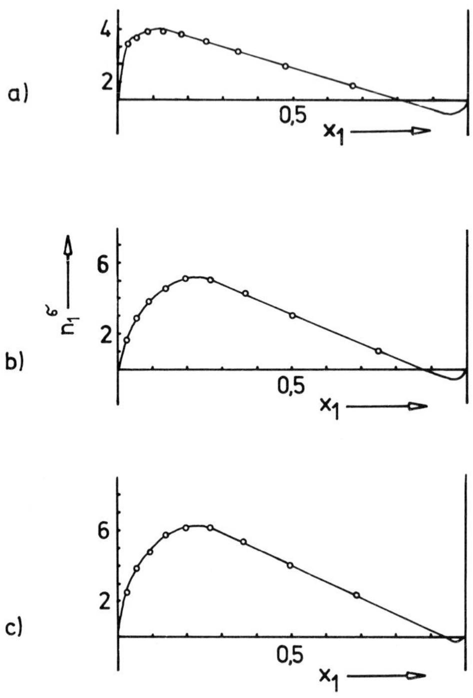

Fig. 4. Surface excess isotherms for DMSO (1)/H_2O (2) mixtures on surfaces with tetradecylammonium ions. χ_1: mole fraction DMSO, n_1^σ. surface excess (mmoles/g). (a) vermiculites, (b) montmorillonite (Germany), (c) montmorillonite (Wyoming)

Adsorption from Binary Liquids

Competitive adsorption from DMSO/H_2O mixtures on alkylammonium vermiculites gives type-IV-isotherms (Fig. 4, Shay-Nagy classification (5)). The isotherms show substantial linear sections which are indicative of extended plateaux in the individual isotherms. A constant ratio DMSO/H_2O in the interlayers persists through most of the mole fraction range of DMSO in solution (x_1(DMSO) = 0.1 - 0.95). The molar ratio DMSO/H_2O slightly increases with the alkyl chain length from 3.7 (n = 12) to 5.4 (n = 18). The independence of the interlayer composition of the mole fraction in solution clearly proves that distinct DMSO/H_2O clusters are built up between the chains.

The alkylammonium montmorillonites adsorb small amounts of water, but high amounts of DMSO. In contact with 1:1-DMSO/H_2O the spacings for $n < 13$ or 15 (Fig. 3) lie between the spacings under water and under DMSO. For longer chains the spacings of DMSO are approximated. It might follow that DMSO/H_2O-mixtures are intercalated at $n < 13$ or 15 and solely DMSO at longer chains. The surface excess isotherms, however, clearly establish that DMSO is preferentially adsorbed and that water is nearly completely excluded for all alkylammonium derivatives (Table 1).

The perceptible water exclusion may result from hydrophobic interactions with the alkyl chains. The spacings under DMSO/H_2O indicate V1(90°) or even its expanded form. If the chains stand upright, 30 percent of the internal surface of vermiculite is covered. In the case of montmorillonites the alkylammonium ions occupy 15 percent of the surface only, but water is more strongly excluded. The silicate surface undoubtedly acts as a hydrophobic surface. The hydrophilicity usually observed is apparently related to the presence of inorganic gegen ions. Thus, the surface supports the hydrophobic effect of the alkyl chains.

TABLE 1

Competitive Adsorption from DMSO/Water The silicate surface carries alkylammonium ions $C_nH_{2n+1}NH_3^+$ in distances increasing from vermiculite ⟶ montmorillonite. The amounts of adsorbed DMSO and H_2O are n_1° and n_2°, (mmoles/g).

n	vermiculite		beidellite		montm. (Germany)		montm. (USA)	
	n_1°	n_2°	n_1°	n_2°	n_1°	n_2°	n_1°	n_2°
12	4.4	1.2			8.3	0.2	8.9	0.3
14	4.7	1.1	6.9	2.3	10.9	0.3	8.8	0.5
16	5.0	1.0	7.5	1.4	9.4	0.5	9.7	0.0
18	5.4	1.0			9.4	1.2	10.2	0.2

Salt Effects

The proceeding results present the decisive role of liquid structuring on the lattice expansion. Salts affecting the liquid structure should influence the behavior of the liquids on the surface. In fact, manifold changes of the spacings were observed, if salts were added to the liquids.

In aqueous solution the spacings of alkylammonium beidellites generally increase with the salt concentration (Fig. 5). The maximal changes are presented in Figure 6. For comparison, the structure breaking and structure making effects of the anions ("Hofmeister series") are illustrated by the clouding points (turbidity points) of a polyethylene glycol in water. Perchlorate gives the highest clouding point and has the highest "structure temperature". It is the most effective structure breaking anion (6-8). The monovalent anions increase the spacing in order of the Hofmeister series, perchlorate gives the highest spacing.

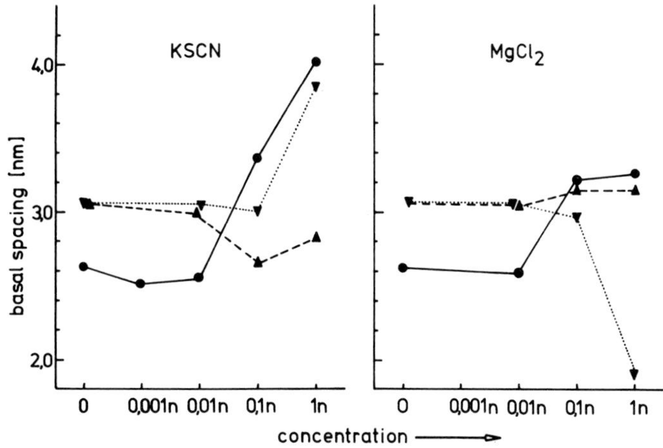

Fig. 5. Basal spacing changes of tetradecylammonium beidellite under solutions of KSCN and $MgCl_2$ in water (●), ethanol (▲) and DMFA (▼) (highest concentration: 1n or saturated).

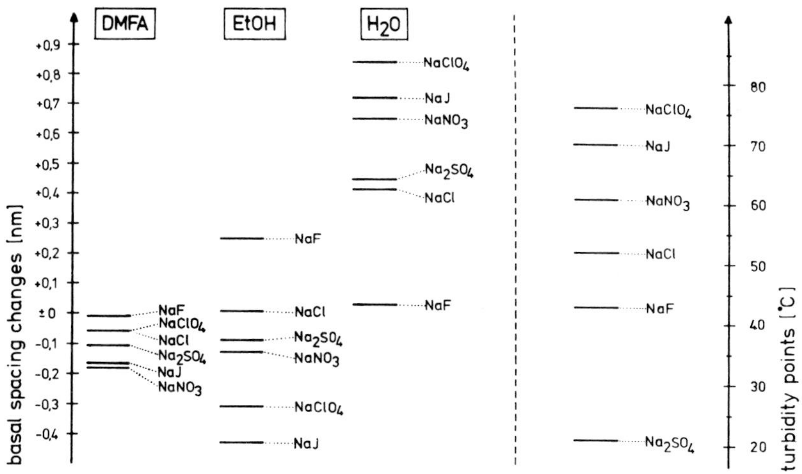

Fig. 6. Basal spacing changes of tetradecylammonium beidellite in sodium salt solutions. On the right: the turbidity points of ethoxylated p-iso octylphenol (6,7).

TABLE 2

Effect of salts on the DMSO Adsorption by Tetradecylammonium Montmorillonite (Wyoming, USA)

salt	Basal spacings (nm) conc.: 1n*)	0.1n	0.01n	0.001n
Na_2SO_4	3.27	3.21	3.21	3.21
NaF	3.21	3.21	3.21	3.21
LiF	3.16	3.21	3.21	3.26
NaCl	1.86	3.21	3.21	3.27
$MgCl_2$	1.86	1.86	3.54	3.21
$BaCl_2$	1.92	1.86	3.21	3.21
$NaNO_3$	1.88	1.88	3.46	3.21
NaI	1.92	1.86	3.27	3.21
$NaClO_4$	1.92	1.86	3.40	3.27
NaSCN	1.80	1.88	3.21	3.27
KI	1.86	1.86	1.86	3.33

*) or saturated

In organic liquids addition of salts increases or decreases the basal spacings (Fig. 5). Very simply behaves tetradecylammonium montmorillonite in DMSO-salt solutions (table 2). Strongly structure promoting salts (Na_2SO_4, NaF, LiF) do not affect the spacings. Structure breaking salts (from NaCl → KI) impede the intercalation of DMSO above a critical concentration. The very low lattice expansion then shows that only some "isolated" DMSO molecules are intercalated. Potassium iodide impedes the DMSO-adsorption even in 0.01 m solution! The salts are effective in concentrations distinctly below those of other systems. The swelling of the alkylammonium silicates responds very sensitively to salt effects.

In ethanol the spacings follow the Hofmeister series. Sodium iodide and perchlorate as strongly breaking salts (highest structure temperature) give the lowest basal spacings. The salt effects in every solvent do not agree with the Hofmeister series. In DMFA the basal spacing changes are small and not in the order of the Hofmeister series.

Introduced is the classification into "structure breaking" and "structure promoting" ions for organic liquids in the same sense as for water. The term "cluster" should imply that the liquid molecules between the chains assume a distinct organization (6,9). Breaking ions are ions which destroy or disorder this organization.

Structure promoting effects in organic solvents are well established (10-13). Engel and Hertz (14) concluded from nuclear magnetic resonance studies and Abraham et al. (13) from entropy calculations that breaking effects are less pronounced and probably restricted to distinct solvents, for instance glycerol and ethylene glycol . Our experiments can be best interpreted by assuming that structure breaking effects are also operative in organic liquids.

The interesting feature of Figure 6 is the pronounced increase of the spacings when structure breaking ions are added to water, and the decrease when they are added to ethanol and DMSO. The volumes of the clusters are relatively small. When structure breaking ions are incorporated, then:

— the clusters are destroyed, or

— if the clusters cannot be destroyed, the volume of the intercalated phase is increased until clusters and domains with increased disorder around the breaking ions simultaneously can exist. In organic solvents the first mechanism is operative: the breaking ions destroy the clusters and the number of intercalated guest molecules is largely reduced. In aqueous solutions the hydrophobic interaction strongly stabilizes the clusters and the breaking ions are intercalated together with additional water molecules.

The structuring of the liquids on the surface then turns out to result from a delicate balance between the intermolecular interactions in the liquid (liquid structure), the structure promoting or breaking effects of the ions and the influence of the alkyl chains (hydrophobic interactions). The latter depends on the length and packing density of the chains and the type of charge distribution in the surface.

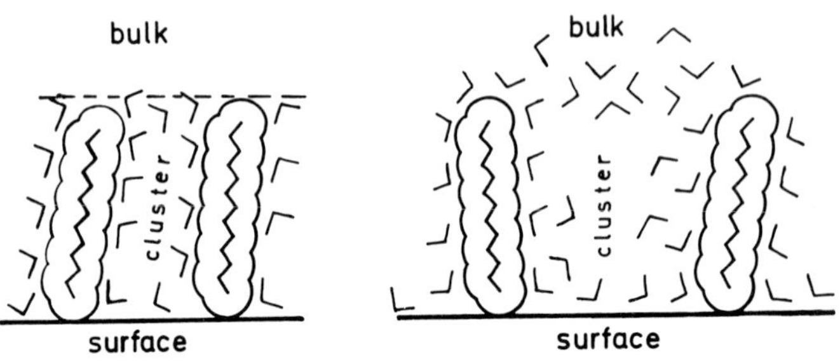

Fig. 7. Clusters of water molecules between alkyl chains on silicate surfaces: a) chains at small distances, strongly organized clusters: hydrophobic surface b) chains at larger distances, "open" clusters: hydrophilic surface.

4. CONCLUSION

Why is a surface with long alkyl chains hydrophobic? The experiments of Weiss (1) prove that not required is the complete coverage of the surface with alkyl chains. The clusters between the chains on the external surfaces are in contact with the bulk phase. One might speculate that the surface phase behaves hydrophobically when clusters by influence of the chains are organized so strongly that a discontinuity is created in the bulk water. This requires sufficiently close-packed chains exceeding a critical length. If the chains are arranged more distantly from each other, the strong aggregation of the water molecules around the chains is lost at some distance from the chains. In this region the molecules assume a more random orientation which destroys the discontinuity in the bulk phase and the surface becomes hydrophilic (Fig.7). To maintain the regular orientation within the whole cluster and the discontinuity to the bulk phase the chain length must be increased. Thus, the hydrophobicity is related to the formation of strongly organized clusters between the chains.

REFERENCES

1. Weiss, A. (1966). *Kolloid Z. Z. Polymere,* 211, 94-97.
2. Lagaly, G. (1981). *Clay Min.* 16, 1-21.
3. Lagaly, G. (1981). *Naturwiss.* 68, 82-88.
4. Lagaly, G., and Witter, R. (1982). *Ber. Bunsenges. physik. Chem.* 86, 74-80.
5. Kipling, J.J. (1965). "Adsorption from Solutions of Non-Electrolytes." Academic Press, London, New York.
6. Luck, W.A.P. (1976). *In* "The hydrogen Bond. III. Dynamics, Thermodynamics and Special Systems." P. Schuster, G. Zundel, C. Sandorfy (eds) North-Holland Publ. Co. Amsterdam, 1368-1423.
7. Luck, W.A.P. (1978). *Progr. Colloid Interf. Sci.,* 65, 6-28.
8. Luck, W.A.P., Schiöberg, D., and Siemann, U. (1980), *J.C.S. Faraday II,* 76, 136-147.
9. Wicke, E. (1966). *Angew. Chem.* 78, 1-19.
10. Hertz, H.G. (1970). *Angew. Chem.* 82, 91-106.
11. Ebert, G. and Wendorff, J. (1970). *Ber. Bunsenges. phys. Chem.* 74, 1072-1077.
12. Kruus, P. and Poppe, B.E. (1979). *Can. J. Chem.* 57, 538-551.
13. Abraham, M.H., Liszi, J., and Papp, E. (1982) *J. Chem. Soc., Faraday Trans. I,* 78, 197-211.
14. Engel, G. and Hertz, H.G. (1968). *Ber. Bunsenges. phys. Chem.* 72, 808-834.

Factors Influencing The Adsorption From Solutions

Bun-ichi Tamamushi

Nezu Chemical Institute
Musashi University, Nerimaku
Tokyo, 176, Japan

1. INTRODUCTION

The adsorption from solutions generally takes place at any interface of liquid/gas, liquid/liquid or liquid/solid. If we arbitrarily confine the phenomenon to the case of liquid/solid interface, we still deal with the system of at least three components, i.e. adsorptive (a single solute), adsorbent(solid) and solvent(liquid medium), where the interaction forces are not only adsorptive/adsorbent, adsorptive/solvent, solvent/adsorbent, but are also adsorptive/adsorptive. The phenomenon under consideration is therefore very complicated and difficult to analyse completely. For the study of the problem it is accordingly desirable to select certain conditions: namely, to select the combination of adsorptive, adsorbent and solvent of definite natures and to find out certain main factors which control experimental data. From such a point of view, some classical and some recent data of the adsorption from solutions at the liquid/solid interfaces will be reviewed in this paper.

2. REVIEW

Many classical works of the adsorption from solutions were carried out with the system: relatively simple inorganic or organic, ionic or non-ionic compounds as adsorptive, charcoal of various origins as adsorbent, and aqueous solutions as liquid medium. For example, in the study of the adsorption of some homologous organic compounds(carbonic acids, urethanes, etc.) from their aqueous solutions onto blood charcoal, a regularity in the order of adsorption amounts, as known TRAUBE's rule, was found, which indicates that the attraction forces between adsorptive molecules and adsorbent surface play an important role(1). It was however later found that this rule does not hold, and that the adsorption order is even reversed, when toluene instead of water as solvent and silica or alumina as adsorbent were applied(2). This indicates that the adsorption depends on the nature of solid surface and of the solvent medium. As well known, polar adsorbents such as alumina or silica, as well as organic solvents of different natures have been successfully applied in the chromatographic adsorption method(3).

In an early study of the present author on the adsorption of some isomeric bisubstituted benzenes (ortho-,meta-,para-nitroanilines,-nitrophenols) from benzene or methanol solutions onto charcoal or alumina, it was found that meta-compounds with greater dipole moments were more strongly adsorbed by polar adsorbent from non-polar benzene solutions than ortho-compounds with smaller dipole moments, while the former compounds (meta) were more weakly adsorbed by charcoal than the latter (ortho) from methanol solutions(4). Such experimental results suggested that the polarity of the adsorbent (solid surface) as well as the polarities of liquid medium and adsorptive molecules are important factors in controlling the adsorption from solutions.

In another study of the present author on the adsorption of some ionic surfactants (long paraffin chain electrolytes); like dodecylammonium chloride, dodecylpyridinium bromide, sodium dodecyl sulphate, from their aqueous solutions onto various adsorbents; alumina, barium sulphate, calcium fluoride, carbon black, ion-exchange resins, it was found that the adsorption isotherms are generally S-shaped and that the second maxima of the adsorption isotherms appear at equilibrium concentrations of the bulk solutions corresponding to the critical micelle concentrations of the respective surfactants(5). It was considered that the adsorption of these amphiphilic compounds would take place in two steps; first, monomolecular adsorption of the LANGMUIR-type, and then, multi-layer adsorption of the BRUNAUER-EMMETT-TELLER-type. At the surface of polar adsorbent, like alumina, polar groups of surfactant molecules will be attracted to the active spots of the surface by the electrostatic attraction forces, forming oriented monomolecular layers and then bi-molecular layers by the VAN DER WAALS attraction forces between non-polar chains of surfactant molecules. It was also remarked that the adsorption of these surfactants on ionic resins took place in two steps; first due to the exchange reaction and then due to VAN DER WAALS adsorption. These experimental results suggested that there appears the role of the interaction forces among adsorptive molecules, in determining the whole process of the adsorption from solutions. Critical reviews on the problem of the adsorption of such surfactant molecules at solution/solid interfaces have been made in some details by the present author as well as by KIPLING in their monographs(6). K.EDA, former collaborator of the present author, in a series of papers published in Japanese(7), reported his studies on the adsorption and desorption processes of paraffin chain compounds at the solution/mercury interface by applying electric differential capacity method, which was particularly developed by GRAHAME(8). It was therefore suggested, from the characteristic curves of C(differential capacity)-E(applied potential),particulary, for paraffin chain anions like dodecyl sulphate ion, that they are adsorbed at the mercury/solution interface, forming bi-molecular layers by a similar mechanism as described above for the case of surfactant ions at the solid/solution interface, provided that the solution concentrations are greater than the critical micelle concentrations of the surfactant ions.

In a more recent work on the molecular alignment of some liquid crystals: N-methoxybenzylidene-p-buthylaniline(MBBA), a mixture of 4-ethoxy-4'-buthylazoxybenzene with 4-ethoxy-4'-ethylbenzene (NP-5A), in the thin films of these compounds sandwiched between two optical glass or quartz plates, whose surfaces were previously covered by hexadecyltrimethylammonium bromide(HTAB) or lecithin, it was confirmed that their molecular alignment is homeotropic, being confirmed qualitatively, by use of a polarization microscope, and confirmed semi-quantitatively, by applying the resonance RAMAN-spectroscopy(9). It was therefore considered that the amphiphilic molecules of HTAB or lecithin will be adsorbed on the clean surface of glass or quartz by the polar-polar attraction forces, forming monomolecular layer, hydrophobic groups being directed outward, and then molecules of liquid crystals will be attracted to this monolayer by the hydrophobic-hydrophobic attraction forces, and thus the homeotropic structure may be realized.

Those results cited above (5.,7.,9.) therefore suggest that the interaction forces between adsorptive/adsorptive play a role in controlling the whole process of the adsorption from solutions.

Finally, the results so far obtained recently on the adsorption of lecithin from solutions will be reported in the following. The systems studied were: I) adsorption of lecithin by alumina or silica from hexane solutions; II) adsorption of lecithin by alumina or silica from aqueous solutions. Materials used: Egg-lecithin, the product of MERCK. Darmstadt alumina, the BROCKMANN's chromatographic alumina of MERCK.,Silica, the special product of Kanto Chem. Co.,Tokyo; Hexane, extra pure sample of Kanto Chem. Co.,Tokyo; and water was twice distilled. The adsorbents: alumina and silica were washed with conductivity water till the conductivity of the washing water was unchanged and then dried carefully. The procedure of the adsorption experiment applied was the usual one carried out at temperature 30°C. The analysis of lecithin in solutions was made by measuring the extinction coefficient of absorption maximum at 280 nm in hexane solutions and at 400 nm in aqueous solutions by applying HITACHI Model 100-10 spectrophotometer. The specific surface areas of the adsorbents were determined by the method of stearic acid adsorption from hexane solution with the result: 41sq. m/g for alumina, and 32.8sq. m/g for silica, respectively. These values are generally smaller than those determined by N_2-adsorption method(for example, 78.8sq. m/g for the same sample of alumina).
In this study, however, the former values were

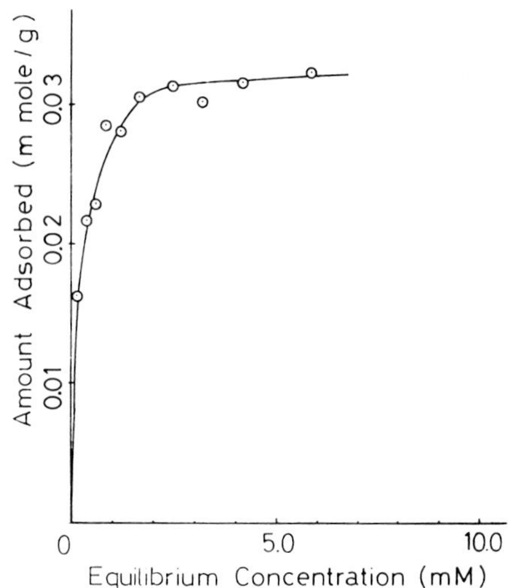

Fig. 1. Adsorption isotherm for lecithin at alumina/hexane interface at 30°C.

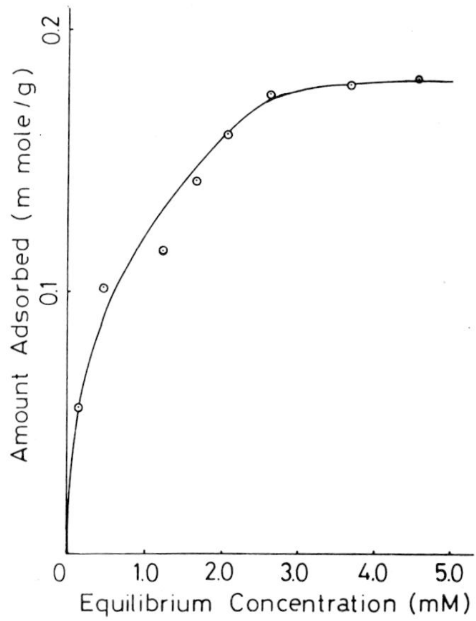

Fig. 2. Adsorption isotherm for lecithin at silica/hexane interface at 30°C.

used, considering that the adsorption experiment was done also in solutions. Adsorption isotherms obtained are shown in Figures 1--2 for alumina-, silica-hexane solution systems; and in Figures 3-4 for alumina-, silica-aqueous solution systems, respectively. As noticed in Fig. 1 and 2, the adsorption isotherms obtained for hexane solutions are the type of LANGMUIR-isotherm, while those for aqueous solutions shown in Fig. 3 and 4, are the type of BRUNAUER-EMMETT-TELLER(BET)-isotherm. It is also noticed that the adsorption amount reaches saturation in the case of Fig. 1,2, and the first maximum point in the case of Fig. 3,4 at about 3-5 mM of the bulk solutions.

If we calculate the molecular areas of the adsorbed lecithin, by combining these maximum amounts of adsorption with the values of specific surface areas for alumina and silica above obtained, we obtain the following approximate values for each case: 140sq. Å for the system; alumina/hexane solution; 30sq. Å for the system; silica/hexane solution; 1350sq. Å for the system; alumina/aqueous solution; and 270sq. Å for the system; silica/aqueous solution; respectively. From the former two values for the adsorption from hexane solutions, it seems probable to assume the monomolecularity of the adsorption layer required by the LANGMUIR-equation, which actually fits the isotherms illustrated in Fig. 1 and 2. The above obtained values of the molecular area for the system: alumina/ and silica/aqueous solutions are, however, too large for assuming the mutual interaction forces among adsorbed molecules to build up bi- or multi-layers at these interfaces. Very small amounts of adsorption at those interfaces might be due to the adsorption of solvent (water) (10). In order to make a further discussion of the problem, it is necessary to have more accurate experimental data for those systems involving aqueous lecithin solutions. It is also desirable to develop a theory of adsorption which explains the BET-type adsorption isotherms at the solid/solution interface. As well known, the BET-theory has been originally developed for the gas/solid interface, but not for the solution/solid interface(11).

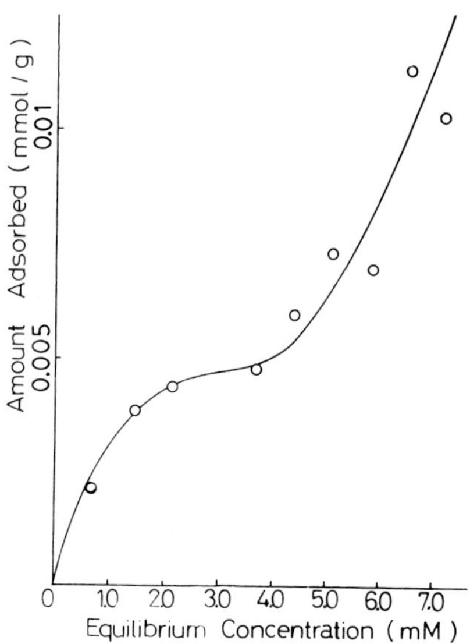

Fig. 3. Adsorption isotherm for lecithin at alumina/aqueous solution interface at 30°C.

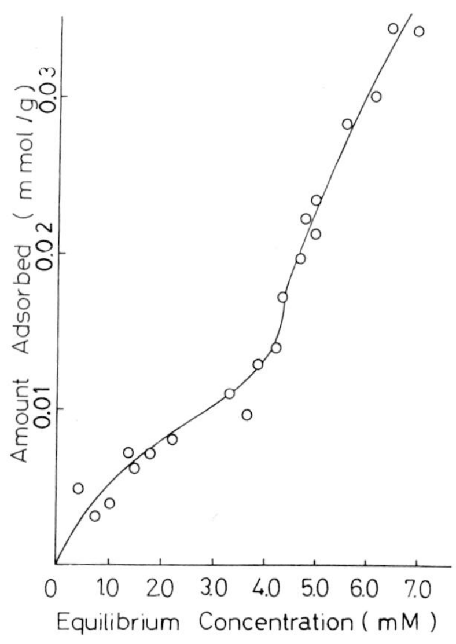

Fig. 4. Adsorption isotherm for lecithin at silica/aqueous solution interface at 30°C.

Acknowledgment

The author's thanks are due to Dr. N.Watanabe, MM. T.Ogawa and K.Onishi for their technical assistance in preparing this paper.

References

1. Freundlich,H.(1930)."Kapillarchemie".Bd.1,pp. 258. Acad.Verlag, Leipzig.
2. Holms,H.N. & Mckelvey.(1928). *J.Phys.Chem.* 32, 1522.
3. Zochmeister,L. & Cholnoky,V.(1938)."Chromatographische Adsorptionsmethode". Springer, Wien.
4. Tamamushi,B. (1931). *Bull.Chem.Soc.Japan* 6,76; (1941),*Sci.Pap.Inst.Phys.Chem.Research* 38,446. (Tokyo).
5. Tamamushi,B. & Tamaki,K. (1957)."Proc.2nd.Int. Cong.Surface Activity". Vol.3,pp. 449. Butterworth,London; (1959). *Trans.Farad.Soc.* 55, 1007-1013.
6. Shinoda,K.,Nakagawa,T.,Tamamushi,B. & Isemura, T. (1963)."Colloidal Surfactants". Chap.3. Acad.Press, New York & London.; Kipling,J.J. (1965)."Adsorption from Solutions of Non-electrolytes". pp. 268. Acad.Press, London & New York.
7. Eda,K. (1959). *J.Chem.Soc.Japan* 80, 343,347, 349,461,465,708; (1960).*ibid.* 81, 689,875.; ref. Shinoda,K.,Nakagawa,T.,Tamamushi,B. & Isemura,T. (1963)."Colloidal Surfactants". Chap.3. Acad.Press, New York & London.
8. Grahame,D.C. (1947). *Chem.Revs.* 41,441.
9. Yamada,H.,Fukumura,K. & Tamamushi,B. (1980). *Bull.Chem.Soc.Japan* 53,3054.
10. Rideal,E.K., (1930)."An Introduction to Surface Chemistry". 2nd Ed.,Chap.4. Univ.Press, Cambridge.
11. Brunauer,S. (1945)."The Adsorption of Gases and Vapours, I Physical Adsorption". Chap.6. Univ.Press, Oxford.

ADSORPTION OF DECAN-1-OL FROM HEPTANE AT THE SOLUTION/GRAPHITE INTERFACE

G.H. Findenegg, C. Koch and M. Liphard

Institute of Physical Chemistry
Ruhr-University Bochum
4630 Bochum, West Germany

1. INTRODUCTION

In this work the adsorption of higher *n*-alkanols from *n*-heptane onto graphitized carbon black (g.c.b.) is studied by measuring the surface excess amount and enthalpy of displacement over a range of concentrations (up to 3 mole percent) and a range of temperatures (15 - 35°C). The isotherms of these systems exhibit a pronounced *step* indicating a strongly cooperative adsorption mechanism which leads to a close-packed monolayer of alkanol molecules oriented with their chain axis parallel to the graphite basal plane. This unusual adsorption behaviour in dilute non-aqueous solutions is related to an order-disorder transition of the pure higher alkanols at the liquid/graphite interface at temperatures 40 - 50 K above their melting point which has been attributed to the *melting* process of a monolayer of alkanol molecules (1). Several interesting aspects of this interfacial phase transition, particularly its influence on the adsorption from binary mixtures, have been studied in Everett's laboratory during the past decade (2 - 4). The present work is concerned with the region of dilute solutions in which the stepwise adsorption behaviour can be treated on the basis of a simple model.

2. EXPERIMENTAL METHODS AND MATERIALS

Enthalpies of displacement were determined by flow calorimetry using the flow unit of an LKB Type 2107 microcalorimeter. The reaction cell was loaded with about 0.2 g of adsorbent and pure solvent was passed through the cell at flow rates between 10 and 20 ml/hr. The displacement reaction was initiated by replacing the solvent with a solution of the desired final concentration, either in a *single step* or by the *cumulative method, i.e.*, a series of small concentration steps. For the present systems these two methods did not generally yield the same result, as will be discussed below.

Surface excess amounts were determined by the solution depletion method using a closed circuit apparatus which enables solutions of known initial composition to be circulated through the system. Changes in solution composition were measured by a Vibrating Tube Densitometer DMA 02 by Paar. This method is particularly well-suited for the determination of changes in adsorption with temperature and with small changes in solution composition.

Materials. Vulcan 3G, an oil furnace black supplied by the Cabot Corporation and graphitized at 2800°C, was taken from the same lot and pretreated in the same way as in previous studies (5, 6). Its specific surface area was 68 m^2/g.

Decan-1-ol (Aldrich Chemicals) had a purity of > 99 % (GC). *n*-Heptane (J.T. Baker, > 99 %) was distilled over a column before use.

3. RESULTS

Isotherms of the specific surface excess amount $n_2^{\sigma(n)}$ of the system (heptane + decanol)/g.c.b. at five experimental temperatures are shown in Fig. 1. They were analysed on the basis of a *monolayer model* by assuming that the adsorbed layer is confined to a single layer of molecules oriented parallel to the solid surface. In this case the fraction of surface covered by solute is given by

$$\phi_2^a = \frac{n_2^{\sigma(n)}}{n_{2,o}^a} \frac{1}{x_1^1 + rx_2^1} + \phi_2^1 \tag{1}$$

where $\phi_2^1 = rx_2^1/(x_1^1 + rx_2^1)$ is the volume fraction of solute in the bulk solution. The specific monolayer capacity $n_{2,o}^a$ was derived from the surface excess isotherms as outlined elsewhere (6). The resulting value (166 μmol/g) corresponds to a cross-sectional area of decanol molecules, $a_{2,o} = 0.68\ nm^2$, in close agreement with a value derived from a geometrical model proposed by Groszek (7) according to which each $-CH_2-$ group of an unbranched chain molecule occupies an area of one hexagon of the graphite basal plane

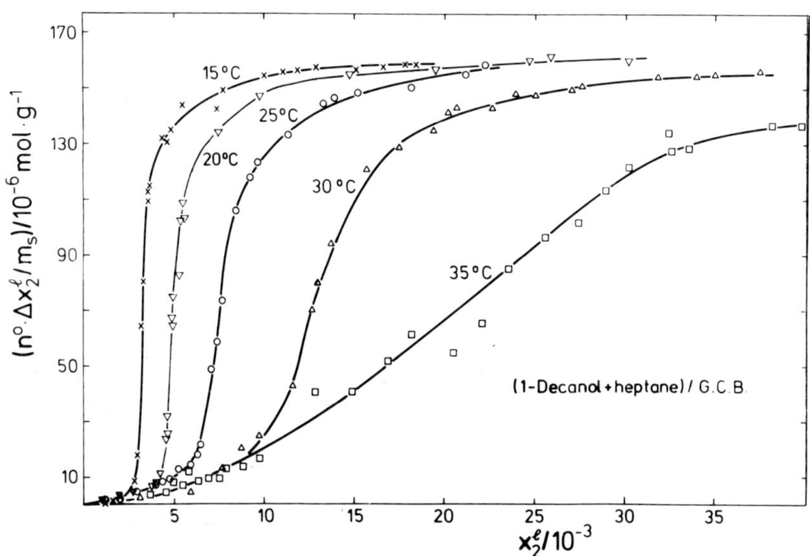

Fig. 1. Adsorption isotherms at various temperatures of the system (heptane + decan-1-ol)/graphitized carbon black.

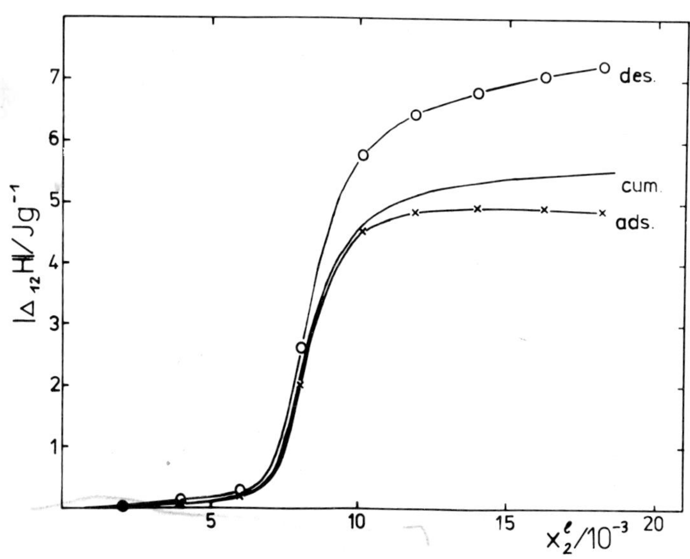

Fig. 2. Enthalpy of displacement isotherm (25°C) measured by the single step adsorption (ads.), single step desorption (des.) and by the cumulative method (cum.).

(0.0524 nm^2) while $-CH_3$ (and -OH) covers twice that area. For the mutual displacement factor this model yields $r = a_{2,o}/a_{1,o} = 13/9$.

Isotherms of the specific enthalpy of displacement $\Delta_{12}H$ obtained by single step adsorption (ads.), cumulative adsorption (cum.) and by single step desorption (des.) (i.e., elution with pure solvent) under otherwise similar experimental conditions are shown in Fig.2. The large differences between these results are caused by dilution effects in the liquid phase, due to the large (positive) excess enthalpy of the heptane + decanol system: (a) partial mixing occurs when the liquid in the calorimeter cell is replaced by a solution of different concentration; this effect can be made negligibly small in the cumulative method; (b) adsorption from solution involves an intrinsic dilution of the bulk solution. Whereas in batch measurements this effect can be accounted for by a suitable correction term, such a correction cannot be applied in an unambiguous way in

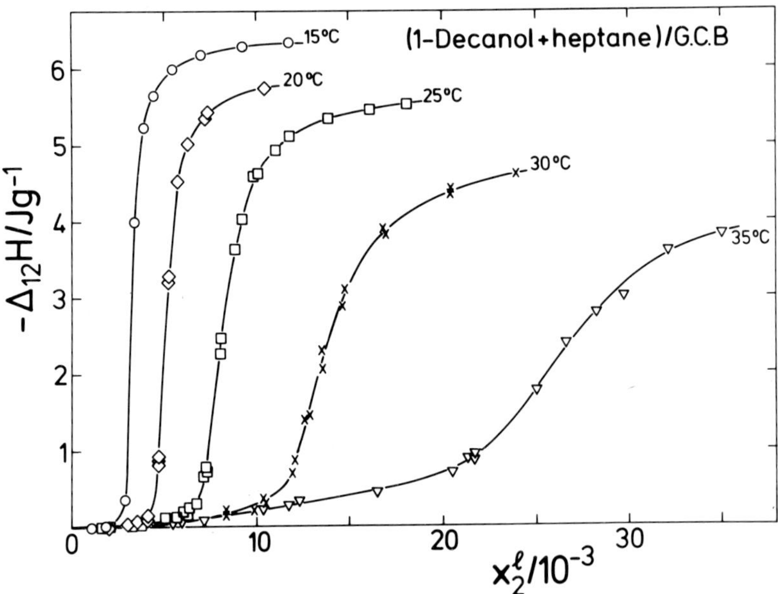

Fig. 3. Enthalpy of displacement isotherms of the system (heptane + decan-1-ol)/graphitized carbon black obtained by the cumulative method for the same temperatures as the adsorption isotherms in Fig.1.

flow adsorption experiments. For these reasons the isotherms obtained by single step adsorption and desorption experiments cannot be used for a quantitative analysis. Results derived by the cumulative method from sufficiently small concentration steps can be used to determine the differential molar enthalpy of displacement, as outlined below.

Fig.3 shows isotherms of $\Delta_{12}H$ (cumulative method) for all experimental temperatures and in Fig.4 $\Delta_{12}H$ is plotted against the surface coverage ϕ_2^a as obtained from the surface excess at given concentrations of the bulk solution. The slope of this plot yields the differential molar enthalpy of displacement at given temperature and composition of the adsorbed layer:

$$\Delta_{12}H_2(T,\phi_2^a) = \frac{\partial \Delta_{12}H}{\partial n_2^a} \, . \tag{2}$$

Numerical results for $\Delta_{12}H_2$ obtained from the graphs in Fig.4 are given in Table 1. The uncertainty of these values is ±5 kJ/mol or less.

4. DISCUSSION

Monolayer Phase Transition

The sharp increase of the surface concentration of decanol within a narrow concentration range exhibited by the isotherms up to about 25°C suggests a phase change in the adsorbed layer. Below we present a simple treatment of this phenomenon based on a three phase equilibrium of two adsorbed monolayer phases (denoted by superscripts ' and ") and the bulk solution (l). For the vertical section of the isotherms the displacement reaction of solvent (1) by solute (2) can be written as

$$r(1)' + (2)^l = r(1)^l + (2)'' \tag{3}$$

and the condition of phase equilibrium is

$$\ln(a_2^l/a_2'') - r\ln(a_1^l/a_1') = (\Delta_a\mu_2''^* - r\Delta_a\mu_1'^*)/RT, \tag{4}$$

where $\Delta_a\mu_2''^* = \mu_2''^* - \mu_2^{l*}$ and $\Delta_a\mu_1'^* = \mu_1'^* - \mu_1^{l*}$

represent differences of the standard chemical potentials. At sufficiently low temperatures the step of the isotherm occurs at very low bulk concentration of solute (i.e., $a_1^l \simeq 1$).

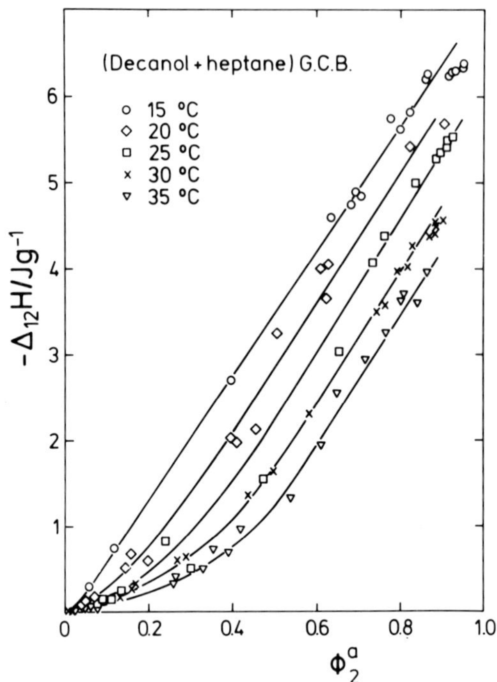

Fig. 4. Enthalpy of displacement plotted against the fraction of surface covered by decanol.

TABLE 1

Differential molar enthalpy of displacement $\Delta_{12}H_2$/kJ mol^{-1}

T/°C	ϕ_2^a: 0	0.2	0.4	0.6	0.8
15	-	43	43	43	43
20	8	33	44	46	46
25	8	22	39	46	46
30	8	16	31	46	46
35	8	10	23	46	46

We assume that in this regime the monolayer phases consist of nearly pure solvent ($a_1' \simeq 1$) and pure alkanol ($a_2'' \simeq 1$) respectively. In this case the activity a_2^l at which the isotherm exhibits the step varies with temperature according to the relation

$$R\,\frac{d\ln a_2^l}{d(1/T)} = \Delta_a H_2''^* - r\Delta_a H_1'^* \equiv \Delta_{12}H_2^* \,, \tag{5}$$

where $\Delta_a H_1'^*$ and $\Delta_a H_2''^*$ represent the standard molar enthalpy differences. The monolayer model relates these quantities to the specific enthalpies of wetting of the pure liquids, $\Delta_a h_i^*$:

$$\begin{aligned} \Delta_a H_1'^* &\equiv H_1'^* - H_1^{l*} = \Delta_w h_1^*/n_{1,o} \\ \Delta_a H_2''^* &\equiv H_2'' - H_2^{l*} = \Delta_w h_2^*/n_{2,o}. \end{aligned} \tag{6}$$

The above thermodynamic model was tested by comparing the value of $\Delta_{12}H_2^*$ as derived from the adsorption isotherms by eqn.(5) with the corresponding quantity obtained from the enthalpies of wetting by eqn.(6). The activities a_2 for eqn.(5) were determined from the mole fractions at which the isotherms exhibit their point of inflection and the corresponding activity coefficients; these were taken from the work of Wieczorek(8) on *hexane* + alkanol systems. The error introduced by equating the activity coefficients of alkanols in mixtures with heptane and with hexane will be small as in both cases deviations from ideality are mainly due to the association equilibrium of the alcohol. Fig.5 shows the resulting plot of $\ln a_2^l$ *vs.* 1/T of heptane + decanol and heptane + dodecanol. The points at $\ln a_2^l = 0$ represent the order-disorder transition temperature of the pure alkanols at the liquid/graphite interface (1). For heptane + decanol linear regression of the data of the solutions in the temperature range 288 - 308 K yields the result $\Delta_{12}H_2^* = -28$ kJ mol^{-1}. This value is to be compared with that obtained from the areal enthalpies of wetting near 298 K given by Brown *et al.* (3), *viz.*-122 mJ m^{-2} (heptane) and -194 mJ m^{-2} (decanol), which gives

$$\Delta_{12}H_2^* = (\Delta_w h_2^* - \Delta_w h_1^*)/n_{2,o} = -29.5 \text{ kJ mol}^{-1}.$$

The close agreement of the two results is surprising in view of the fact that the underlying assumption of mutual immiscibility of the two components in the coexisting monolayer phases seems not justified (except, perhaps, at

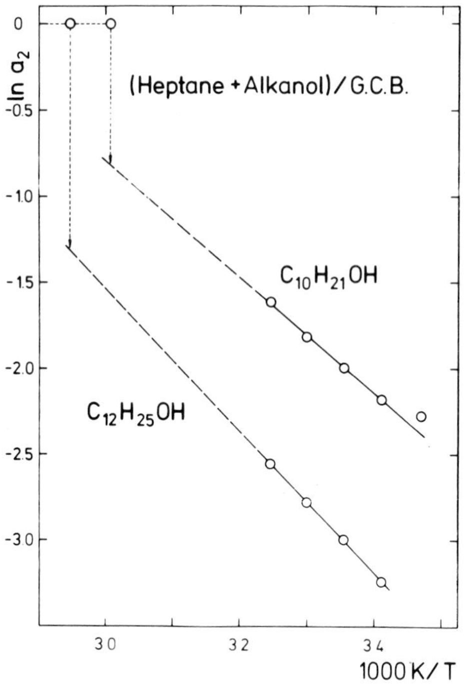

Fig. 5. Determination of standard molar enthalpy of displacement from adsorption isotherms by eqn.(5).

the lowest temperatures). Furthermore, above 25°C the phase change in the adsorbed layer is clearly not a first order transition, as the step of the isotherms spreads over an increasingly wide concentration range of the solution.

Differential molar enthalpy of displacement

The differential molar enthalpy of the exchange reaction (3) is given by

$$\begin{aligned} \Delta_{12}H_2 &= (H_2'' - H_2^1) - r(H_1' - H_1^1) \\ &= \Delta_{12}H_2^o - (H_2^E - rH_1^E), \end{aligned} \tag{7}$$

where

$$\Delta_{12}H_2^o \equiv (H_2'' - H_2^{1*}) - r(H_1' - H_1^{1*}).$$

Here, H_i^p represents the partial molar enthalpy of component *i* in phase *p*, H_i^{p*} is the molar enthalpy of *i* in the pure phase *p*, and H_i^{E1} is the excess partial molar enthalpy of *i* in the liquid solution. If the two surface phases constitute pure components 1 and 2, respectively, as in the previous section, $\Delta_{12}H_2^o$ becomes identical with $\Delta_{12}H_2^*$ and thus independent of the composition of the surface layer. When the coexisting surface phases contain both components, $\Delta_{12}H_2^o$ will differ from $\Delta_{12}H_2^*$ but will still be independent of surface layer composition. On the other hand, in ranges of complete miscibility of the two components in the surface layer (i.e., only *one* surface phase denoted by index *a*) H_1^a and H_2^a and thus $\Delta_{12}H_2^o$ become functions of the surface coverage ϕ_2^a.

$\Delta_{12}H_2^o$ was derived from $\Delta_{12}H_2$ (Table 1) and the molar excess enthalpy H_m^E of the binary liquid systems, using data for 30°C by Woijcicka and Kalinowska (9); for dilute solutions ($x_2^1 \leqslant 0.02$) these data can be represented by the relation

$$H_m^E = x_2^1 H_2^{E\infty} (1 - 23.8\ x_2^1)$$

with $H_2^{E\infty} = 24.0$ kJ mol^{-1}. In eqn.(7) the term containing the excess enthalpies may be approximated for dilute solutions by setting $r = 1$, and from the above relation we obtain

$$H_2^E - H_1^E = H_2^{E\infty}(1 - 47.5\ x_2^1).$$

This expression applies to a temperature of 30°C but will also be used for a qualitative discussion of $\Delta_{12}H_2^o$ at lower temperatures.

At 15°C the surface excess isotherm exhibits a sharp step at $x_2^1 = 3.35 \cdot 10^{-3}$ and $\Delta_{12}H_2$ has a constant value (-43 ± 5 kJ mol^{-1}) independent of ϕ_2^a. As the variation of ϕ_2^a occurs within a narrow range of x_2^1 the correction term for non-ideality of the liquid solution will also be constant ($H_2^E - H_1^E$ = 20 kJ mol^{-1} at x_2^1 = 0.00335 at 30°C) and from eqn.(7) we thus obtain $\Delta_{12}H_2^o$ = -23 kJ mol^{-1}. This value is reasonably close to the standard molar enthalpy of displacement $\Delta_{12}H_2^*$ = -28 kJ mol^{-1}, as to be expected for two coexistent monolayer phases consisting of pure components 1 and 2, respectively.

At 30°C the step in the surface excess isotherm extends over a wide range in x_2^1 and $\Delta_{12}H_2$ strongly increases in magnitude with ϕ_2^a. The corresponding dependence of $\Delta_{12}H_2^o$ on surface coverage is summarized in Table 2.

TABLE 2

Analysis of the differential molar enthalpy of displacement at 30°C (Enthalpies in kJ mol^{-1})

ϕ_2^a	$x_2^1/10^{-2}$	$\Delta_{12}H_2$	$H_2^E - H_1^E$	$\Delta_{12}H_2^o$
0	0	- 8	24	16
0.2	1.08	-16	12	- 4
0.4	1.25	-31	10	-21
0.6	1.41	-46	8	-38
0.8	1.80	-46	4	-42

Unlike the situation at 15°C, $\Delta_{12}H_2^o$ exhibits a pronounced variation with ϕ_2^a as to be expected for complete miscibility in the surface monolayer. The strong increase in magnitude of $\Delta_{12}H_2^o$ with increasing ϕ_2^a can be attributed to strong lateral interactions of the alkanol molecules, due to hydrogen bond formation and chain packing effects. Incidentally, note that the enthalpy change for a transfer of alkanol molecules from infinite dilution in the liquid state to the adsorbed layer at a given surface concentration ϕ_2^a is given by $\Delta_{12}H_2 - H_2^{E\infty}$; at $\phi_2^a = 0$ and higher temperatures this quantity has a weak negative value of -8 kJ mol^{-1}. Thus adsorption of single decanol molecules from heptane onto graphite is energetically not particularly favourable. The preferential adsorption of the alkanol in these systems is largely due to the strong solute-solute attractive interactions in the adsorbed monoalyer. A full account of our results, including those of the system heptane + dodecanol, will be published (10).

REFERENCES

1. Findenegg, G.H. (1972, 1973). *J.Chem.Soc.Faraday Trans.1.* 68, 1799-1806; 69, 1069-1078.
2. Everett, D.H. (1975). *Israel J.Chem.* 14, 267-277.
3. Brown, C.E., Everett, D.H., Powell, A.V. and Thorne, P.E. (1975). *Faraday Discussions Chem.Soc.* 59, 97-108.
4. Everett, D.H. (1978). *Progr. Colloid Polymer Sci.* 65, 103-117.

5. Liphard, M., Glanz, P., Pilarski, G. and Findenegg, G.H. (1980). *Progr. Colloid Polymer Sci.* 67, 131-140.
6. Kern, H.E. and Findenegg, G.H. (1980). *J.Colloid Interface Sci.* 75, 346-356.
7. Groszek, A.J. (1970). *Proc.Roy.Soc.London Ser.A* 314, 473-498.
8. Wieczorek, S.A. (1978, 1979). *J.Chem.Thermodynamics* 10, 187-194; 11, 239-245.
9. Woijcicka, M.K. and Kalinowska, B. (1975). *Bull.Acad.Pol. Sci.* 23, 759.
10. Bien-Vogelsang, U., Findenegg, G.H., Koch, C. and Liphard, M. (1982). *J.Chem.Soc.Faraday Trans.1*, to be submitted.

INFRARED STUDIES OF ADSORPTION AT THE SOLID OXIDE/LIQUID HYDROCARBON INTERFACE

C.H. Rochester*, S.N.W. Cross, J. Graham,
R. Rudham, D.-A. Trebilco and G.H. Yong

**Department of Chemistry, University of Dundee*
Dundee DD1 4HN, U.K.

Department of Chemistry, University of Nottingham,
Nottingham NG7 2RD U.K.

INTRODUCTION

For spectroscopic reasons most early infrared spectroscopic studies of adsorption behaviour *in situ* at the solid/liquid interface involved carbon tetrachloride as the predominant component of the liquid phase (Rochester, 1976). Systems containing hydrocarbon liquids are of considerable interest (Everett, 1973) and therefore the present work was undertaken in order to establish whether useful information about surface-adsorbate interactions could be gained from infrared spectroscopic study of powdered oxides immersed in hydrocarbon liquids.

EXPERIMENTAL

Infrared Cells

Oxides were studied in the form of compressed discs of 25 mm diameter mounted in one of four infrared cells of different design. The cells were glassblown to a conventional vacuum system capable of maintaining a dynamic vacuum of *ca.* 10^{-4} N m^{-2}, fitted with greasefree taps, and containing a section for the purification, storage and admission to the cells of liquids or solutions.

Spectra of silica or rutile immersed in hydrocarbon mixtures or of silica in solutions of substituted anisoles , benzenes or nitrobenzenes in heptane were recorded using a cell with silica optical windows glassblown to the cell body

(Rochester and Trebilco, 1977a). The optical pathlength was 0.5 mm. This cell had the advantage that an oxide disc positioned between the windows could be heat treated *in situ* before the admission of liquid phase. No movement of the disc was necessary between thermal activation and immersion in liquid. However, a disadvantage was that the silica windows were only transparent in the spectral region above 2100 cm^{-1}. A similar cell incorporating a liquid circulation system facilitated stirring and temperature control of the liquid phase (Rochester and Yong, 1980) and was used in studies of the adsorption of propionitrile on silica immersed in 2,2,4-trimethylpentane, toluene or mixtures of the two hydrocarbons.

Cells with fluorite windows were used when it was necessary to record the spectral region below 2100 cm^{-1}. Discs could be moved vertically between an upper cell section surrounded by a furnace (Griffiths *et al.*, 1974) and a lower section to which the windows were fixed with Araldite. For study of the adsorption of ethyl acetate on silica immersed in carbon tetrachloride the lower section was constructed from pyrex glass with the windows 3 mm apart (Buckland *et al.*, 1980). For the study of adsorption of cyclohexanone on silica immersed in 2,2,4-trimethylpentane or 1,4-dimethylbenzene + 2,2,4-trimethylpentane mixtures a variable path-length cell was used which enabled very short path-lengths to be attained during spectroscopic examination of the adsorption systems (Buckland *et al.*, 1978).

Materials and procedure

Degussa aerosil silica had a surface area of 176 m^2 g^{-1}. Rutile (code CL/D 338, Tioxide International Ltd) with a surface area of 30.3 m^2 g^{-1} was freed from surface chloride ions by alternate soxhlet extractions and heat treatments in air (Jackson and Parfitt, 1971). Components of the liquid phases were purified by standard methods.

Silica discs were heated in a vacuum before being cooled and immersed in liquids or solutions. Temperatures during the evacuation stage are given in brackets in the figure legends. Rutile discs were heated in oxygen (13 kN m^{-2}, 20 h, 673 K), equilibrated with saturated water vapour at room temperature and finally evacuated at 400 K (1 h) before admission of liquids to the cell. Spectra of discs in series of solutions of increasing concentration of one component were generally obtained by recording the spectrum after each addition of a series of aliquots of a concentrated solution of that component to the disc initially immersed in a liquid or liquid mixture without any of the component present (Griffiths *et al.*, 1974; Buckland *et al.*, 1978). An exception was for silica immersed in mixtures of hydrocarbons when after each spectrum was recorded the cell was drained and evacuated before the admission of a solution of different composition (Rochester and Trebilco, 1977a).

RESULTS AND DISCUSSION

The perturbation of isolated silanol groups on silica

Spectra of silica immersed in heptane + toluene mixtures contained infrared bands at 3706 and 3611 cm^{-1} (Fig. 1) due to the OH-stretching vibrations of isolated surface silanol groups perturbed by interactions with heptane or toluene molecules, respectively. The series of spectra for different compositions of the liquid phase exhibited an

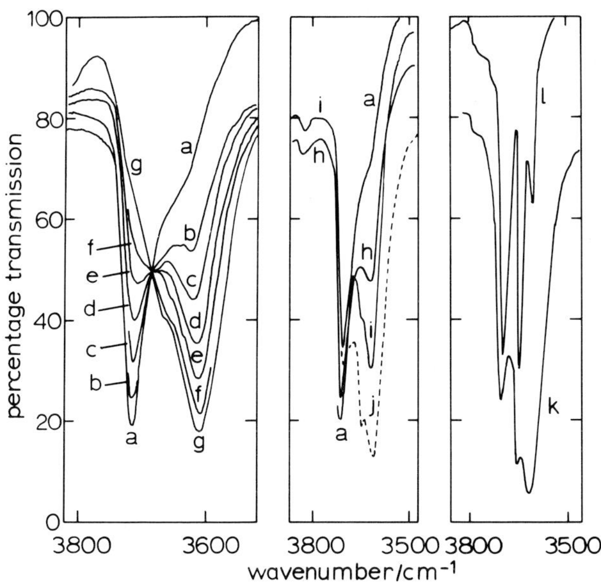

Fig. 1. Spectra of silica (1023 K) immersed in (a) heptane, (b)-(f) heptane + toluene mixtures with mole fractions of toluene equal to (b) 0.067, (c) 0.132, (d) 0.194, (e) 0.313 and (f) 0.580, (g) toluene, (h)-(j) heptane + benzene mixtures with mole fractions of benzene equal to (h) 0.080, (i) 0.226 and (j) 0.622, and (k) benzene. (l) Spectrum of benzene.

isosbestic point except at high mole fractions of toluene for which a weak band at 3660 cm^{-1} in the spectrum of toluene contributed to the intensity of absorption. Both liquids were transparent at 3706 and 3611 cm^{-1} and absorbance values at the two maxima were linearly related. Hence the fractions of silanol groups perturbed by each component of the liquid phase may be evaluated from the absorbance data (Rochester and Trebilco, 1977a). Spectra of silica immersed in heptane + benzene mixtures contained significant contributions from bands in the spectrum of benzene (Fig. 1). However, after appropriate corrections a linear relationship is confirmed between absorbance values at 3706 cm^{-1} and 3621 cm^{-1}. The latter band is ascribed to silanol groups perturbed by interactions with adsorbed benzene molecules.

Fig. 2 shows spectra of silica immersed in solutions of propionitrile in 2,2,4-trimethylpentane. Hydrogen bonding interactions between silanol groups and the cyano groups in

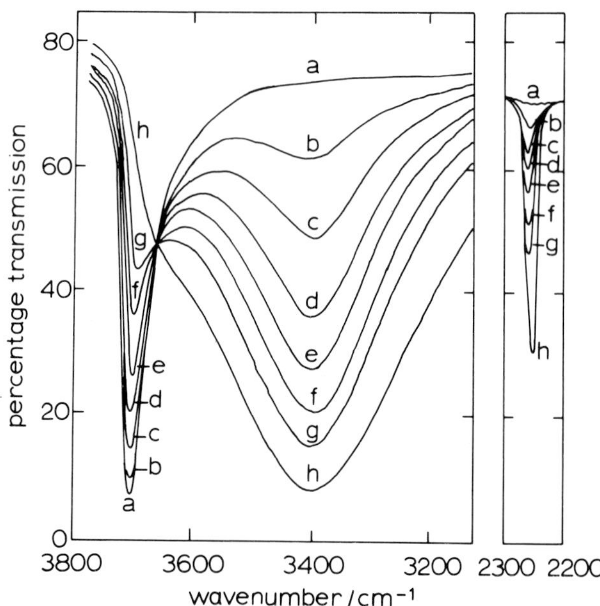

Fig. 2. Spectra of silica (873 K) immersed in (a) 2,2,4-trimethylpentane, (b)-(h) propionitrile + 2,2,4-trimethylpentane mixtures with mole fractions, x_P, of propionitrile given by $10^3 x_P$ equal to (b) 0.16, (c) 0.30, (d) 0.59, (e) 1.16, (f) 2.10, (g) 3.96 and (h) 45.5.

adsorbed molecules led to the appearance of maxima at 3395 and 2260 cm^{-1} due to perturbed silanol and cyano groups, respectively. Neglecting activity coefficients, the equilibrium constant K for the adsorption reaction

$$\rightarrow SiOH + EtCN \rightleftharpoons \rightarrow SiOH...NCEt$$

will be given by the equation

$$K = F_P/[(1-F_P)x_P] \quad (1)$$

where x_P is the mole fraction of propionitrile and F_P is the fraction of silanol groups perturbed by adsorptive interactions. The linearity of a plot of (x_P/F_P) against x_P (Fig. 3a) and the slope of unity supports the applicability of equation (1) and confirms the reliability of the absorbance data. Hence K equals 849 at 298 K. Fig. 3c shows a similar plot which gives K equal to 105 for the adsorption of propionitrile on to isolated silanol groups on silica immersed in toluene.

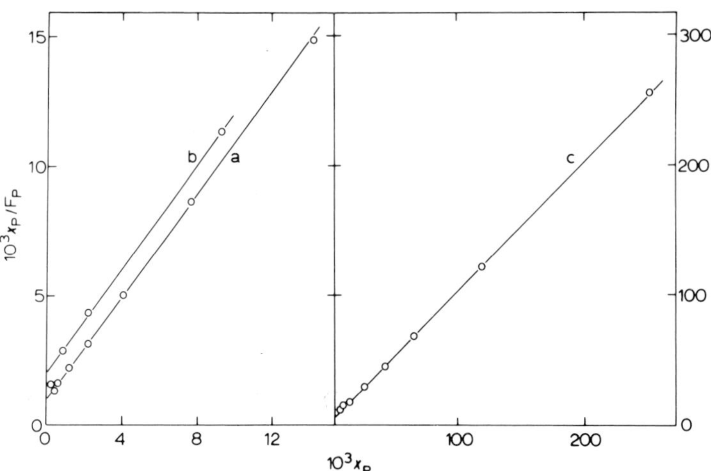

Fig. 3. Test of equation (1) for the adsorption of propionitrile on silica (873 K) in (a) 2,2,4-trimethylpentane, (b) toluene + 2,2,4-trimethylpentane mixtures with a mole fraction ratio (x_M/x_T) of (1/3.22) and (c) toluene.

Adsorption from three component mixtures

Interactions between isolated surface silanol groups on silica and the individual components of three component

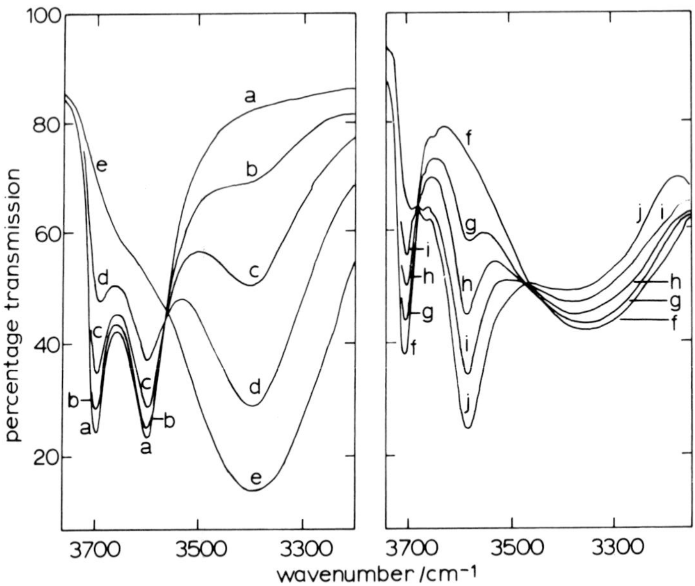

Fig. 4. Spectra of silica (873 K) immersed in propionitrile + toluene + 2,2,4-trimethylpentane mixtures with a constant mole fraction ratio (x_M/x_T) of (1/9.19) and mole fractions of propionitrile given by 10^3x_P equal to (a) 0, (b) 0.41, (c) 1.06, (d) 3.45 and (e) 19.5. Spectra of silica (1073 K) immersed in cyclohexanone + 1,4-dimethylbenzene + 2,2,4-trimethylpentane mixtures with a constant mole fraction ratio ($x_T/10^4x_C$) of (1/3.30) and mole fractions x_D of 1,4-dimethylbenzene equal to (f) 0, (g) 0.016, (h) 0.055, (i) 0.109 and (j) 0.246.

liquid mixtures may be distinguished using infrared spectroscopy provided the strengths of interaction for the three components are appreciably different (Rochester and Trebilco, 1977b). Fig. 4 shows results for silica immersed in mixtures of propionitrile or cyclohexanone with 2,2,4-trimethylpentane and an aromatic hydrocarbon. The differences $\Delta\nu_{OH}$ between the position of the band maximum (3750 cm^{-1}) due to unperturbed silanol groups on silica in vacuum and the band positions for perturbed silanol groups reflect the relative strengths of the surface-adsorbate interactions. Hence propionitrile or cyclohexanone, which gave the biggest shifts $\Delta\nu_{OH}$, were more strongly adsorbed than toluene or 1,4-dimethylbenzene, which gave intermediate shifts but were more strongly adsorbed than 2,2,4-trimethylpentane for which $\Delta\nu_{OH}$ equalled only 45 cm^{-1}. The fractions of isolated silanol groups perturbed by each component of the liquid mixtures may be calculated from absorbance values at the absorption maxima (Buckland *et al.*, 1978; Rochester and Yong, 1980). Fig. 3b illustrates the test of equation (1) for silica in propionitrile + toluene + 2,2,4-trimethylpentane mixtures containing a fixed mole fraction ratio of toluene/2,2,4-trimethylpentane of (1/3.22). The intercept gave K equal to 532 for the adsorption of propionitrile at 298 K. Absorbance values at the maxima due to silanol groups perturbed by toluene and 2,2,4-trimethylpentane were linearly related indicating that the relative numbers of silanol groups perturbed by the two hydrocarbons were, within experimental error, independent of the adsorption of propionitrile up to *ca.* 0.8 coverage of silanol group sites.

Adsorption involving two types of site

Ethyl acetate formed hydrogen bonds with isolated silanol groups on silica to give infrared bands at 3440 and 1712 cm^{-1} assigned to the stretching vibrations of perturbed silanol and carbonyl groups, respectively (Fig. 5b-f). A band at 1747 cm^{-1} was due to carbonyl stretching vibrations of ethyl acetate in solution. Spectra of silica which had been preheated at 453 K contained bands at 3660 and 3550 cm^{-1} which may be ascribed to two types of adjacent interacting surface hydroxyl groups (Cross and Rochester, 1979; Rochester and Trebilco, 1979). The subsequent adsorption of ethyl acetate from solution led to the appearance of a new band at 1688 cm^{-1} (Fig. 5h-k) ascribed to carbonyl groups each simultaneously perturbed by two hydrogen bonds from a pair of adjacent silanol groups.

Measurements of the intensities of the maxima at 1712 and 1688 cm^{-1} allow the extents of adsorption of ethyl acetate on to single hydroxyl group sites and pair sites to be separately monitored. The results showed that adsorption on to single sites was unaffected by the presence of adjacent hydroxyl groups on the silica surface (Cross and Rochester, 1979). Ethyl acetate was more strongly adsorbed on to pair sites than on to single sites.

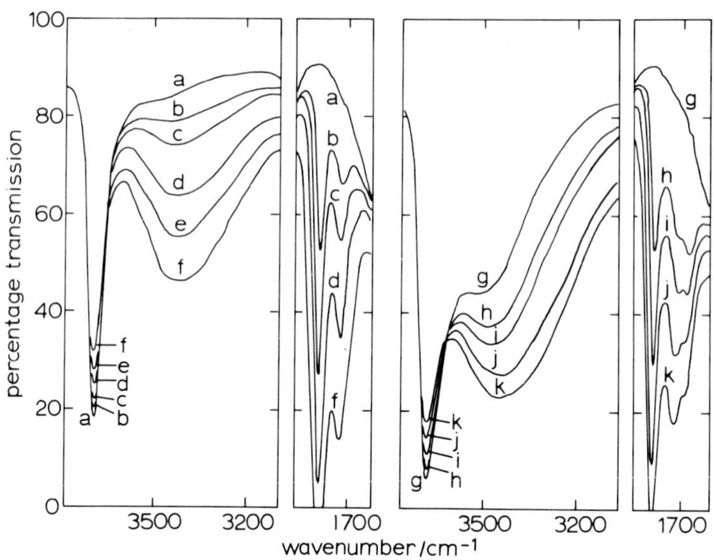

Fig. 5. Spectrum of silica after heat treatment at either 863 K [(a)-(f)] or 453 K [(g)-(k)], cooling and immersion in (a) and (g) carbon tetrachloride, (b)-(f) and (h)-(k) solutions containing increasing concentrations of ethyl acetate in carbon tetrachloride.

Spectra of silica immersed in solutions of cyclohexanone in 2,2,4-trimethylpentane similarly exhibited maxima at *ca.* 1700 and 1677 cm^{-1} due to carbonyl groups perturbed by interaction with one or two silanol groups, respectively (Buckland *et al.*, 1978; Rochester and Trebilco, 1979). The fractions F_{3705} of isolated silanol groups perturbed by cyclohexanone molecules are compared in Fig. 6A, for silica preheated at 1073 and 423 K, with the corresponding fractional coverages F_{1702} and F_{1698} of single silanol group adsorption sites. Subscripts give the spectral positions from which the absorbance data were deduced for use in the calculations. The values of f_{1698} and f_{1677} (Fig. 6B) represent, with an assumption concerning the equality of extinction coefficients, the fractions of the total number of perturbed cyclohexanone molecules which were adsorbed on to single and pair sites, respectively, on silica preheated at 423 K. Adsorption on to pair sites predominated at low cyclohexanone concentrations. However, at complete surface coverage *ca.* 62% of the total number of adsorbed cyclohexanone molecules were bonded to single silanol groups whereas only *ca.* 38% were bonded to adjacent pairs. The total implied concentration of surface hydroxyl groups would be *ca.* 4.5 nm^{-2} providing it can be assumed that the surface concentrations of single sites and of isolated silanol groups were equal, the latter being *ca.* 2 nm^{-2} (Marshall and Rochester, 1975), and that all surface hydroxyl groups were perturbed by cyclohexanone molecules at the maximum attainable surface coverage.

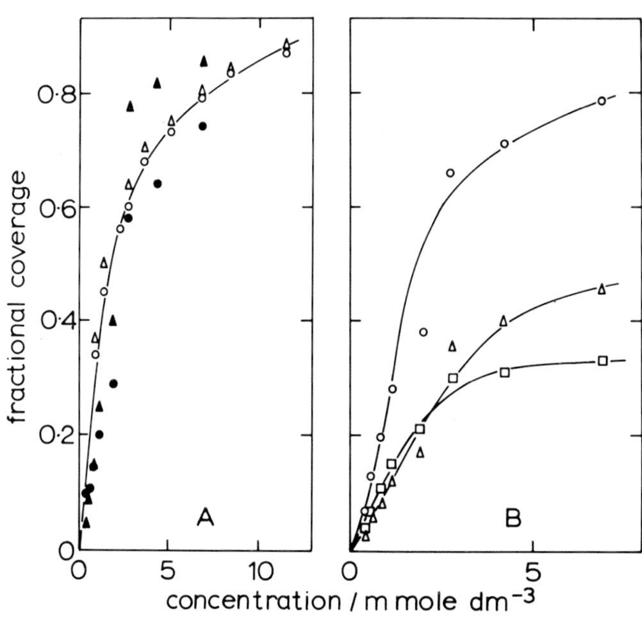

Fig. 6. A. Fractions ● F_{3705} *and* ▲ F_{1698} *for the adsorption of cyclohexanone on silica (423 K) immersed in 2,2,4-trimethylpentane and corresponding fractions* ○ F_{3705} *and* △ F_{1702} *for silica preheated at 1073 K. B. Fractions* △ f_{1698}, □ f_{1677} *and* ○ $(f_{1698} + f_{1677})$ *for the adsorption of cyclohexanone on silica (423 K) immersed in 2,2,4-trimethylpentane.*

Spectra of rutile in vacuum showed two maxima at 3655 and 3410 cm^{-1} (Fig. 7a) due to two types of surface hydroxyl group (Griffiths and Rochester, 1977). The maximum at 3655 cm^{-1} shifted to 3605 cm^{-1} when rutile was immersed in heptane but hydroxyl groups responsible for the band at 3410 cm^{-1}

were unaffected by the presence of liquid phase (Fig. 7b). The band at 3410 cm^{-1} was similarly retained in spectra of rutile in toluene (Fig. 7c). Hydroxyl groups giving the maximum at 3655 cm^{-1} formed hydrogen bonds with adsorbed toluene molecules and their resultant perturbation led to the appearance of a broad maximum at 3540 cm^{-1}. The contrasting behaviour of the two types of hydroxyl group on rutile was further evidenced by spectra of rutile immersed in heptane + 1,4-dimethylbenzene mixtures (Fig. 7d,e). One explanation would be that the groups giving the band at 3410 cm^{-1} existed at sub-surface lattice sites and were not influenced by hydrocarbon liquids in contact with the

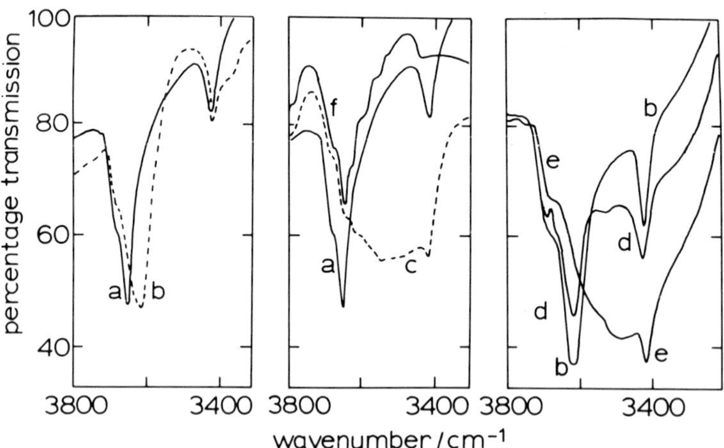

Fig. 7. Spectra of rutile in (a) vacuum, (b) heptane, (c) toluene, (d) and (e) heptane + 1,4-dimethylbenzene mixtures with mole fractions of 1,4-dimethylbenzene of 0.0036 and 0.19, respectively. Spectra (d) and (e), and (b) in the same series, were recorded with a cell containing liquid mixture in the reference beam. (f) Spectrum of toluene.

surface (Buckland *et al.*, 1981; Graham *et al.*, 1981) despite the fact that they undergo rapid H/D isotope exchange in the presence of deuterium oxide vapour (Griffiths and Rochester, 1977).

Distinguishable types of interaction involving particular adsorption sites

Spectra of silica immersed in solutions of 4-methylanisole in heptane contained bands at 3605 and 3390 cm^{-1} (Fig. 8) due to isolated surface silanol groups perturbed by hydrogen bonding interactions with the aromatic nuclei or methoxy groups, respectively, in adsorbed 4-methylanisole molecules. The bands were at 3580 and 3365 cm^{-1} for silica immersed in liquid 4-methylanisole (Fig. 8i). Similar spectra of mixed solutions of triphenylsilanol and 4-methylanisole in carbon tetrachloride resulted from the concurrent formation of two 1:1-addition complexes containing hydrogen bonds between silanol groups and either aromatic nuclei or methoxy groups (Acosta Saracual and Rochester, 1982). By analogy the present results are there-

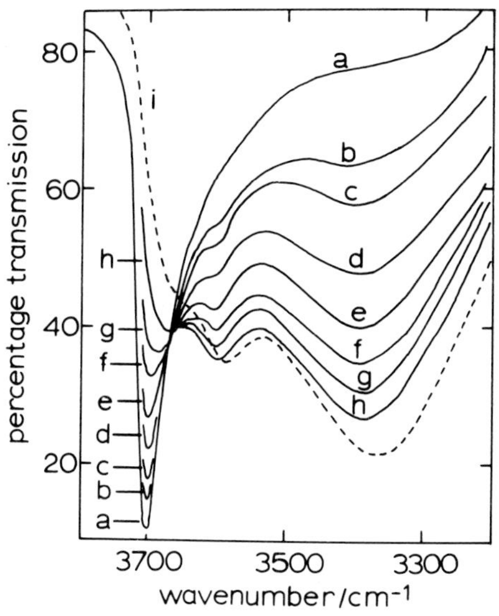

Fig. 8. Spectra of silica (873 K) immersed in (a) heptane, (b)-(h) solutions in heptane of 4-methylanisole at concentrations of (b) 2.4, (c) 4.9, (d) 9.7, (e) 24, (f) 49, (g) 97 and (h) 244 mmol dm^{-3}, and (i) 4-methylanisole.

fore primarily ascribed to two distinguishable modes of adsorption in which each 4-methylanisole molecule was bonded to a surface silanol group via either its aromatic nucleus or its methoxy group. The spectra do not preclude the possibility, however, that some adsorbed molecules each interacted with a pair of isolated silanol groups to form simultaneously two hydrogen bonds, one to the aromatic nucleus and the other to the methoxy group.

The spectroscopic shift $\Delta\nu_{OH}$ (measured from 3705 cm^{-1}) which accompanied the formation of hydrogen bonds between isolated silanol groups and the aromatic nuclei in 4-methylanisole is plotted in Fig. 9c, together with corresponding results for other substituted anisoles and aromatic hydrocarbons (Rochester and Trebilco, 1978a), substituted phenols (Rochester and Trebilco, 1978b), ethyl benzoate (Cross and Rochester, 1981) and benzonitrile (Rochester and Yong, 1982) against the sum of the para σ-constants for all the substituents in the benzene rings (Barlin and Perrin, 1966). Increasing electron density in the benzene ring favoured a stronger surface-adsorbate interaction which led to a greater perturbation of silanol groups and a bigger shift

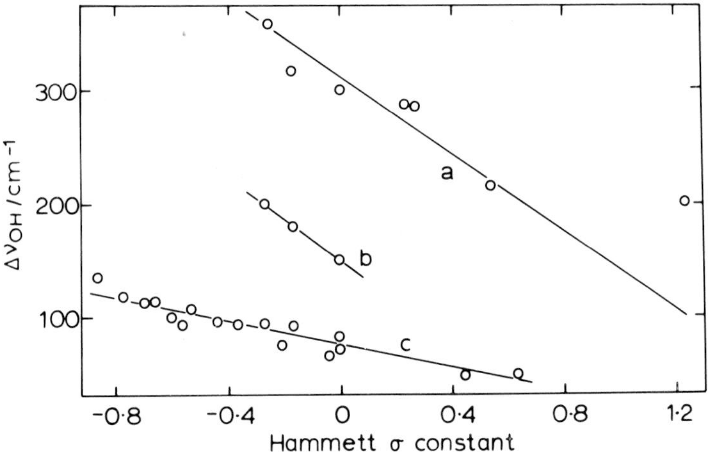

Fig. 9. Correlations between $\Delta\nu_{OH}$ and Hammett σ-constants for the adsorption of substituted (a) anisoles, (b) nitrobenzenes and (c) benzenes onto isolated surface silanol groups on silica (873 K) at the solid/liquid interface.

$\Delta\nu_{OH}$. Electron withdrawing substituents in anisole weakened the interactions between silanol groups and methoxy groups and gave a smaller spectroscopic shift for the formation of hydrogen bonds with methoxy groups (Fig. 9a). The result for 4-nitro substitution (σ = 1.24) was not consistent with the data for other substituents but led to a bigger shift value than expected. However, comparison with the shifts observed for the adsorption of nitrobenzene and nitrotoluene on silica in heptane (Fig. 9b; $\sigma = -0.27$ for 4-methoxy subsitution in nitrobenzene) confirmed that the predominant mode of adsorption of 4-nitroanisole involved hydrogen bonding interactions between surface silanol groups and nitro groups in adsorbed molecules.

CONCLUSION

The present results illustrate how infrared spectroscopy may be used to characterize surface-adsorbate interactions at the solid oxide/liquid hydrocarbon interface and to gain quantitative estimates of the relative extents of adsorption of the components of liquid mixtures on to specific surface sites. The involvement of more than one type of site in the total adsorption process may be monitored and interactions involving different functional groups in adsorbate molecules may be distinguished. Other advantages of infrared spectroscopy in this context have been emphasized elsewhere (Rochester, 1980a,b).

ACKNOWLEDGEMENTS

We thank the British Council, SERC and Tioxide International Ltd for financial assistance, and A.D. Buckland and K. Wigfield for the construction of infrared cells.

REFERENCES

Acosta Saracual, A.R. and Rochester, C.H. (1982). *J. Chem. Soc. Faraday Trans. 1*, in press.

Barlin, G.B. and Perrin, D.D. (1966). *Quarterly Revs.* 20, 75.

Buckland, A.D., Rochester, C.H., Trebilco, D.-A. and Wigfield, K. (1978). *J. Chem. Soc. Faraday Trans. 1* 74, 2393.

Buckland, A.D., Rochester, C.H. and Topham, S.A. (1980). *J. Chem. Soc. Faraday Trans. 1* 76, 302.

Buckland, A.D., Graham, J., Rudham, R. and Rochester, C.H. (1981). *J. Chem. Soc. Faraday Trans. 1* 77, 2845.

Cross, S.N.W. and Rochester, C.H. (1979). *J. Chem. Soc. Faraday Trans. 1* 75, 2865.

Cross, S.N.W. and Rochester, C.H. (1981). *J. Chem. Soc. Faraday Trans. 1* 77, 1027.

Everett, D.H. (1973). *Chem. Soc. S.P.R., Colloid Science* 1, 49.
Graham, J., Rochester, C.H. and Rudham, R. (1981). *J. Chem. Soc. Faraday Trans. 1* 77, 2735.
Griffiths, D.M., Marshall, K. and Rochester, C.H. (1974). *J. Chem. Soc. Faraday Trans. 1* 70, 400.
Griffiths, D.M. and Rochester, C.H. (1977). *J. Chem. Soc. Faraday Trans. 1* 73, 1510.
Jackson, P. and Parfitt, G.D. (1971). *Trans. Faraday Soc.* 67, 2469.
Marshall, K. and Rochester, C.H. (1975). *J. Chem. Soc. Faraday Trans. 1* 71, 2478.
Rochester, C.H. (1976). *Powder Technology* 13, 157.
Rochester, C.H. (1980a). *Progr. Colloid Polymer Sci.* 67, 7.
Rochester, C.H. (1980b). *Adv. Colloid Interface Sci.* 12, 43.
Rochester, C.H. and Trebilco, D.-A. (1977a). *J. Chem. Soc. Faraday Trans. 1* 73, 883.
Rochester, C.H. and Trebilco, D.-A. (1977b). *J. Chem. Soc. Chem. Comm.*, 621.
Rochester, C.H. and Trebilco, D.-A. (1978a). *J. Chem. Soc. Faraday Trans. 1* 74, 1125.
Rochester, C.H. and Trebilco, D.-A. (1978b). *J. Chem. Soc. Faraday Trans. 1* 74, 1137.
Rochester, C.H. and Trebilco, D.-A. (1979). *J. Chem. Soc. Faraday Trans. 1* 75, 2211.
Rochester, C.H. and Yong, G.H. (1980). *J. Chem. Soc. Faraday Trans. 1* 76, 1466.
Rochester, C.H. and Yong, G.H. (1982). *Chem. and Ind.*, 101.

HEATS OF IMMERSION OF MICROPOROUS SOLIDS

A. C. Zettlemoyer, P. Pendleton and F. J. Micale

Center for Surface and Coatings Research
Sinclair Laboratory #7, Lehigh University
Bethlehem, PA 18015, U.S.A.

1. INTRODUCTION

Adsorption isotherms of inert adsorbates by microporous adsorbents are typified by an enhanced adsorption. This enhancement was first recognized by de Boer and Custers (1) as a result of "overlapping force fields" in pores of molecular dimensions. Gurfein et al. (2) and Everett and Powl (3) have made extensive calculations of the adsorption energy in cylindrical capillaries. Recent advances in the understanding of microporosity and the interpretation of Type I isotherms have been discussed in an excellent review by Nicholson and Sing (4).

The quantity of published results concerning the measurement of the heat of adsorption within a micropore appears to be very limited (4). Rouquerol et al. (5,6) using a refined Tian-Calvet calorimeter studied the variation of the differential enthalpy of adsorption as a function of surface pre-coverage of nitrogen (5) and argon (6) by microporous and mesoporous silicas. Their results, however, were somewhat inconclusive as to any effect of micropores. On the other hand, Diano (7) has shown that the net differential heat of adsorption of n-hexane on a microporous carbon black calculated from measurements of the heat of immersion in liquid hexane was, at the lowest uptake, approximately 3.5 times higher than the heat of adsorption on an open surface.

The microporous polymer-based carbon, Carbosieve-S, has received considerable attention with respect to its adsorptive properties (8) and its surface topography (9). Both Stoeckli et al. (8) and Fryer (9) concluded that Carbosieve-S contains a narrow micropore size distribution centered around 0.5 nm. Since this sample appears to approach the qualifications of a "well defined" microporous system, the present study has been undertaken to attempt to define both qualitatively and quantitatively the mechanism of adsorption within such micropores.

2. EXPERIMENTAL

Materials and Experimental Methods

The argon used was Union Carbide "zero-gas" grade; it was dried by slow passage through an activated molecular sieve zeolite maintained at 77 K.

The water was distilled and de-ionized. Both methanol and 2-propanol were Fisher Certified A.C.S. grade containing <0.05% water, and used without further purification.

The adsorbent, Carbosieve-S, a thermally prepared polymer-based carbon, was supplied by Supelco Inc., as 120/140 mesh and used as received. The specific surface area of Carbosieve-S was determined volumetrically, at 77°K, using argon. The equilibrium pressure was measured with a 1000 torr capacitive differential manometer (Datametrics Inc.) which has a sensitivity of 2×10^{-3} torr.

The vapor adsorption isotherms were measured on a quartz spring (Wordon Quartz Co.) with a sensitivity of 10 μg g^{-1}. The equilibrium vapor pressure was measured with a 100 torr capacitive differential manometer (Datametrics) which had a sensitivity of 2×10^{-4} torr. These isotherms were all measured at 25 ± 0.05^{o}C.

The heats of immersion were measured in a thermistor type adiabatic calorimeter (10). The temperature changes in the calorimeter vessel were followed by measuring the change in resistance of the thermistor in a Wheatstone bridge and recording the off-balance of the bridge after amplification by a Hewlett-Packard DC Null Voltmeter 419A. The workable sensitivity was 1×10^{-4} ^{o}C with a background noise of 2×10^{-4} ^{o}C in water. For the organics, the sensitivity was also 1×10^{-4} ^{o}C but with a background noise of 5×10^{-4} ^{o}C. The sources of error in this calorimeter, i.e. instrument sensitivity, reproducibility of heat of bulb breaking, and reproducibility of heat capacity determination led to a cumulative error at $\pm$0.2J for each heat of immersion measurement in water; the cumulative error was $\pm$0.5 J for the organics.

Sample activation conditions were 25°C at 5×10^{-6} torr for 16 hr. For calorimetry, the pre-weighed samples were activated, allowed to equilibrate with the vapor, left at equilibrium for 15 min., then sealed.

A well characterized, non-porous and monodispersed polystyrene (11) was used as the standard for an α_s-analysis of the Carbosieve-S/vapor adsorption isotherm. We believe this

PS to be a good choice as a standard for low energy porous materials.

3. RESULTS

Sorptive Properties of Carbosieve-S

The sorptive properties of Carbosieve-S expressed in various formats are summarized in Tables 1 and 2. The argon adsorption isotherm, shown by Fig. 1, has Type I character with a B-point at 341 cc(STP)g^{-1}; i.e., an equivalent surface area of 1540 m^2g^{-1}. The cross-sectional surface area of argon was taken as 0.168 nm^2 (12). The α_s-analysis, using polystyrene as the standard, suggested a micropore volume of 0.434 cc(liq)g^{-1} and an external surface area of 14 m^2g^{-1}. An excellent agreement was found between the B-point equivalent surface area and the total equivalent surface area from α_s-analysis.

TABLE 1

Molecular Sieve Properties of Carbosieve-S

Adsorbate	B-pt, (mg/g)	Σ_B (m^2/g)	Mol. dia. (nm)	Gurvitsch Vol. (cc(liq)/g)
Argon	341*	1540	0.29	0.450
Water	341	1198	0.35	0.346
Methanol	295	1194	0.46	0.396
2-Propanol	294	974	0.58	0.387

*units, cc(STP)/g

TABLE 2

Sorptive Properties of Carbosieve-S

Adsorbate	Σ_B (m^2/g)	Σ_{pore} (m^2/g)	Σ_{ext} (m^2/g)	Σ_{TOTAL} (m^2/g)	Pore Vol. (cc(liq)/g)
Argon	1540	1533	14	1547	0.434
Water	1198	1145	75	1220	0.326
Methanol	1194	1140	37	1177	0.356
2-Propanol	974	962	6	968	0.370

A Type V character was found for both water and methanol adsorption, Figs. 2 and 3, respectively. Virtually no

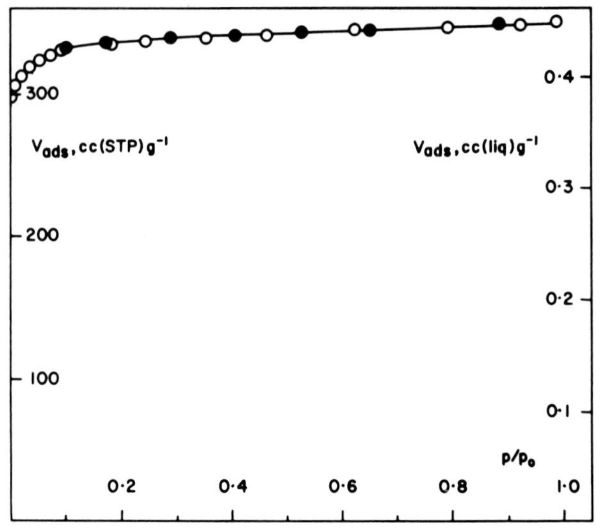

Fig. 1. Adsorption isotherm of argon on Carbosieve at -196°C; ○*, adsorption;* ●*, desorption.*

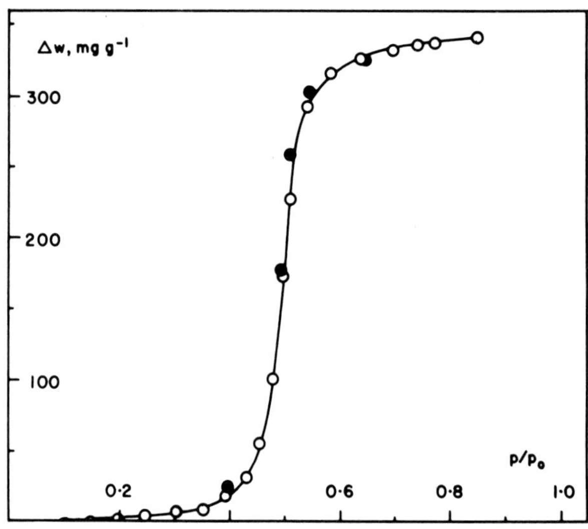

Fig. 2. Adsorption isotherm of water on Carbosieve at 25°C; ○*, adsorption;* ●*, desorption.*

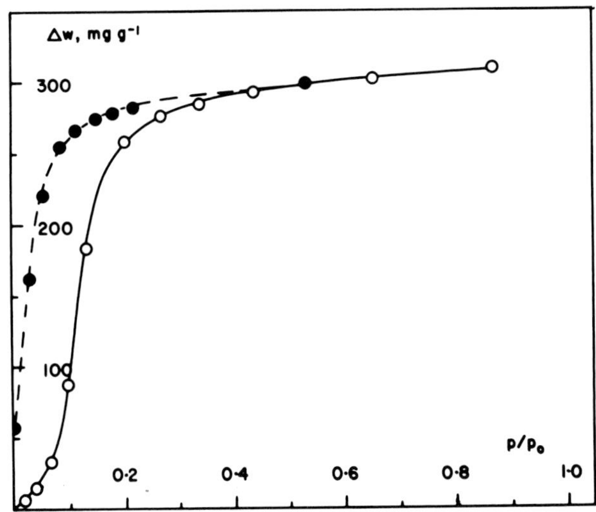

Fig. 3. Adsorption isotherm of methanol on Carbosieve at 25°C; O, *adsorption;* ●, *desorption.*

hysteresis is evident in the case of water; however, a very significant hysteresis cycle was found for methanol. A value at 342 mg g^{-1} was chosen as the B-point for water equivalent "area" analysis. However, in the low pressure range, <0.2 P_o, a small "knee" was observed and a B-point was determined at 1.1 mg g^{-1}, equivalent to the adsorption of 0.024 water molecules nm^{-2}. Conductometric titration yielded approximately 0.02 acid groups nm^{-2}.

The methanol B-point was taken as 295 mg g^{-1}. Close agreement was found between the B-point equivalent surface area and the total equivalent surface area from α_S-analysis.

Fig. 4 shows Type I adsorption of 2-propanol. A B-point of 294 mg g^{-1} results in a lower equivalent surface area, 974 m^2 g^{-1}, than those areas for argon, water and methanol. Hysteresis was observed at low pressures (a not unusual characteristic of microporous solids), with approximately 15% of the B-point adsorption value being retained after evacuation at 25°C overnight. The sample had to be heated to 200°C for 6 hours to desorb all the 2-propanol. Again, close agreement was found between the B-point equivalent area and that from the α_s-analysis. This close agreement between B-point and α_s-analysis, even for water, confirms a reasonably accurate, albeit fortuitous, choice of standard in polystyrene.

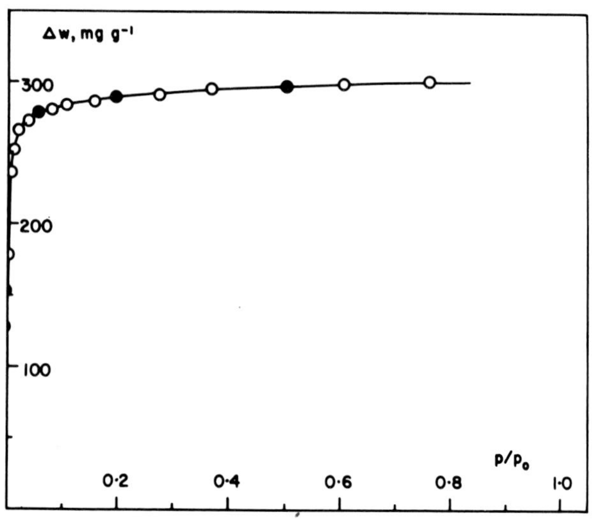

Fig. 4. Adsorption isotherm of 2-propanol on Carbosieve at 25°C; ○, adsorption; ●, desorption.

Heats of Immersion

The heat of immersion of the solid into the liquid is related to the integral heat ΔH_{ads} of adsorption from the vapor by:

$$q_o - q_i = \Gamma(\Delta H_{ads} - \Delta H_L) \,, \qquad (1)$$

where the immersional heats q_o and q_i refer to the bare solid and film covered solid at coverage Γ, respectively, with ΔH_L the heat of liquefaction.

Fig. 5 shows the heat of immersion of Carbosieve as a function of pre-coverage of adsorbed water. The heat of immersion decreases to a plateau after approximately 25% of the micropore volume filled. The heat decreases again after 80% micropore volume filling.

In the case of methanol pre-coverage, shown in Fig. 6, a plateau of heat of immersion was observed at a pre-coverage equivalent to 20% of the micropore volume filled. This plateau decreases again at 70% until approximately 93% volume filling, after which a second plateau was observed.

Fig. 7 shows that the heat of immersion of Carbosieve as a function of pre-coverage of 2-propanol gradually decreases up to 34% volume filling. Pre-coverage from 34 to 65% causes a decrease from 100 to 30 Jg^{-1} where a plateau similar

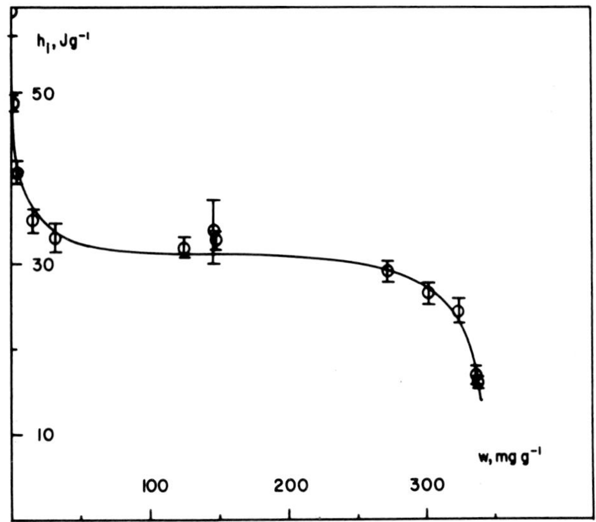

Fig. 5. Heat of immersion of Carbosieve in water as a function of pre-coverage at 25°C.

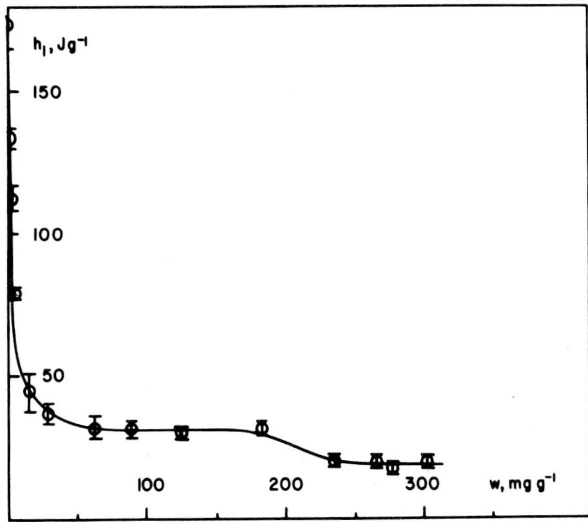

Fig. 6. Heat of immersion of Carbosieve in methanol as a function of pre-coverage at 25°C.

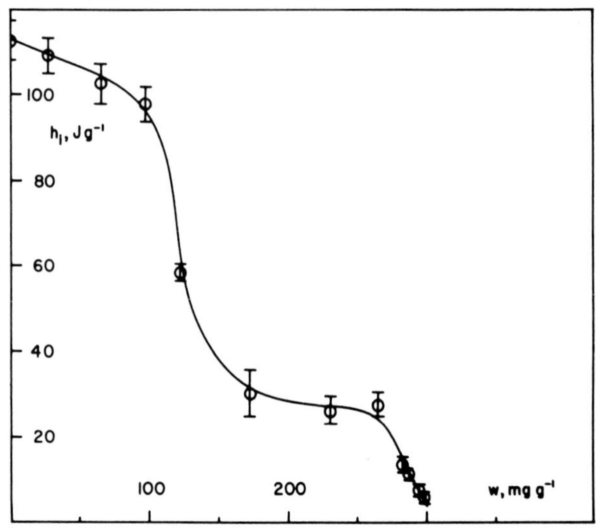

Fig. 7. Heat of immersion of Carbosieve in 2-propanol as a function of pre-coverage at 25°C.

to those found with water and methanol occurs. At approximately 90% micropore filling, the energy decreases once more.

These three curves were plotted as heats of immersion with units J g^{-1} vs. mg g^{-1} rather than the more usual units of mJ m^{-2} vs. mg g^{-1} since area is a debatable dimension in the discussion of micropores. Each point on all three curves is an average of 3 separate measurements.

Heats of Adsorption

The slope of the (q_i,Γ) curves (Fig. 5 - 7) gives the net differential heat of adsorption according to Eq. 1. The extent of the micropore filling, θ, has been assumed unity at the micropore volume calculated from the intercept of the α_s-plots for the respective adsorbates. Results shown in Figs. 8, 9, and 10 should be considered approximate in view of errors in graphical differentiation; however, the estimated error (±3%) is not expected to change the shape of any of the curves.

The differential heat of adsorption of water on Carbosieve is shown in Fig. 8. The initial heat, equivalent to the adsorption of 1 mg g^{-1}, is 186 kJ $mole^{-1}$. This value decreases to approximately 82 kJ $mole^{-1}$ for the adsorption of 2 mg g^{-1}. The adsorbate approaches liquid-like behavior

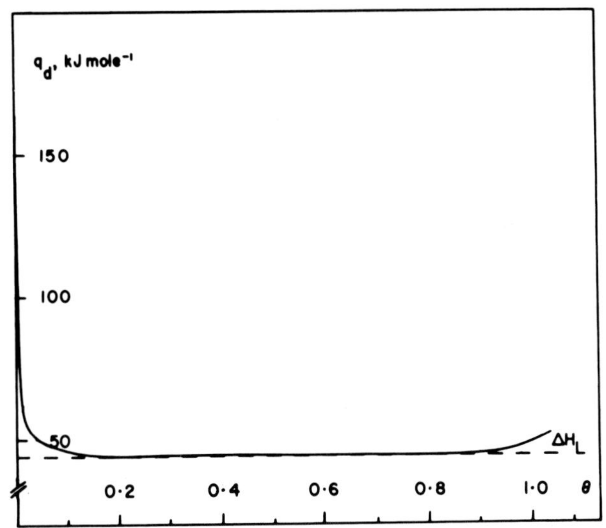

Fig. 8. Differential heat of adsorption of water on Carbosieve at 25°C.

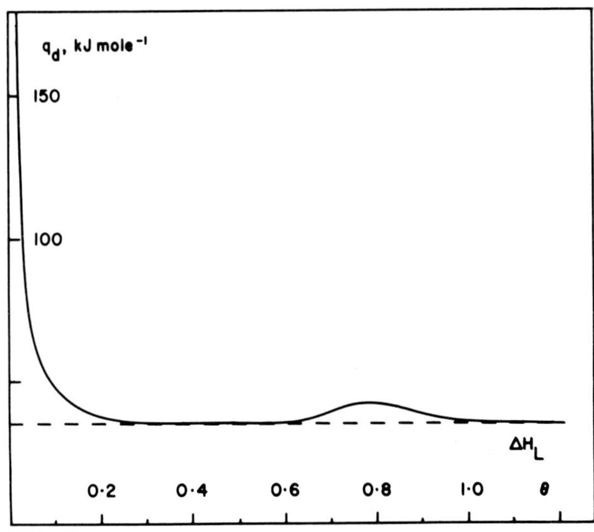

Fig. 9. Differential heat of adsorption of methanol on Carbosieve at 25°C.

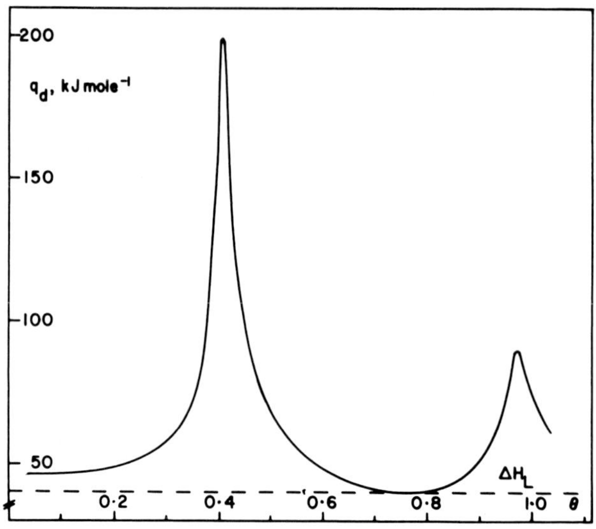

Fig. 10. Differential heat of adsorption of 2-propanol on Carbosieve at 25°C.

after approximately 23% volume filling (75 mg g^{-1}). A small increase in enthalpy is seen as the micropore filling process approaches completion.

Fig. 9 shows the differential heat of adsorption of methanol by Carbosieve. The heat of adsorption at very low coverages is not included in Fig. 9, but is identified in Table 3. Inclusion of these results would cause a contraction of the differential heat axis and a loss in definition of the small heat changes observed close to the equivalent of the micropores being filled. At approximately 30% micropore filling, the differential heat of adsorption implies liquid-like behavior. A maximum occurs close to the equivalent of the micropores being filled.

TABLE 3

Differential Heat of Adsorption of Methanol by Carbosieve

W (mg/g)	θ	q_d (kJ/mole)
1	0.004	884
2	0.008	835
3	0.016	508

Fig. 10 has two maxima due to the shape of the heat of immersion curve in Fig. 7. The first maximum occurs at approximately 40% micropore filling with the second occurring at 95% filling. The differential heat of adsorption of 2-propanol on Carbosieve-S displays two regions of near liquid-like behavior: up to approximately 20% filling and after approximately 60% filling.

4. DISCUSSION

Sorptive Properties of Carbosieve-S

Carbosieve-S has been shown by Stoeckli (8) and Fryer (9) to contain pores of diameter approximately 0.5 nm. This value is, of course, of molecular dimensions commensurate with those of the adsorbate molecules included in this study. The enhanced volume uptake of the adsorbates indicates micropore filling. The position of the large uptake, with respect to the relative pressure, strongly indicates, in each case, the effect of the surface on the adsorbate. The water adsorption (Fig. 2) is greatly affected by the hydrophobicity of the external surface and the micropores, resulting in a Type V isotherm. The shape of the curve suggests that hydrogen bonded structures are not viable within the constricted environment of the hydrophobic micropore. The isotherms of the adsorbates water, methanol and 2-propanol (Fig. 2, 3 and 4, respectively) clearly demonstrate a change in interaction between the surface and adsorbate, as the adsorbate contributes more van der Waals forces.

The mechanism of adsorption of water, methanol and 2-propanol on well characterized polystyrene has been discussed by Micale et al. (11). They found a close agreement between the number of water molecules adsorbed at the B-point and the number of titratable surface acid groups. The volume uptake at point-B of both methanol and 2-propanol was found to exceed greatly the number of acid groups when that number was below 1.0 acid group nm^{-2}. They concluded that the predominant mechanism for 2-propanol adsorption was via the van der Waals forces between the propyl backbone and the polystyrene surface.

There is a close agreement between the low pressure water B-point (on Carbosieve-S) and the number of titratable surface acid groups; thus it is not an unreasonable assumption that the low pressure water adsorption occurs at these localized acid sites. These sites must also exist within the micropores since α_s-analysis of the argon isotherm suggests only 1% of the specific equivalent surface area is external area; the specific equivalent surface area was used to calculate the number of acid groups both from B-point and titration analysis. Thus, a proposed mechanism for water adsorption (or probably any molecule whose primary mode of adsorption is via polar interactions) is localized adsorption until a sufficient hydrostatic pressure exists to compress the adsorbate into liquid-like behavior in the micropore.

Methanol (Fig. 3) also shows some S-shaped character. In the light of the above discussion, the mechanism of adsorption is not only via the alcohol moiety, but must also occur as a van der Waals dispersion interaction via the methyl group.

The mechanism for 2-propanol adsorption (Fig. 4) is obviously different from that of water and methanol, leading to a Type I isotherm. The enhanced adsorption is attributed to a strong van der Waals dispersion interaction between the propyl chain and the carbonaceous surface. Within the confines of a slit-shaped micropore, however, the interaction of the alcohol group with any acid groups would also contribute to an enhanced adsorption. Liquid-like volume filling behavior is assumed, even at very low relative pressures ($p/p_o > 0.001$), because of the general shape of the curve.

Molecular sieve effects are also a characteristic feature of microporous solids. This effect has been clearly demonstrated in Table 1 by the decreasing B-point equivalent surface area with increasing adsorbate molecular diameter. The "Gurvitsch Rule" is quite closely obeyed, the average volume being 0.376 ml g^{-1} (±7%). The Gurvitsch volume of the methanol is higher than that for 2-propanol because the high relative pressure range of the methanol isotherm has a slight Type II character.

Our criterion for the choice of a non-porous standard for α_s-analysis was that the adsorption process on the standard surface should be as close to, if not exactly the same as, that on the external surface of the test sample. Polystyrene was chosen as the standard because it had been well characterized (11), was monodispersed, displayed similar adsorptive properties to Carbosieve-S and was readily available. The α_s-plots of the various adsorbates in Table 2 all displayed linearity in the high relative pressure range, with a large deviation towards the x-axis for α_s-values <1.0; indicative of the presence of micropores only (13). The pore volume calculated from the argon analysis is regarded as the maximum pore volume available to those molecules considered in this study. An interesting and initially somewhat intriguing result of the α_s-analysis is that the pore volumes, expressed as a volume of liquid, increase with increasing molecular diameter. The most likely reason for this increase is the increasing dispersion contribution, and the relatively decreasing polar contribution to the adsorption process within the constricted environment of an assumed slit-like micropore of carbonaceous origin.

Heats of Adsorption on Carbosieve

The mechanism of adsorption of both water and methanol discussed above is confirmed by the heat of immersion measurements and the resulting differential heats of adsorption, q_d. A large heat exchange occurs at low surface coverage caused by localized adsorption by the acid groups. The very high q_d values for methanol at low coverages are caused by adsorption by acid groups within the micropore followed by, or accompanied by, van der Waals interaction between the methyl group and the carbon surface. The large difference in heat between methanol and 2-propanol is also caused by the difference in molecular diameter. According to Gurfein et al. (2), the maximum in the ratio of the energy of interaction of a molecule in a micropore to that on the external surface occurs when the ratio of the diameter of the pore to the diameter of the adsorbate molecule is 1·1:1. The molecular diameter of methanol is 0.46 nm. Thus, the maximum energy difference will occur in a pore of diameter 0.51 nm. Carbosieve has been shown to contain pores of diameter approximately 0.5 nm (9). Gurfein's calculations suggest that this maximum (micropore) interaction energy, at the ratio of 1·1:1, should be 3.37 times greater than that on an external surface. The interpretation of these calculations is somewhat clouded by the fact that adsorption by the external surface occurs simultaneous to that in the micropore.

The error in graphical differentiation of the heats of immersion of water and methanol at low coverage is less than 5%. Irrespective of this error, the general shape is of major consequence. A qualitative interpretation of these curves suggests a high energy of interaction at low coverage proceeding to liquid-like behavior by 20% volume filling of the micropores. Energetics suggest that micropore adsorption occurs as a liquid-like volume filling, confirming the usual practice to express micropore volumes as a volume of liquid. The maximum which occurs at approximately 80% filling for methanol and near 100% filling for water is caused by compression of the adsorbate, somewhat similar to the maximum caused by lateral interactions near monolayer coverage observed for nitrogen adsorption by graphitized carbon black (14). A second suggestion for this maximum is that the molecules are clustered around the "mouth" of the micropore and are still affected by the overlapping force fields from the pore walls. The position of the maximum for water is most likely an underestimate due to the difficulties in determining an accurate α_s-plot, and hence micropore volume, for a Type V isotherm.

The differential heat of adsorption of 2-propanol by Carbosieve clearly implies a different mechanism of adsorption than that for water or methanol.

In their calculations, Gurfein et al. (2) assumed that the repulsive energies were such that a molecule could adsorb in a pore smaller than itself. For this condition, the ratio of the diameter of the pore to the diameter of the molecule is obviously less than unity. In this study the pore diameter distribution is assumed to center at 0.5 nm and the adsorbate diameter is 0.58 nm.; i.e. the ratio is 0.86:1. Their calculations suggested that the ratio of energy of adsorption in the micropore to the energy of adsorption on the external surface is zero when the diameter ratio is 0.93:1. When this ratio is less than 0.93:1, the condition for micropore filling is no more favorable than for adsorption on the external surface. Adsorption of 2-propanol by polystyrene suggested strongly that the adsorption of the alkyl moiety is the predominant interaction (11). For the adsorption of 2-propanol by Carbosieve-S, the "mouth" of the micropore is most probably acting as a constriction to adsorption within the pore. Thus, for 2-propanol to exhibit liquid-like behavior at such a low coverage implies adsorption clustering around the micropore mouth until a sufficient pressure is generated to force adsorption within the pore, i.e. akin to the "ink-bottle" effect observed sometimes in mesopore adsorption. Liquid-like behavior of the adsorbate then follows. The second maximum is again explained in terms of compression of the molecules close to the completion of the pore filling.

5. ACKNOWLEDGEMENTS

The authors wish to thank Mr. E. Hekal and Miss D. Lammey for their assistance with the heat of immersion measurements.

6. REFERENCES

1. de Boer, J.H. and Custers, J.F.H. (1934). *Z. Phys. Chem.* (B) 25, 225.
2. Gurfein, N.S., Dobychin, D.P. and Koplienko, L.S. (1970). *Zhur. fiz. Khim.* 44, 741 (Russ. J. Phys. Chem. 44, 411).
3. Everett, D.H. and Powl, J.C. (1976), *J.C.S. Faraday I* 72 619.

4. Nicholson, D. and Sing, K.S.W. (1973). *In* "Colloid Science" Specialist Periodical Reports (Ed. D. H. Everett). Vol. 3, pp. 10-31, The Chemical Society, London.
5. Grillet, Y., Rouquerol, F. and Rouquerol, J. (1976). *Revue Générale de Thermique* 171, 237.
6. Rouquerol, J., Rouquerol, F., Peres, C., Grillet, Y. and Boudellal, M. (1979). *In* "Characterization of Porous Solids" (Eds S.J. Gregg, K.S.W. Sing and H.F. Stoeckli). pp. 107-116. Soc. Chem. Ind., London.
7. Diano, W. (1969) Ph.D. Thesis, Exeter University.
8. Stoeckli, H.F., Houriet, J.Ph., Perret, A. and Huber, U. (1979). *In* "Characterization of Porous Solids" (Eds S.J. Gregg, K.S.W. Sing and H.F. Stoeckli). pp. 31-39. Soc. Chem. Ind., London.
9. Fryer, J. R., ibid. pp. 41-52.
10. Topic, M., Micale, F.J., Leidhesier, Jr., H. and Zettlemoyer, A.C. (1979). *Rev. Sci. Instrum.* 45, 487.
11. Micale, F.J., Vanderhoff, J.W. and El-Aasser, M.S. (1980). *A.C.S. Organic Coatings Preprints*, p. 38. Amer. Chem. Soc., Houston.
12. Micale, F.J. Unpublished Results.
13. Sing, K.S.W. (1970). *In* "Surface Area Determination" (Eds. D.H. Everett and R.H. Ottewill). pp. 25-34. Butterworths, London.
14. Joyner, L.G. and Emmett, P.H. (1948). *J. Amer. Chem. Soc.* 70, 2353.

DIRECT MEASUREMENTS OF MIXED SURFACTANT CONCENTRATIONS AT THE GAS-LIQUID INTERFACE AND THEIR RELATION TO MICELLAR COMPOSITIONS

G. G. Jayson, G. Thompson

Department of Chemistry & Biochemistry
Liverpool Polytechnic
Byrom Street
Liverpool L3 3AF

M. Hull, A. L. Smith

Unilever Research Laboratories
Port Sunlight
Wirral, Merseyside

ABSTRACT

Direct measurements have been made of adsorption at the nitrogen-solution interface of sodium dodecyl sulphate (SDS), mono-octyltetraethylene glycol and their mixtures using a radiotracer technique.

Changes in the surface tension of aqueous solutions of the mixtures have been compared with those calculated from surface excess concentrations, giving good agreement via the Gibbs' adsorption equation.

For the mixed surfactant system, micellar compositions were calculated from critical micelle concentration data (CMC) and surface compositions from the surface tension data of the single surfactants, on the basis of regular solution mixing, and compared with experimental surface compositions. Both the calculated micellar and experimental surface compositions showed a change from anionic rich to nonionic rich as the fraction of SDS was increased, with the cross-over point moving towards higher bulk SDS content as electrolyte (NaCl) was added.

INTRODUCTION

A number of workers(1-5) have reported the surface properties of mixtures of homogeneous surfactants such as those of anionic-anionic, nonionic-nonionic systems; however, there are relatively few papers dealing with binary heterogeneous surfactant solutions which are known to exhibit non-ideal mixing(6-9). In none of these have individual surface excess concentrations been determined. In this work a radiotracer technique has been used to measure directly the surface excess concentrations of surfactants adsorbed at the nitrogen-solution interface from mixed surfactant solutions at $25^{o}C$. The system studied was that of sodium dodecyl sulphate (SDS) and mono-octyltetraethylene glycol (C_8E_4), chosen because of their similar critical micelle concentrations. The measured surface compositions have been compared with surface tension measurements and with micellar compositions calculated from CMC data.

EXPERIMENTAL

The tritiated surfactants, SDS and C_8E_4 were synthesised using the unsaturated acids as the starting materials, i.e. 2-dodecenoic acid and 3-octenoic acid. Tritiation of the double bond was carried out at the Radiochemical Centre, Amersham, and the subsequent synthetic routes are described elsewhere(10). The specific radio-activities of the final purified products were 1.08 $TB_q mol^{-1}$ and 29 $GB_q mol^{-1}$ for SDS and C_8E_4 respectively. The non radio-labelled surfactants were prepared using identical procedures to those used in synthesising the radio-labelled compounds. In no case was there a minimum in the surface tension - concentration curve.

Water used in the experiments was triply distilled. Sodium chloride (Analar grade) and ammonium sulphate (Analar grade) were used without further purification, except that the saturated solution of the latter (used for calibration purposes) was treated with activated charcoal prior to use.

The technique used to measure the individual surface excess concentrations in the mixed surfactant system was that of sheet scintillation counting and is described elsewhere(11). The reported data are based on equilibrium values which were taken when two consecutive count rates were constant within the normal statistics of counting. All measurements were made at 25°C.

The surface tension measurements were obtained using an enclosed Wilhelmy plate apparatus.

TREATMENT OF DATA

A method by which surface compositions can be calculated from surface tension measurements on single surfactants and their mixtures has been published(12,13). For systems showing non-ideal mixing this method involves a regular solution treatment for the activity coefficients in the surface in the binary mixture. The surface composition is given as:

$$x_s f_1 = C_1^m/\phi_{1,0}(\gamma)$$
$$(1-x_s) f_2 = C_2^m/\phi_{2,0}(\gamma) \qquad (1)$$

where x_s is the mole fraction of surfactant (1) in the surface and C_1^m and C_2^m are the monomer activities of surfactant (1) and (2) respectively in solution. $\phi_{1,0}(\gamma)$ and $\phi_{2,0}(\gamma)$ are surface tension - concentration (activity) functions for the pure components. These take the form:

$$\phi = \exp\left[k^1 - k^{11}\gamma\right]$$

the surface activity coefficients are given as:

$$f_1 = \exp \beta_s (1-x_s)^2 \;;\; f_2 = \exp \beta_s x_s^2$$

where β_s is the surface interaction parameter.

At constant surface tension where $\gamma mixt = \gamma_{1,0} = \gamma_{2,0}$ $\phi_{1,0}(\gamma)$ and $\phi_{2,0}(\gamma)$ can be calculated. Equation (1) is then solved numerically to yield x_s and β_s.

Above the CMC the monomer activities are calculated using a similar treatment(14).

$$x_m f_1 = C_1^m / C_1$$

$$(1-x_m) f_2 = C_2^m / C_2 \qquad (2)$$

where C_1 and C_2 are the CMCs of surfactants (1) and (2) respectively and x_m is the mole fraction of surfactant (1) in the mixed micelle. The micellar activity coefficients are given by:

$$f_1 = \exp \beta_m (1-x_m)^2 \quad ; \quad f_2 = \exp \beta_m x_m^2$$

where β_m is a mixed micellar interaction parameter.

At any concentration of total surfactant (C) above the CMC the micellar mole fraction (x_m) is given by:

$$x_m = \frac{-(C-\Delta) \pm \sqrt{(C-\Delta)^2 + 4\alpha C\Delta}}{2\Delta} \qquad (3)$$

where $\Delta = C_2 f_2 - C_1 f_1$ and α is the mole fraction of surfactant (1) in bulk solution. Iterative solution of equation (3) yield x_m which from equation (2) gives C_1^m and C_2^m. These values are used in equation (1) to calculate x_s.

RESULTS AND DISCUSSION

The SDS-C_8E_4 mixed surfactant system, as expected, shows marked non-ideality for the CMC as a function of composition, which is shown in Fig. 1, where the dotted lines are calculated from ideal mixing in the micelles(5). This non-ideal mixing in the micelles can be looked upon as arising from the shielding of the electrical repulsions between the SDS head groups by the ethylene oxide groups of the nonionic. As the ionic strength is increased the electrical free energy of the system is decreased giving less deviation from ideal behaviour.

The measured individual surface excess concentrations for the SDS-C_8E_4 system is shown in Figs. 2 and 4 with no added electrolyte and with 10^{-1} mol dm^{-3} constant added sodium chloride respectively. From the data in Fig. 2 the surface tension changes for the mixture can be obtained

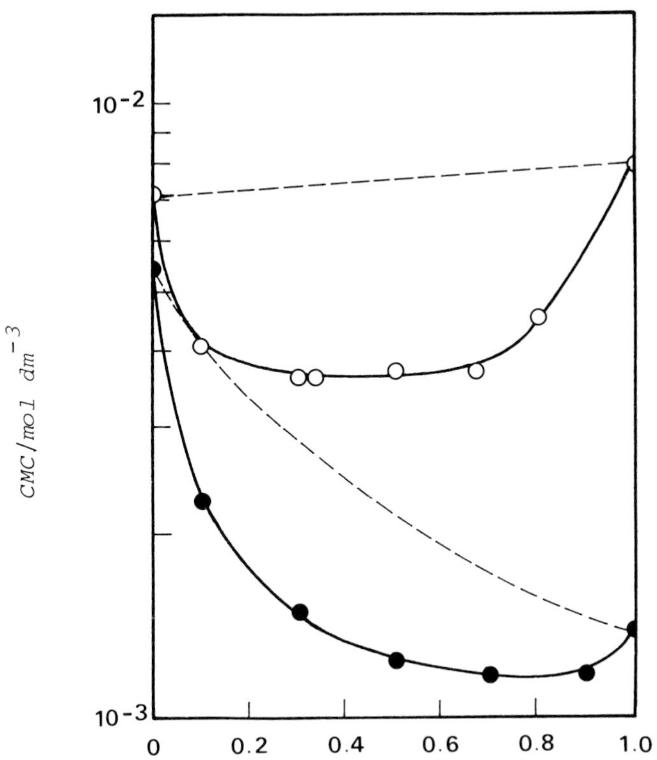

Fig. 1 - Critical micelle concentrations for SDS/C_8E_4 in the absence (O) and presence (●) of 0.1 mol dm^{-3} NaCl. The solid lines are non-ideal values fitted to a best value of the mixing parameter β. The dotted lines are ideal values.

based on the measured surface tension at 0.33 SDS mole fraction by means of the Gibb's adsorption equation.

$$-d\gamma = RT\left[\Gamma_1 d\ln a_1 + \Gamma_2 d\ln a_2\right]$$

where a_1 and a_2 are the activities of surfactants (1) and (2), and Γ_1 and Γ_2 are the surface excess concentrations of surfactants (1) and (2). The calculated surface tension changes are shown in Fig. 3 where the solid line shows the

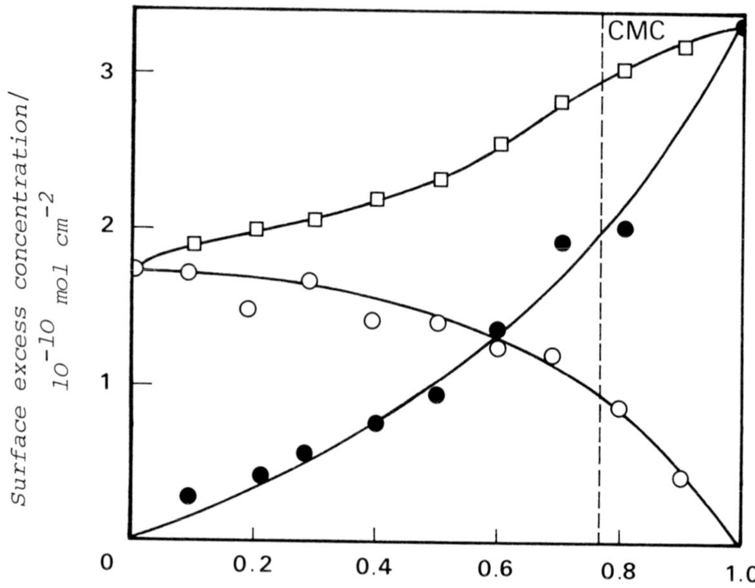

Fig. 2 - Variation of the surface excess concentrations of SDS (●), C_8E_4 (o) and total (□), at the nitrogen solution interface, with the mole fraction of surfactant in the bulk solution, and a constant nonionic concentration of 1 x 10^{-3} mol dm^{-3}

calculated surface tension. In this system the nonionic was present at a constant value of 1 x 10^{-3} mol dm^{-3} and hence $d\ln a_2 = 0$. The results obtained give confidence in the adsorption data obtained by the radiotracer results.

The effect of added electrolyte on the mixed surfactant system is shown in Fig. 4 where the experimental conditions were identical to those for Fig. 2 except that a constant level of 0.1 mol dm^{-3} sodium chloride was used. In this case the mixed CMC is markedly lowered by the electrolyte such that for most of the composition range the system is micellar. The slight depression in the total surface excess concentration measured in the micellar region is not mirrored by an increase in surface tension, which is

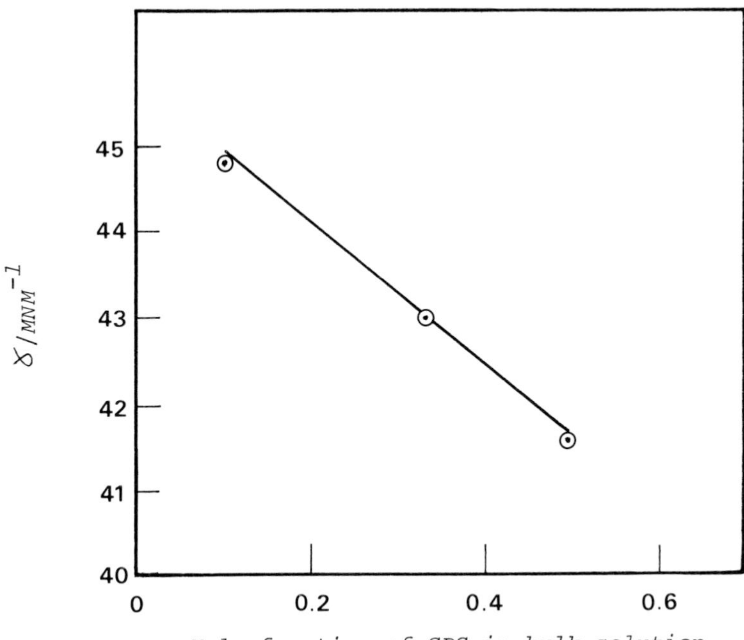

Fig. 3 - The surface tension of mixtures of SDS and C_8E_4 at a constant C_8E_4 concentration of 1×10^{-3} mol dm^{-3}

virtually constant throughout the SDS mole fraction range 0.3-0.9. Hence the decrease in the total surface excess concentration is attributable to changes in the monomer activities of the two surfactants in the micellar solutions.

Figs. 5 and 6 show changes in surface composition for the systems having a constant nonionic concentration of 1×10^{-3} mol dm^{-3} C_8E_4 and in the absence and presence of 0.1 mol dm^{-3} sodium chloride respectively. The adsorption is seen to favour the anionic at low SDS mole fractions and the nonionic at high mole fractions. The point of equal surface and bulk compositions moves towards higher SDS mole ratios as electrolyte is added, ie. as the electrical free energy contribution from the charged surfactant is reduced.

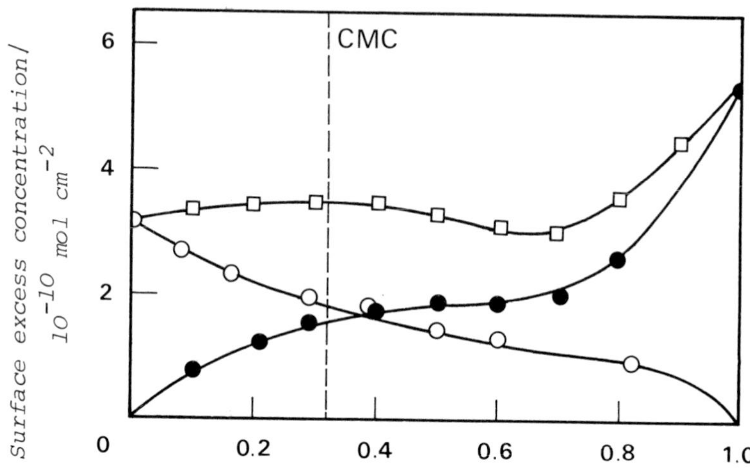

Fig. 4 - Variation of the surface excess concentrations of SDS (●), C_8E_4 (o), and total (□), at the nitrogen solution interface, with the mole fraction of surfactant in the bulk solution, and a constant concentration of C_8E_4 of 1×10^{-3} mol dm^{-3} and NaCl concentration of 0.1 x mol dm^{-3}

The surface tension data and micellar data for the mixed SDS-C_8E_4 have been used to calculate the surface composition of these systems. This is shown in Figs. 5 and 6. The calculated and the experimentally measured surface compositions are qualitatively similar. However, in the low electrolyte case (Fig. 5) the absolute agreement between the calculated and measured composition is not good. The calculations give rise to a surface interaction parameter (β_S) which is numerically smaller than that required to give full agreement with the experimental data. In the micellar region, in the presence of 0.1 mol dm^{-3} sodium chloride, the calculated and measured surface compositions are in good agreement. However, submicellar, the regular solution model yields a surface interaction parameter more negative than indicated by the experimental data. In Fig. 6 the micellar composition calculated from CMC data was used to calculate the surface composition of the mixed surfactants and the calculated and measured surface compositions are in good agreement.

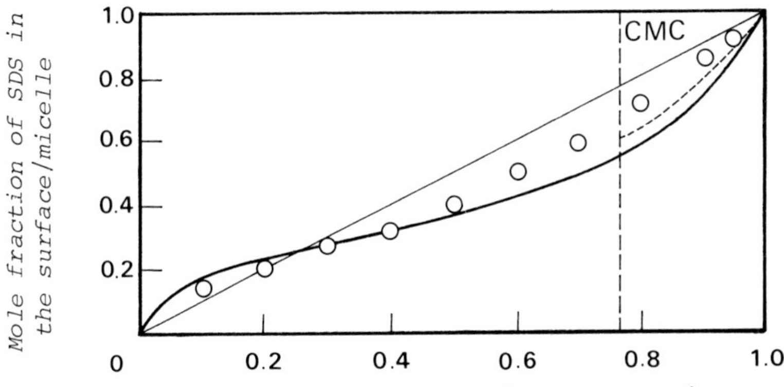

Fig. 5 - Variation of the measured surface (o), and calculated surface (-) and micellar (---) compositions with mole fraction of SDS in bulk solution for SDS:C_8E_4 with constant C_8E_4 concentration of 1×10^{-3} mol dm^{-3}

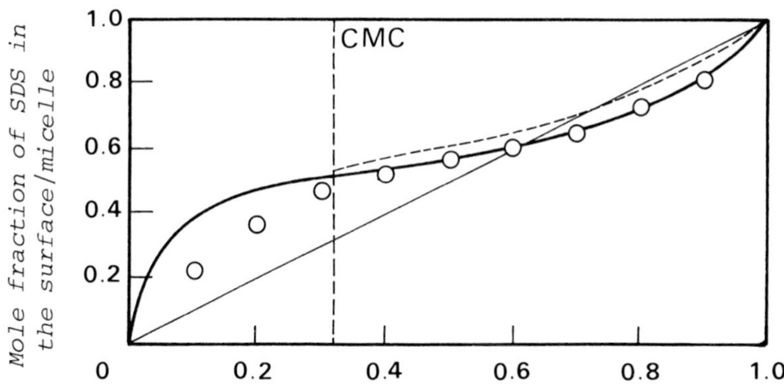

Fig. 6 - Variation of the measured surface (o), and calculated surface (-) and micellar (---) compositions with the mole fraction of SDS in the bulk solution for SDS:C_8E_4 with a constant C_8E_4 concentration of 1×10^{-3} mol dm^{-3} in the presence of 0.1 mol dm^{-3} NaCl

G.T. would like to thank the SERC for a CASE award and Unilever PLC for support.

REFERENCES

1. Mysels, K.J. and Otter, R.J. (1961). *J. Colloid Sci.*, 16, 462.

2. Shedlovsky, L., Jacob, C.W. and Epstein, M.B. (1963). *J. Physical Chem.*, 67, 2075.

3. Moroi, Y., Motomura, K. and Matuura, R. (1974). *J. Colloid Sci.*, 46, 111.

4. Clint, J.H. (1975). *J. Chem. Soc. Faraday Trans.*, 71, 1327.

5. Lange, H. and Beck, K.H. (1973). *Kolloid Z uz Polymere*, 251, 424.

6. Corkill, J.M., Goodman, J.F. and Tate, J.R. (1964). *Trans. Faraday Soc.*, 60, 986.

7. Funasaki, N. (1978). *J. Colloid and Interface Sci.*, 67, 384.

8. Funasaki, N. and Hada, S. (1979). *J. Physical Chem.*, 83, 19, 2471.

9. Schick, M.J. and Manning, D.J. (1966). *J. Am. Oil Chem. Soc.*, 43, 133.

10. Thompson, G. (1980). *PhD Thesis*, Liverpool Polytechnic, England.

11. Muramatsu, M. (1973). *"Surface and Colloid Science"* ed Matijevic, E., Wilay Interscience, New York, Vol 6, p. 101.

12. Garrett, P.R. (1976). *J. Chem. Soc. Faraday Trans. I*, 72, 2174.

13. Ingram, B.T. (1980). *Colloid and Polymer Sci.*, 258, 2, 191.

14. Rubingh, D. (1978). *"Solution Chemistry of Surfactants"* ed Mittal, K.L., Vol 1, p. 337 Published.

DETERMINATION OF THE ADSORPTION OF SURFACE ACTIVE AGENTS ON LATEX PARTICLES BY SMALL ANGLE NEUTRON SCATTERING

N.M. Harris[†], R.H. Ottewill[*] and J.W. White[**]

Present address: [†]*Saint-Gobain Recherche, Aubervilliers, France*

[*]*School of Chemistry, University of Bristol*

[**]*Physical Chemistry Laboratory, University of Oxford*

INTRODUCTION

A unique feature of a neutron beam is its ability to distinguish between hydrogen and deuterium since these atoms have quite different coherent scattering cross-sections (1). Moreover, since H_2O and D_2O differ substantially in neutron scattering length density it is frequently possible to find a mixture of these which balances the neutron scattering length density of a particle but not its adsorbed layer (2). In this way the properties of the adsorbed layer can be examined by small angle neutron scattering in the presence of the particle. If small colloidal particles are used a monolayer of simple surface active agent molecules can occupy a very considerable volume with respect to that of the particle and it is possible to determine the population density and thickness of such layers in a suitable scattering experiment. For such studies it is convenient to prepare the adsorbent in the form of small spherical particles, e.g. a polymer latex, since the scattering from isotropic spherical particles of this type can be treated exactly and has been examined in some detail both theoretically and experimentally (3,4,5). Adsorption onto a spherical particle can then be treated by a concentric sphere model if a uniform distribution of adsorbate is assumed (4,6). A model of a particle with an adsorbed layer is discussed in this paper and it is shown that several experimental approaches can be developed from it. One of these has been qualitatively demonstrated previously using as the basic system the adsorption of d_{23}-dodecanoate ions onto polystyrene latex particles (2,7). Our experiments have continued with this system and we now report a more detailed study using two methods of approach to analyse the experimental small angle neutron scattering data. A complete adsorption isotherm has been obtained and an examination made of the area per adsorbed molecule on the polystyrene surface as a function of pH.

Our interest in this method of approach was stimulated by the fact that there are many systems in colloid science where a molecular layer is covalently linked to the surface (8) or where removal of the adsorbed layer would lead to the complete destabilisation of the colloidal dispersion (9). For these systems conventional methods of starting with the bare materials and then adding the adsorbate cannot be used to investigate the properties of the layer. In these circumstances small angle neutron scattering can be applied directly to the total system in situ. Even for more conventional systems the task of removing the particles by mechanical means, e.g. centrifugation, in order to analyse the supernatant is avoided and the adsorbed layer is examined directly on the substrate.

This paper investigates the feasibility of using neutron scattering to study the adsorption of surface active molecules on spherical particles.

EXPERIMENTAL

Materials

The distilled water used was doubly distilled, the final distillation being carried out in a quartz still.

The ammonium chloride and ammonium hydroxide were reagent grade materials.

d_{23}-dodecanoic acid was obtained from Merck, Sharpe and Dohme, Canada, Limited, with an isotopic purity of 98%.

The polystyrene latex was prepared by a seeded growth procedure similar to that previously described (10). Potassium persulphate was used as the initiator and sodium octadecyl sulphate as the stabilizing surface active agent. The latter together with oxidation products from the polymerisation was removed by extensive dialysis against distilled water before the material was used for neutron scattering experiments. Electron microscope examination of this latex gave a number average diameter of 456 ± 30 Å with a standard deviation on the mean of 11%.

Adsorption Isotherm Measurements were made mainly on latices at a concentration of 0.2% w/v prepared in an 8% D_2O: 92% H_2O mixture. Samples were prepared at various total concentrations of dodecanoate up to 1.6×10^{-2} mol dm^{-3}. The critical micelle concentration of sodium dodecanoate is listed as 1.82×10^{-2} mol dm^{-3} at 25°C in 2.18×10^{-2} mol dm^{-3} sodium chloride solution (11,12). The pH of the systems was controlled by the use of ammonium hydroxide/ammonium chloride buffer solutions. Most studies were carried out at pH 8.1 ± 0.1. The total ionic strength of the systems was maintained at an initial value of 2×10^{-2} by the addition of sodium chloride. The scattering length density of the buffer solutions was taken to be that of water (-0.56×10^{10} cm^{-2}).

Scattering data were obtained on the basic latex, i.e. containing bare particles, on the samples containing d_{23}-dodecanoate and on blank solutions of dodecanoate.

Small Angle Neutron Scattering (SANS) Measurements The SANS measurements were made at the Institut Laue Langevin (I.L.L.) Grenoble, using the neutron diffractometer D11A. The instrument was used at a sample-detector distance of 10 m using a neutron beam of wavelength 10 Å; The full width at half height of the distribution of wavelength, $\delta\lambda/\lambda$ was 9%. This gave an observable range of scattering vector, Q, from 2×10^{-3} Å^{-1} to 2×10^{-2} Å^{-1}. The counting time per sample required to give adequate counting statistics was typically 20 min. All the measurements were made at room temperature, 22 ± 1°C. The samples were contained in optical-quality quartz cells with a path length of 1 mm.

The experimental data were obtained as counts on a two-dimensional detector, the plane of which was perpendicular to the incident neutron beam. Standard I.L.L. computer programs (13) were used to process the basic data to give I(Q) as a function of the scattering vector, Q.
The solvent background was subtracted from the scattering patterns of each of the samples.

The subtraction of the solution backgrounds required an estimate of the decrease of the d_{23}-dodecanoate concentration on adsorption to be made. The background for each sample was estimated by interpolation between backgrounds measured at intervals of dodecanoate concentration. Below 10^{-2} mol dm^{-3} dodecanoate the difference between solution and solvent was small. At 1.6×10^{-2} mol dm^{-2} a significant difference was observed particularly at the lower Q values; this suggested that some micellar aggregation could have occurred even though the system was expected to be below the critical micelle concentration.

THEORY

Homogeneous Spherical Particles

For a system containing N_p homogeneous particles per unit volume, each of volume V_p then, I(Q), the intensity of neutrons scattered at a particular scattering vector, Q, is given by (1,3)

$$I(Q) = K(\rho_p - \rho_s)^2 \ N_p V_p^2 \ . \ P(Q) \tag{1}$$

where K = an instrument constant, ρ_p = the neutron scattering length density of the particle and ρ_s = the scattering length density of the dispersion medium. The scattering vector Q is given by,

$$Q = \frac{4\pi}{\lambda} \ . \ \sin(\theta/2) \tag{2}$$

with λ = the wavelength of the incident neutrons and θ = the scattering angle. The volume of the particles is given by,

$$V_p = \frac{4}{3}\pi R^3 \tag{3}$$

where R = the radius of the particles. P(Q) is the particle form factor which for spherical particles is given by,

$$P(Q) = \left(3(\sin QR - QR \ . \ \cos QR)/(QR)^3\right)^2 \tag{4}$$

For small values of QR (QR << 1) equation (4) can also be written in the form,

$$P(Q) = \left[1 - Q^2R^2/10 + \ldots\ldots\right]^2 \tag{5}$$

Substitution of (5) into (1) leads to,

$$I(Q) = K(\rho_p - \rho_s)^2 \ N_p V_p^2 \left(1 - \frac{Q^2R^2}{10} \ \ldots\ldots\right)^2 \tag{6}$$

so that for Q = 0,

$$I(0) = K(\rho_p - \rho_s)^2 \ N_p V_p^2 \quad \ldots\ldots \tag{7}$$

It follows therefore that for $\rho_p = \rho_s$, then I(Q) and I(0) = 0, and hence this forms the basis of a method for the determination of the neutron scattering length density of the particle (1,2). A number of values of scattering length density are given in Table 1.

TABLE 1

Neutron Scattering Length Densities

Material	$\rho/10^{10}$ cm^{-2}
Water, H_2O	- 0.56
Heavy Water, D_2O	6.40
h-polystyrene	1.41
d_{23}-dodecanoic acid	6.43

It was shown by Guinier (3) that equation (6) leads directly to,

$$I(Q) = I(0)\ \exp(-\ Q^2 Rg^2/3) \tag{8}$$

with Rg = the radius of gyration of the particle. For a single homogeneous sphere,

$$Rg^2 = \frac{3}{5}\ R^2 \tag{9}$$

Rewriting equation (9) as,

$$\ln\ I(Q) = \ln\ I(0) - R_g^2 \cdot \frac{Q^2}{3} \tag{9a}$$

indicates immediately that a plot of ln I(Q) against Q^2 should be linear thus enabling ln I(0) and hence I(0) to be obtained by extrapolation and R_g to be obtained from the gradient.

Homogenous Particles with an Adsorbed Layer Figure 1 illustrates the situation of a homogeneous spherical particle, radius R_1, and neutron scattering length density ρ_p surrounded by an adsorbed layer of neutron scattering length density, ρ_L, such that the overall diameter of the particle is R_2.

The intensity of scattering for this situation is given by,

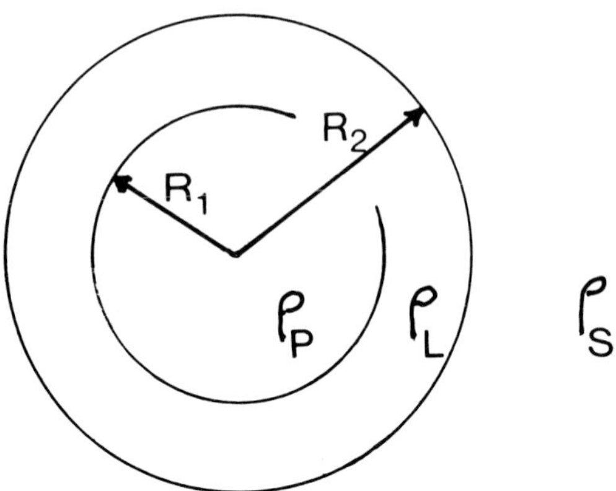

Fig.1 Model particle showing basic parameters used.

$$I(Q) = K N_p \cdot \frac{16\pi^2}{9} \left[(\rho_L - \rho_S) \left\{ 3R_2^3 \left(\frac{\sin QR_2 - QR_2 \cdot \cos QR_2}{Q^3 R_2^3} \right) - 3R_1^3 \left(\frac{\sin QR_1 - QR_1 \cdot \cos QR_1}{Q^3 R_1^3} \right) \right\} + (\rho_p - \rho_s) 3R_1^3 \left(\frac{\sin QR_1 - QR_1 \cdot \cos QR_1}{Q^3 R_1^3} \right) \right]^2 \quad (10)$$

This equation gives immediately a method of analysing the experimental data. However, for QR < 1 this reduces to,

$$I(Q) = K N_p \left[(\rho_L - \rho_S) V_L + (\rho_p - \rho_s) V_p - \frac{Q^2}{10} \left\{ (\rho_L - \rho_S)(V_T R_2^2 - V_p R_1^2) + (\rho_p - \rho_s) V_p R_1^2 \right\} \right]^2 \quad (11)$$

where V_L = the volume of the adsorbed layer,
V_p = the volume of the core particle,
so that the total volume of the composite particle is given by

$$V_T = V_p + V_L \tag{12}$$

Thus we obtain,

$$(\rho_L - \rho_S) V_L + (\rho_p - \rho_s) V_p = V_T \left(\frac{\rho_L V_L + \rho_p V_p}{V_T} - \rho_s \right) \tag{13}$$

For the composite particle a mean scattering length, ρ_m, can be defined as,

$$\rho_m = \frac{\rho_L V_L + \rho_p V_p}{V_T} \tag{14}$$

and the difference between the mean scattering length of the particle and the dispersion media by,

$$\bar{\rho} = \rho_m - \rho_s \tag{15}$$

whence,

$$(\rho_L - \rho_S) V_L + (\rho_p - \rho_s) V_p = \bar{\rho} \cdot V_T \tag{16}$$

Consequently equation (11) can be rewritten as,

$$I(Q) = K N_p \bar{\rho}^2 V_T^2 \left[1 - \frac{Q^2}{10 \bar{\rho} V_T} \left\{ (\rho_L - \rho_S)(V_T R_2^2 - V_p R_1^2) + \frac{Q^2}{10 \bar{\rho} V_T} (\rho_p - \rho_s) V_p R_1^2 \right\} \cdots \right]^2 \tag{17}$$

For Q = 0 this gives,

$$I(0) = K N_p (\bar{\rho} V_T)^2 \tag{18}$$

After multiplying out the term in the bracket in equation (17) we also obtain,

$$I(Q) = I(0) \left[1 - \frac{Q^2}{5 \bar{\rho} V_T} \left\{ (\rho_L - \rho_S)(V_T R_2^2 - V_p R_1^2) \right\} + \frac{Q^2}{5 \bar{\rho} V_T} \left\{ (\rho_p - \rho_s) V_p R_1^2 \right\} \right] \cdots \tag{19}$$

By analogy with the homogeneous particle we can define a root mean square radius $\bar{R}$, such that

$$\bar{R}^2 = \frac{1}{\bar{\rho} V_T} \left\{ (\rho_L - \rho_S)(V_T R_2^2 - V_p R_1^2) + (\rho_p - \rho_s) V_p R_1^2 \right\} \quad (20)$$

and a mean radius of gyration $\bar{R}g$, such that

$$\bar{R}g^2 = \frac{3}{5} \bar{R}^2 \quad (21)$$

and hence

$$I(Q) = I(0) \exp(-Q^2 \bar{R}g^2 / 3) \quad (22)$$

which shows that a Guinier plot can be utilised also for the composite particle and an analysis made in terms of $\bar{R}g$ (4). An alternative is to use equation (22) as a means of extrapolating the experimental data to $Q = 0$ in order to obtain I(0). The latter approach was used in the present work.

RESULTS AND DISCUSSION

Adsorption Determination from Intensity Measurements over a Range of Q Values

The results of S.A.N.S. measurements on the polystyrene latex in the presence and absence of sodium dodecanoate are shown in Figure 2. These were carried out using a dispersion medium composed of 8% D_2O (w/v) and 92% H_2O (w/v), a mixture which gives zero scattering length for the dispersion medium, i.e. $\rho_s = 0$. As can be seen distinct changes in the scattering curves were obtained as the concentration of sodium dodecanoate in the system was increased.

For the condition, $\rho_s = 0$, the scattering by the bare particles is given by equation (1) in the form,

$$I(Q) = K \rho_p^2 \phi V_p P(Q) \quad (24)$$

In order to apply this equation the volume fraction of the latex $\phi = N_p V_p$ was determined analytically and ρ_p was taken as 1.41×10^{10} cm^{-2} (see Table 1). In order to allow for the polydispersity of the latex a zeroth order logarithmic distribution of particle size was assumed since this has been established to give a good fit in light scattering experiments (14).

This has the form,

$$p(R) = \frac{\exp\left[-\frac{(\ln R - \ln R_m)^2}{2\sigma_o^2}\right]}{(2\pi)^{\frac{1}{2}} \sigma_o R_m \exp(\sigma_o^2/2)} \qquad (25)$$

with R_m = the modal radius of the particles and σ_o = a parameter which gives a measure of the width and skewness of the distribution. The relationship between R_N, the number average value of R, and R_m is given by,

$$\ln R_N = \ln R_m + 1.5\,\sigma_o^2 \qquad (26)$$

and the standard deviation σ by

$$\sigma = R_m \left[\exp(4\sigma_o^2) - \exp(3\,\sigma_o^2)\right]^{\frac{1}{2}} \qquad (27)$$

Equation (25) was found to be particularly suitable for computer processing in order to compare theoretical curves with experimental data. For the bare particles, using a non-linear least squares method, equations (4) and (24) were used with various values of R_N and σ to find the best fit to the experimental values. This is shown as curve a in Figure 2 for R_N = 205 Å and σ = 16%. The fitting procedure was found to be more sensitive to variation of R_N than σ.

Once the size and standard deviation of the bare latex particles had been established the experimental data on the coated particles were examined using equation (10) with the condition $\rho_s = 0$. Several approaches to this problem are possible (4) but the one found most useful can be illustrated in the following way. For $\rho_s = 0$ at low Q values equation (10) becomes,

$$I(Q) = K\,N_p\left[\rho_L V_L\left(1 - \frac{Q^2R_2^2}{10}\,\ldots\right) + \rho_p V_p\left(1 - \frac{Q^2R_1^2}{10}\,\ldots\right) - \frac{Q^2\rho_L V_p}{10}(R_2^2 - R_1^2)\,\ldots\right]^2 \qquad (28)$$

Since N_p, R_1, σ and ρ_p are known the variables to be fitted are ρ_L and R_2, or $R_2 - R_1$ with ρ_L = the scattering length density of d_{23}- dodecanoate. The adsorbed layer parameter in equation (28) is essentially the term $\rho_L\ V_L$. V_L may be

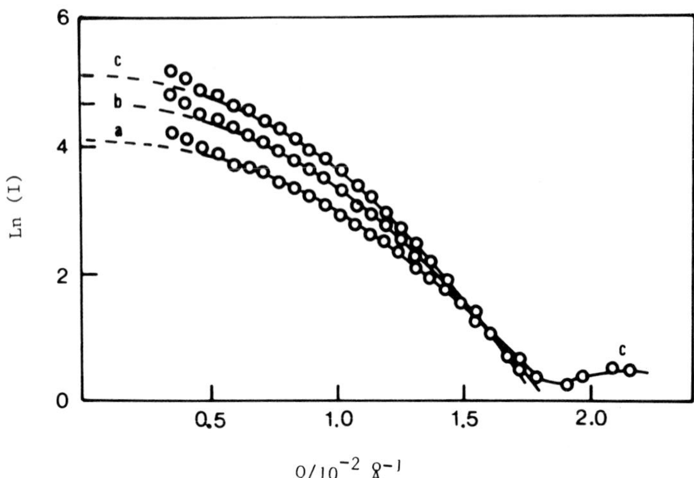

Fig.2 Examples of scattering data obtained. a) bare latex; b) latex in 4 x 10^{-3} mol dm^{-3} total dodecanoate concentration; c) latex in 1.2 x 10^{-2} mol dm^{-3} total dodecanoate concentration. ——, *fits of concentric shell model.*

written as,

$$V_L = (R_2 - R_1) \cdot S_p \tag{29}$$

where $(R_2 - R_1)$ = the thickness of the adsorbed layer and S_p = the area of the particle i.e. $4\pi R_1^2$. Moreover,

$$\rho_L = b_L/V_{LM} \tag{30}$$

where b_L = the neutron scattering length of the adsorbed molecule, d_{23}-dodecanoate, and V_{LM} its molecular volume. Hence,

$$\rho_L \, V_L = \left\{ \frac{(R_2 - R_1) \cdot S_p}{V_{LM}} \right\} b_L \tag{31}$$

where the term in curly brackets is the number of molecules adsorbed on the particles, n and thus we obtain,

$$\rho_L V_L = n\, b_L \tag{32}$$

where the right hand side is the product of a number density and a scattering length. This shows clearly that the scattering from the adsorbed layer is directly dependent on the number of molecules adsorbed.

If the area occupied by an adsorbed molecule is taken as A then,

$$A = S_p/n \tag{33}$$

and

$$\rho_L V_L = S_p . b_L/A \tag{34}$$

Since b_L was known the best fit to the experimental curves was obtained over a range of Q values by assuming various values of A. Thus it was possible to obtain the number of molecules of adsorbate directly independently of the thickness of the layer. This is a consequence of the insensitivity of the intensity to thickness in this Q range when the layer thickness is small compared to the particle size (4). Although the scattering is more sensitive to the structure of the adsorbed layer at high values of QR the scattered intensity decreases substantially with QR and good counting statistics are difficult to achieve in this region.

Adsorption Determination from I(0)

From the physical nature of the apparatus used for neutron scattering it is not possible to make direct measurements of scattered intensity at Q = 0. However, it can be seen from Figure 2 that it is possible to obtain this quantity by extrapolation of the data from experimental measurements made at low Q values. In order, to obtain a good extrapolation a linear plot is preferable and it can readily be seen from equations (9a) and (22) that this can be achieved using plots of ln I(Q) against Q^2. The data given in Figure 2 is shown replotted in this form in Figure 3. The apparent linearisation of the Guinier plot outside its anticipated range of validity is clearly apparent, for example, in the results for the bare latex. The good linear fit is partly a consequence of some particle size polydispersity. This allowed representative fits to be obtained away from the smallest Q values where the scattering was very sensitive to the presence of small aggregates.

For analysis of the data at zero Q we obtain from equation (11),

$$I(0) = K\,N_p\left[(\rho_p - \rho_s)\,V_p + (\rho_L - \rho_S)\,V_L\right]^2 \tag{35}$$

which on combining with equation (34) gives,

$$I(0) = K\,N_p\left[(\rho_p - \rho_s)\,V_p + (b_L - \rho_s\,V_{LM})\,\frac{S_p}{A}\right]^2 \tag{36}$$

and for $\rho_s = 0$

$$I(0) = K\,N_p\left[\rho_p\,V_p + \frac{b_L \cdot S_p}{A}\right]^2 \tag{37}$$

an equation which allows a direct determination of A.

For polymer latices measurements can be made under identical conditions of volume fraction and instrument settings on the system containing bare particles and on the system containing the particles plus their adsorbed layers. Thus if we designate the scattering from the bare particles as $[I(0)]_p$ and that from the particles plus adsorbed layer as $[I(0)]_L^p$ we obtain for the ratio,

$$I_R = \frac{[I(0)]_L^{\frac{1}{2}}}{[I(0)]_p^{\frac{1}{2}}} = 1 + \frac{b_L \cdot S_p}{A\,V_p\,\rho_p} \tag{38}$$

or putting,

$$I_R^{\,1} = I_R - 1$$

we obtain,

$$A = \frac{b_L}{\rho_p}\left(\frac{S_p}{V_p}\right)\frac{1}{I_R^{\,1}} \tag{39}$$

an equation which provides a straightforward method for the determination of the area per adsorbed molecule on the surface. As written equation (39) is only strictly valid for a monodisperse substrate. The errors introduced by the presence of a particle size distribution can be estimated and are shown to be less than 5% for $\sigma < 15\%$ (4).

An adsorption isotherm for d_{23}-dodecanoate ions on polystyrene at pH 8.1 ± 0.1, in a total 1:1 electrolyte

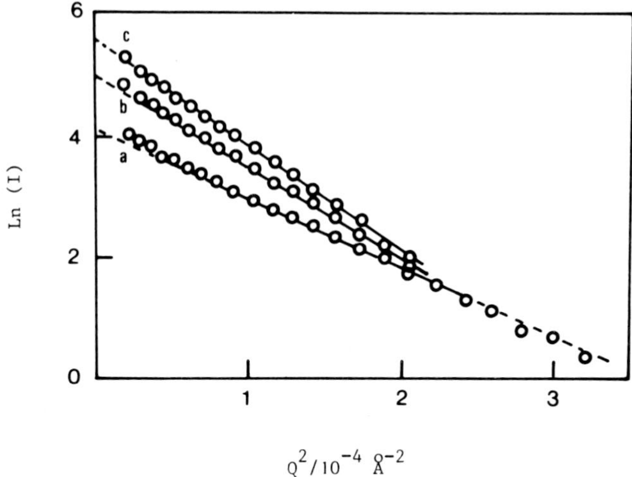

Fig.3 Guinier plots of scattering curves presented in Figure 2. ——, *Guinier fit;* - - - , *extrapolated curves.*

concentration of 2 x 10^{-2} mol dm^{-3} at 22 ± 1°C is shown in Figure 4. In this figure the results obtained by Guinier analysis are compared with those obtained using the concentric shell model. Both methods of analysis are seen to give the same result within experimental error but the latter gives the more systematic approach.

Adsorption of Dodecanoate Ions at pH 8 The adsorption isotherm obtained shows a monotonic increase in the extent of adsorption to a plateau value. The latter is reached just before the critical micelle concentration of the adsorbate and corresponds to an area per adsorbed dodecanoate ion of 28 ± 5 Å^2. The isotherm is similar in shape to that obtained by Connor (15) using radio-labelled h_{23}- dodecanoate at pH 8.0 and in 10^{-3} mol dm^{-3} sodium bromide solution. On the plateau of the isotherm he obtained an area per ion of 45 Å . However, since his ionic strength was a factor of 10 lower than that used in the present work it is difficult to make a direct comparison of the results.

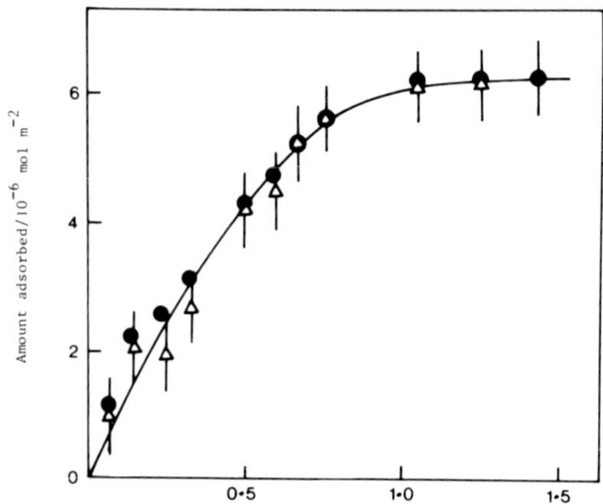

Fig.4 Adsorption isotherm for d_{23}-dodecanoic acid on polystyrene latex at pH 8.1 and 22 ^{o}C with a total electrolyte concentration of 2.2 x 10^{-2} mol dm^{-3}. △, *calculated by fitting a spherical shell model plus error bars;* ●, *points obtained from Guinier analysis.*

Adsorption as a Function of pH The variation of the area occupied per adsorbed dodecanoate ion as a function of pH, taken at the plateau region of the isotherms, is shown in Figure 5. The values given were based on the use of the concentric sphere model to analyse the data. The results demonstrate a pronounced variation of the area per adsorbed ion/molecule with pH. A value of 28 $Å^2$ was obtained for the area occupied at the lowest pH investigated. This suggests that at pH 7.0 and lower pH values the fatty acid molecules form an adsorbed layer on the polystyrene latex particles in which the fatty acid molecules are close-packed and vertically oriented. The experimental values appear to be approaching the cross-sectional area of a monobasic acid molecule of 20.2 $Å^2$.

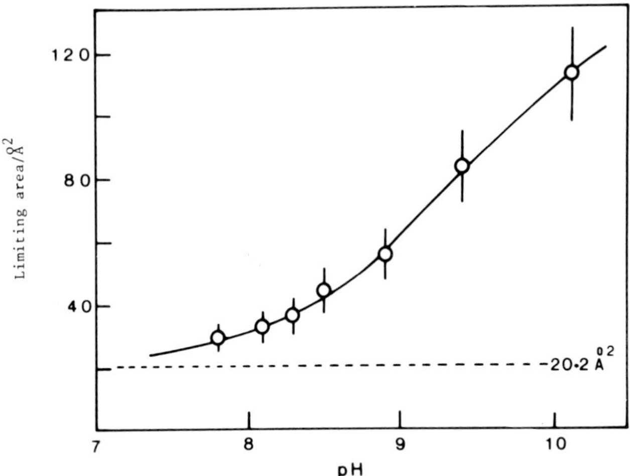

Fig.5 Variation of limiting area per molecule with pH. Total d_{23}*-dodecanoate concentration,* 10^{-2} *mol* dm^{-3} *in a total electrolyte concentration of* 2×10^{-2} *mol* dm^{-3}.

CONCLUSIONS

Model systems have been used to examine the possibility of using small angle neutron scattering to determine the adsorption of small molecules on surfaces. For the situation examined with polystyrene latex and d_{23}-dodecanoate ions the validity of the technique has been demonstrated. In this work it was possible to examine the bare particles initially and then the particles plus their adsorbed layer without separating the particles from the dispersion medium. In more recent work we have successfully used the method to examine dispersions with a permanently attached layer (9) where it was not possible to examine the particles independently. We conclude that the technique of small angle neutron scattering offers considerable versatility of approach in the study of adsorption in that it enables the observer to examine in a direct manner the adsorbed layers on particles in their own environment.

ACKNOWLEDGEMENTS

This work was carried out at the Institut Laue-Langevin, Grenoble and we wish to express our thanks to the Directors for the neutron beam facilities provided. We also wish to thank the S.E.R.C. for their support of this work.

REFERENCES

1. Jacrot, B. (1976). *Rep.Prog.Phys.*, 39, 911.
2. Cebula, D., Thomas, R.K., Harris, N.M., Tabony, J. and White, J.W. (1978). *Faraday Discuss.Chem.Soc.*, 65, 76.
3. Guinier, A. and Fournet, G. (1955). "Small Angle Scattering of X-rays", John Wiley, New York.
4. Harris, N.M. (1980) "Small Angle Neutron Scattering from Colloidal Systems", D.Phil. thesis, Oxford University.
5. Goodwin, J.W., Harris, N.M. and Ottewill, R.H. in press.
6. Ottewill, R.H. (1982). "Chem.Soc.Review Symposium on Colloid Science", Royal Society of Chemistry, London.
7. Harris, N.M., Ottewill, R.H. and White, J.W. (1979). I.L.L. Annual Report, 394, 400.
8. Cebula, D.J., Goodwin, J.W. and Ottewill, R.H. (1981). I.L.L. Annual Report, 318.
9. Cebula, D.J., Marković, I. and Ottewill, R.H. (1981). I.L.L. Annual Report, 323.
10. Ottewill, R.H. and Shaw, J.N. (1967). *Kolloid Z.u.Z. Polymere*, 215, 161.
11. Merrill, R.C. and Getty, R. (1948). *J.Phys.Colloid Chem.*, 52, 74.
12. Mukerjee, P. and Mysels, K.J. (1971). "Critical Micelle Concentrations of Aqueous Surfactant Systems", Nat. Bur.Stand., Washington, D.C.
13. Ghosh, R.E. (1978). "A Computing Guide for Small-Angle Scattering Experiments", Institut Laue-Langevin, Grenoble.
14. Kerker, M. (1969). "The Scattering of Light and Other Electromagnetic Radiation", Academic Press, New York.
15. Connor, P. (1967). "The Interaction of Cationic Surface Active Agents with Polystyrene Latices", Ph.D. thesis, University of Bristol.

THE ADSORPTION OF SURFACE ACTIVE AGENTS ON POLYTETRAFLUOROETHYLENE LATEX PARTICLE SURFACES

Hilary E. Bee†, R.H. Ottewill, D.G. Rance* and Rachel A. Richardson

School of Chemistry, University of Bristol

† *I.C.I. Australia, Melbourne*

* *I.C.I. Petrochemicals and Plastics Division, Welwyn Garden City*

INTRODUCTION

Polytetrafluoroethylene (PTFE) surfaces are unique in the field of surface chemistry in that they have a very low surface energy, 18.5 mN m^{-1}, and can be regarded as being both hydrophobic and oleophobic. Chemically, they are regarded as being rather inert but since PTFE is prepared by a dispersion polymerisation process (1,2), which is free radical initiated, chemical groupings from the initiator are incorporated into the surface of the latex particles. Evidence for these has been obtained in our previous work (3,4,5). In view of this evidence it appears that the surface of a PTFE latex should be regarded as being predominantly composed of $-CF_2-$ groupings with a relatively small number of hydrophilic ionic groupings distributed, probably non-uniformly, over the surface.

This combination of properties makes the surfaces of considerable interest, both academically and industrially, as substrates for the adsorption of surface active agents but surprisingly there seems to be little work reported in the literature on this topic. In this study we have examined the adsorption of anionic, cationic and non-ionic surface active agents onto PTFE latex particles. In the case of anionic surface active agents the adsorption of both hydrocarbon-chain and fluorocarbon-chain carboxylic acids was examined. The latices were extensively cleaned as described previously (3) to remove from the PTFE particle surfaces materials used in their preparation.

EXPERIMENTAL

Materials

All the distilled water used in the experiments was doubly distilled, the second distillation being carried out using an all-Pyrex apparatus.

Ammonium chloride, ammonium nitrate and sodium nitrate were Analar grade materials supplied by B.D.H. Chemicals Limited.

Surface Active Agents The dodecyl hexaoxyethylene glycol monoether ($C_{12}E_6$) was kindly supplied, as a pure sample, by Dr. R. Laughlin of the Procter and Gamble Company, Cincinnati, Ohio. The critical micelle concentration of this material in water was found to be 9.5×10^{-5} mol dm^{-3}.

Unlabelled sodium octanoate and sodium dodecanoate were obtained from B.D.H. Chemicals Limited. The ^{14}C radio-labelled sodium octanoate and dodecanoic acid were supplied by the Radiochemical Centre, Amersham. The former was supplied as a 50 μC sample having a specific activity of 31.2 mC $mmol^{-1}$ and the latter as a 250 μC sample of specific activity 31.2 mC $mmol^{-1}$.

Ammonium perfluorooctanoate (APFO) was a 99% pure sample obtained from Rimar SpA, Vicenza, Italy, ^{14}C labelled APFO was kindly prepared by S.R.D. Oldland, I.C.I. Limited, Plastics Division and was supplied as a 0.4 μC sample of specific activity 14.0 μC $mmol^{-1}$.

Decyltrimethylammonium bromide (DeTAB) was pure material prepared by the interaction of bromodecane with trimethyl-amine followed by recrystallisation of the product from moist acetone. The surface tension against log concentration curve obtained with this material did not exhibit a minimum and had a critical micelle concentration in water of 5.6×10^{-2} mol dm^{-3} at $25^{\circ}C$.

Polytetrafluoroethylene, PTFE, Latices The PTFE latices were kindly supplied by I.C.I. Limited, Plastics Division, and were prepared by dispersion polymerisation. APFO, 0.14% w/w based on monomer, was used as the emulsifier and bis-(β-carboxypropionyl) peroxide was used as the initiator. The polymerisation was carried out at $70^{\circ}C$ in a stainless steel pressure vessel and tetrafluoroethylene monomer was added batchwise at a pressure of 2.6×10^{6} N m^{-2}.

The PTFE latices were stored in an undialysed state until required. Samples were then transferred to a length of well-boiled Visking dialysis tubing and dialysed against a ten fold excess of distilled water which had been adjusted to pH = 10 by the addition of sodium hydroxide. Fifteen exchanges of water were made at pH = 10 and a further fifteen under neutral conditions. All the measurements made with the latices were carried out immediately after dialysis.

Characterization of the PTFE Latices The particle size distribution of the latex particles was obtained by electron microscopy using the sample preparation techniques described previously (6). The magnification range of the instrument was calibrated using a carbon replica of a diffraction grating. Particle size analysis was carried out as described in a previous paper by measuring both the long axis and the short axis of the asymmetric particles (3).

The surface charge density of latex DR2 was determined by conductimetric titration with sodium hydroxide solution in the presence of 10^{-3} mol dm^{-3} sodium chloride.

The basic characteristics of the three latices used are summarized in Table 1.

TABLE 1

Properties of PTFE Latices used

Property	Latex DR2	Latex HB1	Latex RR1
Modal long dimension/nm	130	170	198†
Modal short dimension/nm	75	110	
Aspect ratio	1.7	1.6	-
Specific surface area/ $m^2\ g^{-1}$	22.7 ± 1.0	21.4 ± 1.0	12.6 ± 0.8
Surface charge density/ $\mu C\ cm^{-2}$	0.68 ± 0.02	-	-

† Number average diameter

Determination of Adsorption Isotherms

General Procedure Basically a total sample volume of 10 cm^3 was used. Initially measured quantities of the stock solution of surface active agent, electrolyte solution and distilled water were put into clean 15 cm^3 borosilicate tubes. After adding an aliquot of the concentrated PTFE latex, Teflon sleeved stoppers were inserted and the tubes sealed with Parafilm. Each tube was inverted at least fifty times to ensure thorough mixing and then the dispersions were equilibrated for 24 h in a thermostatted water bath at 25 ± 0.01°C. During the equilibration period the tubes were inverted from time to time to prevent settling of the PTFE particles. After 24 h the solid latex particles were separated from the supernatant by centrifugation using an MSE 18 "High Speed" centrifuge in which the centrifuge bowl was maintained at 25 ± 0.5°C. The clear supernatant was removed for analysis of the surface active agent concentration using either surface tension or radioactive assay.

The pH of the solutions was maintained at 6 for APFO and at pH 9 for the hydrocarbon surface active agents. At these pH values both types of agent were in the form of fully dissociated carboxylic acid salts. The amount of latex used for an experiment was varied so that the depletion of the agent caused by adsorption was of the order of 50%. In this way errors in calculating the extent of adsorption from the initial and final concentrations were minimised.

Analysis of Supernatant by Surface Tension Measurements

Surface tension measurements at the air-water interface were carried out using a du Noüy tensiometer. Initially a calibration curve for the particular surface active agent was determined as a plot of surface tension against the logarithm of the molar concentration of the agent. This established the region in which the surface tension changed in a linear fashion and also gave a determination of the critical micelle concentration (c.m.c.). The solution under investigation was contained in a water-jacketed Pyrex vessel, which was maintained at a temperature of 25 ± 0.1°C. A period of ten minutes was allowed for a solution to reach temperature equilibrium. The platinum ring was flamed and allowed to cool before each surface tension measurement. Measurements were made on each solution until a constant surface tension value was obtained corresponding to adsorption equilibrium at the air-water interface.

Analysis of Supernatant by Radiochemical Assay A small modification in general technique was made when radiolabelled materials were used in that after addition of the solutions to the tubes an initial count rate was determined on the solution of the surface active agent after it had equilibrated with the tube surface. After this the PTFE latex was added and the procedure carried out as described above.

In the radiochemical studies trace amounts of the surface active agents containing ^{14}C on the carboxyl group were used in combination with the unlabelled material: Liquid scintillation counting was used for determination of the concentration of the radio-labelled materials. Initially, a calibration curve was constructed of count rate against known concentrations of surface active agent. This curve was subsequently used to determine the concentration of the supernatants from the adsorption experiments using identical volumes of aqueous solution and scintillator to those used in the calibration experiments.

Radioactive assay was found to be the most sensitive technique for adsorption determinations at low equilibrium concentrations. At equilibrium concentrations above the c.m.c. the aqueous solution was diluted quantitatively to bring the concentration below the c.m.c. prior to surface tension determination. Good agreement was obtained between radiochemical and surface tension determinations.

Analysis of Supernatant by Dye Transfer For studies using the cationic surface active agent, decyltrimethylammonium bromide, its concentration was determined using the dye-transfer technique described by Few and Ottewill (7).

RESULTS

Studies on Latex DR2

Adsorption of APFO The adsorption of APFO onto particles of PTFE latex DR2 was examined at pH 7.0 in 10^{-3} mol dm^{-3} ammonium nitrate solution using radiotracer measurements. Adsorption was also studied at pH 6.0 in 10^{-2} mol dm^{-3} ammonium chloride solution using surface tension measurements to determine the equilibrium concentration of the surface active agent. The adsorption isotherms obtained are shown in Figure 1. The isotherm obtained at pH 7.0 shows initially a low adsorption with a small plateau which corresponds to an area per adsorbed perfluorooctanoate ion of 630 ± 30 Å^2. This is followed by a steep rise in adsorption to a second plateau which corresponds to an area per ion of 83 ± 5 Å^2.

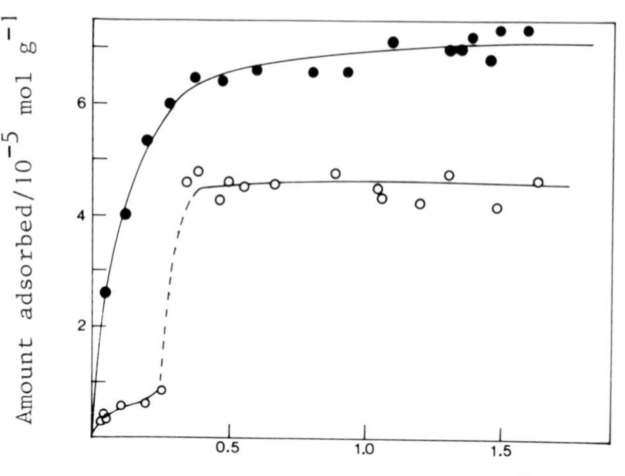

Fig.1. Adsorption of APFO on latex DR2. ○ *, results at pH 7.0 in 10^{-3} mol dm^{-3} ammonium nitrate solution;* ● *, results at pH 6.0 in 10^{-2} mol dm^{-3} ammonium chloride solution.*

At the lower pH and higher ionic strength the amount of material adsorbed increases rapidly from the lowest concentration and rises steadily to an adsorption plateau corresponding to an area per adsorbed ion of 54 ± 4 Å^2. The commencement of essentially constant adsorption density occurs below the critical micelle concentration of 2.26 x 10^{-2} mol dm^{-3}. The adsorption process is clearly aided by the increase in salt concentration as judged by the higher initial slope of the adsorption isotherm and the smaller area occupied per ion at saturation adsorption. Both factors can be attributed to the decreased electrostatic repulsion between head groups and head groups and between head group and surface group which occurs with the increase in ionic strength. At the air-water interface at this salt concentration the limiting area per ion for APFO was found to be 47 Å^2. This compares closely with the figure of 54 ± 4 Å^2 found at the PTFE - solution interface and suggests that in the plateau region essentially a close-packed vertically oriented layer is formed.

Adsorption of Sodium Octanoate Over the sodium octanoate concentration range, 3 x 10^{-5} to 2 x 10^{-2} mol dm^{-3}, at pH 9.0 and in 10^{-3} mol dm^{-3} sodium nitrate solution, it was not possible to measure, within experimental error, any adsorption of octanoate ions.

Sodium Dodecanoate The adsorption of dodecanoate ions was studied at pH 9.0 in 10^{-3} mol dm^{-3} sodium nitrate solution using C^{14} labelled material. The adsorption isotherm obtained is shown in Figure 2. The solubility limit of sodium dodecanoate at this pH was 2 x 10^{-2} mol dm^{-3} which meant that this was the highest initial concentration that was used. At the highest equilibrium concentration attained, 7 x 10^{-3} mol dm^{-3}, the amount adsorbed was 2.6 x 10^{-5} mol g^{-1}, which corresponded to an area per adsorbed ion of 120 Å^2. This is a larger area than the value of 28 ± 5 Å^2 found for the adsorption of dodecanoate ions on polystyrene at pH 8.0 and

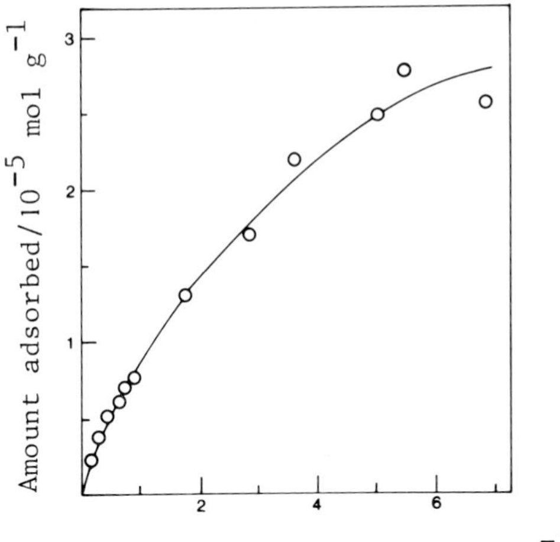

Fig.2. Adsorption of dodecanoate ions on latex DR2 at pH 9.0 in 10^{-3} mol dm^{-3} sodium nitrate solution.

similar electrolyte conditions at an equilibrium concentration of 1.23 x 10^{-2} mol dm^{-1}; this corresponded to a plateau on the adsorption isotherm (8,9). At an equilibrium concentration of 7 x 10^{-3} mol dm^{-3} an area per dodecanoate ion on polystyrene was found of 32 ± 5 Å^2 (8,9). It was clear that in the present work an adsorption plateau was not reached for sodium dodecanoate. Consequently, the area per ion attained on PTFE of 120 Å^2 is substantially larger than the value of 32 Å^2 found on the essentially hydrocarbon substrate polystyrene, under similar conditions (9). In combination with the apparent non-adsorption of octanoate this demonstrates the lack of affinity of hydrocarbon chains for PTFE surfaces. It should be noted, however, that Na^+ ions were used in this study rather than NH_4^+. Thus the possibility of H-bonding which could occur with NH_4^+ was not present.

Studies on Latex HB1

Adsorption of $C_{12}E_6$ The adsorption isotherm obtained for the nonionic surface active agent, $C_{12}E_6$, onto PTFE particles of latex HB1 is shown in Figure 3. The isotherm

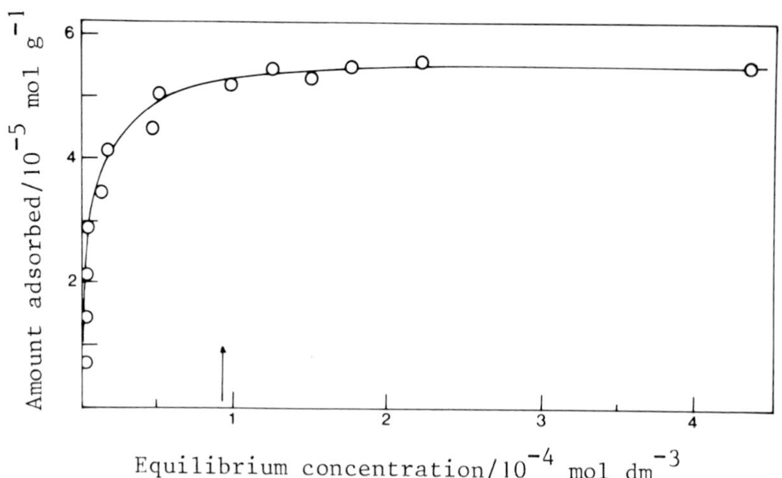

Fig.3. Adsorption of $C_{12}E_6$ from water on latex HB1. ↑, critical micelle concentration of $C_{12}E_6$.

shows continuous adsorption of the nonionic material with the isotherm reaching a plateau close to the critical micelle concentration of the surface active agent. The area per adsorbed molecule in the plateau region was found to be 65 ± 5 Å^2. This compares very closely with the area per molecule at the air-water interface of 60 Å^2 (10) and 55 Å^2 found on graphon (11); a lower value of 40 Å^2 was found on polystyrene latex (12). These values suggest that $C_{12}E_6$ forms a close-packed vertically oriented layer on PTFE.

This result appeared to be somewhat surprising in view of the results obtained with the sodium dodecanoate where there appeared to be a marked reluctance of the $C_{11}H_{23}$- chain to adsorb on PTFE. As a complementary study therefore the influence of $C_{12}E_6$ on the stability of PTFE latices to coagulation by magnesium sulphate was examined. The results obtained are given in Figure 4 in the form of log [critical coagulation concentration for $MgSO_4$] against log [equilibrium concentration of $C_{12}E_6$]. The adsorption data is also

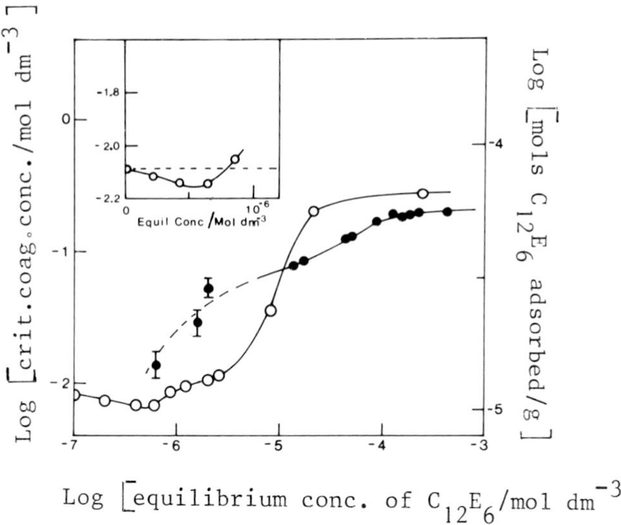

Fig.4 ○, *Log [critical coagulation concentration] against log [equilibrium concentration of $C_{12}E_6$] using $MgSO_4$ and latex HB1.* ●, *adsorption results for $C_{12}E_6$.*

plotted in the same figure. At low equilibrium concentrations of $C_{12}E_6$ the latex shows sensitized coagulation, i.e. a lower concentration of magnesium sulphate is needed to effect coagulation than in the complete absence of $C_{12}E_6$. This result suggests that the particle has become more hydrophobic and hence indicates that the $C_{12}E_6$ molecule has adsorbed via the ethylene oxide head group leaving the $C_{12}H_{25}$ chain exposed in the solution phase. This facilitates further adsorption via hydrocarbon-hydrocarbon co-operative association to provide a monolayer on the surface. Consistent with this hypothesis is the increasing stability of the latex to coagulation by magnesium sulphate. It can be seen from Figure 4 that once monolayer adsorption of $C_{12}E_6$ has occurred the critical coagulation concentration is increased by a factor of 33, a clear indication that the latex has become sterically stabilised by virtue of the ethylene oxide groups being exposed to the aqueous phase. A similar behaviour has been observed with polystyrene latices (12,13).

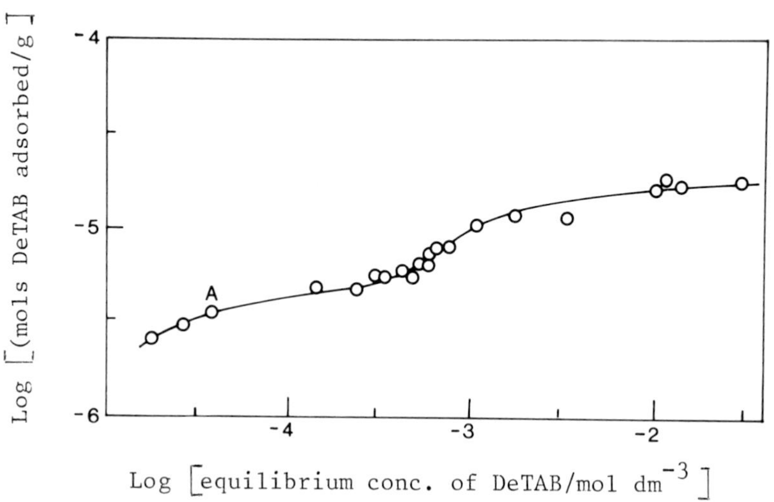

Fig.5 Adsorption of DeTA$^+$ions, from water, on latex RR1

Studies on Latex RR1

Adsorption of DeTAB The adsorption isotherm of DeTAB onto PTFE latex RR1 is shown in logarithmic form in Figure 5. Two distinct plateau regions are observable on the isotherm, one at low concentrations of surface active agent and the other at higher concentrations, approaching the critical micelle concentration of 5.6×10^{-2} mol dm^{-3}. The point A (Figure 5) at the beginning of the first plateau corresponds to an area per adsorbed ion of 632 Å^2. This point was chosen since it corresponded to an equilibrium concentration of DeTAB equivalent to the critical coagulation concentration for this material which suggested that it was the point at which neutralization of the negative charge on the latex surface occurred. The second plateau corresponded to an area per adsorbed ion of 118 Å^2. This is somewhat larger than the value of 65 Å^2 found for the saturation adsorption of $DeTA^+$ on negatively charged polystyrene latices (14).

Complementary studies were also carried out of the coagulation of the PTFE latex by cationic surface active agents. The results are given in Table 2 in which they are compared with the corresponding values obtained using polystyrene latices.

TABLE 2

Coagulation of PTFE Latex by Cationic Surface Active Agents

Agent	Critical Coagulation Concentrations /mol dm^{-3}	
	PTFE Latex	Polystyrene Latex
$C_8H_{17}.\overset{+}{N}Me_3.Br^-$	4.5×10^{-5}	1.59×10^{-4}
$C_{10}H_{17}.\overset{+}{N}Me_3.Br^-$	3.1×10^{-5}	1.99×10^{-5}
$C_{12}H_{17}.\overset{+}{N}Me_3.Br^-$	2.9×10^{-5}	1.29×10^{-6}

An interesting feature of these results is that whereas on polystyrene latices there is a pronounced chain length effect, on PTFE latices the effect of chain length is almost insignificant. The former results reflect the

Traube's rule effect and suggest that the hydrocarbon chain is to a large extent removed from the water on adsorption and lies flat on the surface of the hydrocarbon substrate. In this context the PTFE results suggest that the coagulation values are charge dominated and that although there may be some hydrophobic effect it appears unlikely that the hydrocarbon chains of the surface active agent lie flat on the PTFE surface.

DISCUSSION

The Nature of the PTFE Latex Surface

The preparation of the PTFE latex was carried out using disuccinic acid peroxide (bis-(β-carboxypropionyl peroxide) a material which cleaves homolytically on heating to give two symmetrical free radicals. Rapid decarboxylation then occurs to give the free radical,

$$.CH_2CH_2COOH$$

This radical can then initiate polymerisation of the tetrafluoroethylene leading to the formation of,

$$.(CF_2CF_2)_n\ CH_2CH_2COOH \qquad \text{I}$$

In addition reaction of radicals with the water molecules present, with transfer of H, can lead to the formation of .OH radicals and the subsequent formation of,

$$.(CF_2CF_2)_n.OH$$

This is an unstable perfluoroalkanol which undergoes elimination of hydrogen fluoride to give the acid fluoride. The latter then undergoes rapid hydrolysis to give,

$$.(CF_2CF_2)_n.COOH \qquad \text{II}$$

From this mechanism it seems likely that carboxyl groups of type I and type II will be present on the surface of a PTFE latex particle. The pK_a values for groups of type I and II are ca. 4.2 and 2.0 respectively so that under the conditions used in the adsorption experiments both groupings would have been above their pK_a values. The presence of acidic groups on the surface of PTFE particles has been confirmed experimentally by both microelectrophoresis and titration with alkali (5,15).

One of the latices used in the present work, DR2, was examined by conductometric titration and gave a value for the surface charge density of 0.68 μC cm^{-2}. This corresponds to one carboxyl group per 2353 Å^2. If the area occupied per perfluoroalkyl carboxylate at the air-water interface of 47 Å is assumed as a reasonable value for the PTFE-water interface, then we find that only 2% of the surface is occupied by carboxyl groups and the rest by $-CF_2-$ groups. At the moment the distribution of carboxyl groups on the surface is not known but it is doubtful for a mainly crystalline latex that it is a uniform distribution.

The Role of the Head Group of the Surface Active Agent The three head groups examined in the present work were,

carboxylate, $-C(=O)OH \rightarrow -C(=O)O^-$

ethylene oxide, $-(CH_2CH_2O)_6\ OH$

trimethyl ammonium, $-\overset{+}{N}Me_3$

In view of the inertness of polytetrafluoro ethylene it is very doubtful that there was any interaction of significance between any of the head groups and the $-CF_2-$ group on the surface. Therefore the most likely interaction appears to be that between the head group of the surface active agent and the carboxylate groups on the latex surface.

In the case of APFO the carboxyl group is a strong acid and can be considered to be totally ionised at the pH values used in the experiments. Consequently, repulsion between the carboxylate groups of APFO and the latex surface would be anticipated. In the case of dodecanoate ions the acid is a much weaker acid ($pK_a \sim 5$). Hence, whilst the same arguments can be maintained, there is a possibility of a small percentage of acid molecules being present which could hydrogen bond, by ion-dipole forces, to the carboxylate groups on the latex surface.

In the case of the nonionic surface active agent, $C_{12}E_6$, the coagulation studies provide direct evidence for initial adsorption via the head group. This implies again the formation of hydrogen bonds by ion-dipole or dipole-dipole interactions with the terminal hydroxyl group, viz:-

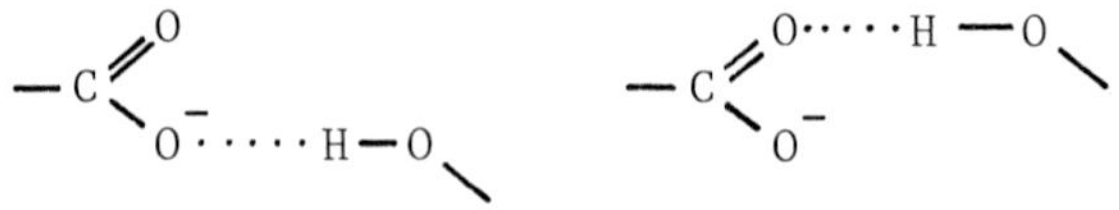

$$-C\overset{\displaystyle /\!\!/ O}{\underset{\displaystyle \searrow O^{-}\cdots\cdots H-O\diagdown}{}} \qquad -C\overset{\displaystyle /\!\!/ O\cdots\cdots H-O\diagdown}{\underset{\displaystyle \searrow O^{-}}{}}$$

or linkage to the ether oxygens via a water molecule,

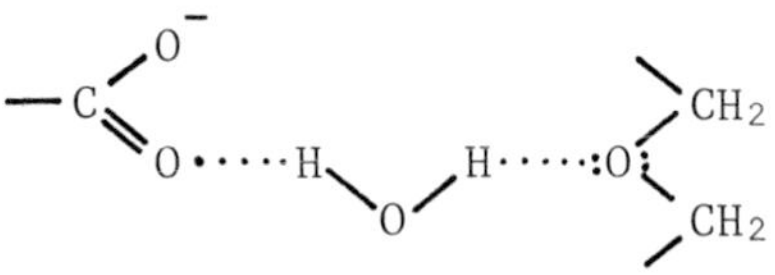

$$-C\overset{\displaystyle \nearrow O^{-}}{\underset{\displaystyle \searrow\!\!\searrow O\cdots\cdots H\diagdown_{O}\diagup H\cdots\cdots :O\langle^{CH_2}_{CH_2}}{}}$$

Whatever the mechanism, and possibly all three can occur, the adsorption via the head group leaves the hydrocarbon chain of the $C_{12}E_6$ in a receptive position for further adsorption via the hydrocarbon tails of the surface active agent.

In the case of the DeTAB the possibility of direct electrostatic interaction occurs with the formation of a salt-linkage,

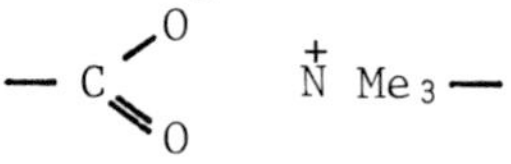

$$-C\overset{\displaystyle \nearrow O^{-}}{\underset{\displaystyle \searrow\!\!\searrow O}{}} \qquad \overset{+}{N}Me_3-$$

at the surface. The implication from the coagulation studies is again that the hydrocarbon chain remains primarily in the aqueous phase thus forming a site for further adsorption of hydrocarbon chains. The initial plateau on the adsorption isotherm (Figure 5) appears to conform to this suggestion. However, the area per adsorbed ion found, 632 Å^2, corresponds to a surface charge density of 2.45 μC cm^{-2}. This is not unreasonable but seems higher than anticipated for a PTFE latex; direct conductometric data were not available for latex RR1.

The Form of the Adsorption Isotherms The data for APFO obtained at pH 6.0 in 10^{-2} mol dm^{-3} ammonium chloride solution and shown in Figure 1 were found to fit a Langmuir type I isotherm and to give a good straight line using the linear form of the Langmuir equation. Since the latter equation assumes that the heat of adsorption remains constant with surface coverage this would be consistent with the view that adsorption occurred onto $-CF_2-$ surface groups via the fluorocarbon chain of the APFO. The area per adsorbed ion in the plateau region of 54 Å^2 is strongly in favour of a vertically oriented monolayer of perfluorooctanoate ions.

The data for the dodecanoate ions gave a good fit to a Freundlich isotherm suggesting variation of the heat of adsorption with coverage. This could be consistent with some initial slight adsorption via the head group with a concentration dependent deionization occurring at the lower pH in the double layer of the particles. However, it is clear from these data and the octanoate studies that there is no strong affinity of the hydrocarbon chain for the PTFE surface.

In the case of the DeTAB data (Figure 5) the isotherm is clearly a two-stage Langmuirian isotherm. This also appears to be the case for the $C_{12}E_6$ but here the absence of an accurate method for assaying low concentrations of $C_{12}E_6$ made the low concentration data somewhat uncertain. However, it is clear for both of these surface active agents that the initial adsorption occurred via the head groups and was followed by subsequent adsorption of hydrocarbon tails to the tails already present on the surface. In the case of $C_{12}E_6$ this seems to lead to a close-packed monolayer and rather less than a monolayer for the DeTAB; possibly this reflects a difference in cross-sectional area if clustering occurs around the head down molecules.

The Mechanism of Adsorption Accurate thermodynamic data do not yet exist for these systems and hence the mechanisms proposed must be speculative. They are largely based on the processes which occur on the micellization of surface active agents where accurate data have been obtained (16).

In the adsorption of APFO the interaction between the head group and the surface will be unfavourable to adsorption. In the case of the chain, however, replacement of a fluorocarbon-water interface (interfacial energy 57.7 mN m^{-1} (17)) by PTFE-fluorocarbon contact, i.e. the C_7F_{15} chain lying on the PTFE surface (interfacial energy $\approx$ 0), will lead to a negative enthalpy of adsorption. Furthermore, the water molecules which were oriented close to the chain in the bulk phase are to a large extent stripped off on adsorption of the chain and return to the bulk phase. An entropy gain is thus obtained. On both counts a favourable free energy of adsorption is obtained from the chain which overcomes the unfavourable effects of the head group. The form of the isotherm for APFO at pH 7.0 in 10^{-3} mol dm^{-3} ammonium nitrate may well indicate an initial horizontal orientation followed by reorientation to a vertical monolayer.

We now examine the effect of the hydrocarbon. In the bulk phase the hydrocarbon-water interface has as interfacial energy of ca. 50 mN m^{-1}. In the adsorbed phase the interface formed between hydrocarbon and PTFE will have a finite interfacial energy, which will be lower than that of the water-hydrocarbon interface. This should lead to a negative enthalpy of adsorption which is smaller than in the fluorocarbon-PTFE case. Similarly stripping water off the hydrocarbon chain on adsorption and allowing the water molecules to return to the bulk phase leads to a positive entropy of adsorption. However, in considering the interactions between hydrocarbon and PTFE it is now essential to consider the difference in size between the CF_2 group and the CH_2 group. The former is somewhat larger than the latter (17) and hence the chains of (CF_2) groups and (CH_2) groups cannot fit into the geometric proximity required to maximise the dispersion interactions between the two groups. It becomes imperative that the molecular configuration of the hydrocarbon chain on the PTFE surface is considered. In terms of the results obtained it must be concluded that arrangement of a hydrocarbon chain on a PTFE surface is entropically unfavourable and that the adsorption of hydrocarbon tails on PTFE surfaces is largely dominated by what is essentially an unfavourable "surface entropy of mixing" term for hydrocarbons and fluorocarbons. In consequence hydrocarbon surface active agents do not adsorb, to any extent, on PTFE surfaces unless motivated to do so by head group interactions.

ACKNOWLEDGEMENTS

We wish to thank the Fluon Group of I.C.I. Plastics Division for generously providing the PTFE latices used in this work and for many useful discussions. We also thank S.E.R.C. and I.C.I. Plastics Division for a CASE Award during the tenure of which part of this work was carried out.

REFERENCES

1. Sperati, C.A. and Starkweather, H.W. (1961). *Fortschr. Hochpolym.Forsch.*, 2, 465.
2. Sherratt, S. (1966). "Kirk-Othmer Encyclopedia of Chemical Technology", 2nd Edition, Interscience, 9, 805.
3. Ottewill, R.H. and Rance, D.G. (1977). *Croatica Chemica Acta*, 50, 65.
4. Ottewill, R.H. and Rance, D.G. (1979). *Croatica Chemica Acta*, 52, 1.
5. Ottewill, R.H. and Rance, D.G. (1982). in press.

6. Goodwin, J.W., Hearn, J., Ho, C.C. and Ottewill, R.H. (1974). *Colloid and Polymer Sci.*, 252, 464.
7. Few, A.V. and Ottewill, R.H. (1956). *J.Colloid Science*, 11, 34.
8. Harris, N.M. (1980). "Small Angle Neutron Scattering from Colloidal Systems", D.Phil., Oxford University.
9. Harris, N.M., Ottewill, R.H. and White, J.W. (1982). This volume p.
10. Lange, H. (1965). *Kolloid Z.u.Z.Polymere*, 201, 131.
11. Corkill, J.M., Goodman, J.F. and Tate, J.R. (1966). *Trans.Faraday Soc.*, 62, 979.
12. Ottewill, R.H. and Walker, T. (1968). *Kolloid Z.u.Z. Polymere*, 227, 108.
13. Ottewill, R.H. and Walker, T. (1974). *J.Chem.Soc. Faraday Trans.I*, 70, 917.
14. Connor, P. and Ottewill, R.H. (1971). *J.Colloid and Interface Science*, 37, 642.
15. Rance, D.G. (1976). "Studies on the Characterisation and Colloid Chemistry of Polytetrafluoroethylene Latices", Ph.D. thesis, Bristol University.
16. Corkill, J.M., Goodman, J.F. and Tate, J.R. (1968). "Hydrogen-Bonded Solvent Systems", p.181.
17. Fowkes, F.M. (1964). *Ind.Eng.Chem.*, 56, 40.

ADSORPTION INFLUENCE ON PARTICLES INTERACTION AND DISPERSE SYSTEMS STABILITY

E.K. Shchukin and V.V. Yaminsky

The Institute of Physical Chemistry
Academy of Sciences of the USSR
Moscow, USSR

The diverse applications of surfactants as stabilizers, plasticising agents, flocculants, detergents, etc., are based on their capacity for adsorption at various interfaces with consequent alteration of the properties of surfaces. The most fundamental effect of adsorption of surfactants is the reduction of interfacial tension which, in the case of equilibrium-reversible adsorption, is described by the Gibbs equation.

With mobile interfaces (liquid/gas or liquid/liquid) this reduction may be measured directly. At a solid (S)/liquid(L) interface adsorption effects on interfacial tension σ_{SL} may also sometimes by easily detected, e.g. by the variation of the contact angle θ according to the Young equation:

$$\Delta^{L} \sigma_{SA} \equiv \sigma_{SA} - \sigma_{SL} = \sigma_{LA} \cos \theta + \Pi \tag{1}$$

where σ_{SA} and σ_{LA} are the surface tensions of the solid and the solid and the liquid at the interface with air and Π is the two-dimensional pressure of the saturated vapour of the liquid adsorbed on the solid surface. Surface tension may also change under non-equilibrium conditions and when adsorption is irreversible. However, no universal methods describing such changes have been proposed and the very idea of surface tension in certain cases becomes ambiguous.

Apart from surface tension, adsorption may change other physico-chemical properties of interfaces, e.g. electric charge, surface viscosity, etc; hence the variety of adsorption-induced effects in disperse systems. For example, the classical DLVO theory |1,2| links variations of stability to those of long-range interaction forces between particles (surfaces in terms of surface charge and the changes of the structure of the electrical double layer during adsorption of ions. The structural-mechanical barrier concept |3,4| attributes the strong stabilizing effect of adsorption layers formed by certain substances, particularly polymers, to their special mechanical properties (strength, viscosity) which prevent the layers from destruction as the particles approach each other and to the surface lyophilicity of the layers which prevents them sticking together. The structural-mechanical barrier concept also includes later notions on the sterical stabilization produced by adsorption of layers of macromolecules |5|; this takes into account osmotic effects and the elasticity of polymer chains.

As long ago as the early 1930's P.A. Rehbinder |6| pointed out that the stability of dispersions of solids in liquids may depend directly on such a fundamental thermodynamical parameter of interfaces as surface tension. However, until very recently these ideas have not been analyzed quantitatively and, in particular, the effect of reversible adsorption of surfactants on the stability of sols and suspensions has not been considered from this standpoint. At the same time a number of recent experimental investigations |7-12| provide strong support to this concept.

According to Derjaguin's thermodynamical theory |13,14|, the force of adhesion N of convex particles with the effective radius $\bar{R}$($\bar{R}$ = R for two equivalent spheres of radius R) is related to the specific (per unit area) free energy F of adhesion of plane-parallel surfaces by an equation

$$N = \Pi R F \tag{2}$$

The quantity F, the depth of the minimum in the F(H) curve (the potential energy of interaction plotted against the distance between the surfaces H), may be expressed in terms of surface tensions of solid particles S_1 and S_2 at the interface with the medium M as follows:

$$F_{S_1MS_2} = \sigma_{S_1M} + \sigma_{S_2M} - \gamma_{S_1MS_2} \tag{3}$$

where γ is the excess free energy (tension) arising from the existence of a boundary between two solids in contact (at the equilibrium distance $H = H_o$, corresponding to the minimum of the potential of interaction); for two identical surfaces $F_{SMS} = 2\sigma_{SM} - \gamma_{SMS}$. The meaning of γ is similar to that of a grain boundary free energy in polycrystalline materials, and even in direct contacts between identical surfaces this quantity may differ from zero.

Proceeding from equation (3) with the subscript L instead of M for surfaces in a liquid and with the subscript A for surfaces in air, we arrive at a correlation between the energies of adhesion in a liquid $F_{S_1LS_2}$ and in air $F_{S_1AS_2}$ in the form,

$$F_{S_1LS_2} = F_{S_1AS_2} - (\Delta^L \sigma_{S_1A} + \Delta^L \sigma_{S_2A}) + \Delta^L \gamma_{S_1AS_2} \quad (4)$$

where $\Delta^L \sigma_{SA} = \sigma_{SA} - \sigma_{SL}$, $\Delta^L \gamma_{S_1AS_2} = \gamma_{S_1AS_2} - \gamma_{S_1LS_2}$ (for two identical surfaces $F_{SLS} = F_{SAS} - 2\Delta^L \sigma_{SA} + \Delta^L \gamma_{SAS}$). With the absolute (the deepest) minimum of potential energy of interaction a condition $\Delta^L \gamma_{S_1AS_2} \geqq 0$ should hold, so that

$$F_{S_1LS_2} \geqq F_{S_1AS_2} - (\Delta^L \sigma_{S_1L} + \Delta^L \sigma_{S_2L}) \quad (5).$$

Indeed, if H_o in a liquid corresponds to an intermolecular distance between two solids in a direct contact and the liquid medium components are absent between the solid surfaces (no adsorption of the molecules of the liquid takes place in the equilibrium contact), the contact does not differ from that in air. Thus $\gamma_{S_1LS_2} = \gamma_{S_1AS_2}$ or $\Delta^L \gamma_{S_1AS_2} = 0$ and equation (5) becomes an equality. The presence of individual adsorbed molecules or a film of the medium in the equilibrium contact means positive adsorption in the gap and decreases the excessive free energy so that $\gamma_{S_1LS_2} < \gamma_{S_1AS_2}$, $\Delta^L \gamma_{S_1AS_2} > 0$ and equation (5) becomes an inequality. The opposite sign of the inequality in (5) is impossible for the absolute minimum of F(H). If it is nevertheless observed, the system is situated in a local (shallow) minimum and a deeper absolute minimum must exist in which condition (5) holds.

If the energy of interaction between the surfaces $F_{S_1AS_2}$ exceeds that of their interaction with the medium $(\Delta^L \sigma_{S_1A} + \Delta^L \sigma_{S_2A})$, the minimum of F(H) may correspond to a direct contact and forces of attraction will act at small distances between the surfaces. At $F_{S_1AS_2} < (\Delta^L \sigma_{S_1A} + \Delta^L \sigma_{S_2A})$ the direct contact becomes definitely impossible, since at $\Delta^L \gamma_{S_1AS_2} = 0$, $F_{S_1LS_2}$ becomes negative. In this case a more or less thick interlayer of the medium stays between the surfaces to cause their repulsion at small distances.

Table 1 from our work with E.A. Amelina |10| presents experimental values of $F_{S_1AS_2}$ and $F_{S_1LS_2}$ obtained according to equation (2) from the measurements of adhesion forces between crossed cylinders ($\bar{R} \sim 1$ mm) of various hydrophilicity (non-modified glass, methylated glass and glass covered with a film of acetylcellulose) in air and in water. It also gives the free energies of wetting of these surfaces with water $\Delta^L \sigma_{SA}$ estimated according to (1), the two-dimensional pressure of the adsorbed vapour being neglected. The table confirms that the condition (5) is indeed observed for all the combinations of two identical or different surfaces studied. In systems where the sum of the energies

TABLE 1

Specific Free Energies of Adhesion for Surfaces of Various Hydrophilicity in Air ($F_{S_1AS_2}$) and in Water ($F_{S_1LS_2}$) in mN m^{-1}

Surfaces $S_1 - S_2$	$\Delta^L \sigma_{S_1A} + \Delta^L \sigma_{S_2A}$	$F_{S_1AS_2}$	$F_{S_1LS_2}$
Mt - Mt	- 40	40	75
Gl - Gl	140	100	$\lesssim 0.01$
Mt - Gl	50	45	5
Ac - Ac	80	30	$\lesssim 0.01$
Ac - Mt	20	30	30
Ac - Gl	110	50	$\lesssim 0.01$

Gl *is (non-modified) glass,* $\Delta^L \sigma_{GlA} \approx 70$ mN m^{-1}

Mt - *methylated glass,* $\Delta^L \sigma_{MtA} \approx 40$ mN m^{-1}

Ac - *acetylcellulose,* $\Delta^L \sigma_{AcA} \approx 40$ mN m^{-1}

of wetting for the two surfaces ($\Delta^L \sigma_{S_1L} + \Delta^L \sigma_{S_2L}$) is far greater than the energy of their adhesion in air $F_{S_1AS_2}$ the equilibrium contact should contain a thermodynamically stable interlayer of the medium. As such surfaces are brought together to form an immediate contact, the effect is repulsion, i.e. the appearance of a positive disjoining pressure (hydrophilic interaction). The energy of interaction in this case has no minimum at all or its depth is negligible and falls below the level of experimental sensitivity. When the sum of the energies of wetting ($\Delta^L \sigma_{S_1A} + \Delta^L \sigma_{S_2A}$) is compatible with or less than the energy of adhesion in air $F_{S_1AS_2}$, the energy of adhesion in the liquid assumes finite values which increase with the difference between $F_{S_1AS_2}$ and ($\Delta^L \sigma S_1A + \Delta^L \sigma_{S_2A}$). in these systems the disjoining pressure at small distances is negative and may give rise to strong adhesion of particles (hydrophobic interaction).

In such terms the effect of surfactants on adhesion of particles may result from the reduction of surface tension σ due to adsorption of the surface active agent on the particle/medium interface as well as from the reduction of tension γ by virtue of its adsorption in the contact between the particles. Considering interaction between surfaces in a solution of a surfactant of concentration C, we shall write the parameters of equation (3) with C, while symbols with a subscript L will refer to the pure liquid (solvent). Then a correlation between the energy of adhesion in the surfactant solution $F_{S^1CS^2}$ with that in the pure solvent $F_{S_1LS_2}$ may be presented as,

$$F_{S_1CS_2} = F_{S_1LS_2} - (\Delta^C \sigma_{S\ L} + \Delta^C \sigma_{S\ L}) + \Delta^C \gamma_{S_1LS_2} \quad (6)$$

where, $\Delta^C \sigma_{SL} = \sigma_{SL} - \sigma_{SC}$

and, $\Delta^C \gamma_{S_1LS_2} = \gamma_{S_1LS_2} - \gamma_{S_1CS_2}$

(for two identical surfaces $F_{SLS_C} - 2\,\Delta^C \sigma_{SL} + \Delta^C \gamma_{SLS}$). By analogy with the above-said, $\Delta^C\gamma = 0$ if adsorption of the surfactant in the gap is zero and $\Delta^C \gamma > 0$ if surface active molecules are present in the equilibrium contact. Therefore a condition

$$F_{S_1CS_2} \geq F_{S_1LS_2} - (\Delta^C \sigma_{S_1L} + \Delta^C \sigma_{S_2L}) \quad (7)$$

should be obeyed for the absolute minimum of F(H).

If ($\Delta^C \sigma_{S_1L} + \Delta^C \sigma_{S_2L}$) is less than $F_{S_1LS_2}$ adsorption in the contact region may be close to zero and equation (7) should become an equality. In this case addition of a surfactant would cause a decrease in the energy of adhesion which would practically be due entirely to the reduction of surface tension at the particle/medium interface, i.e. to the two-dimensional pressure of adsorption layers of the surfactant at the surface of both particles. If the sum of two-dimensional pressures ($\Delta^C \sigma_{S_1L} + \Delta^C \sigma_{S_2L}$) exceeds the energy of adhesion in the pure solvent $F_{S_1LS_2}$, the equilibrium contact would definitely contain molecules of the surfactant (otherwise the energy of adhesion in the surfactant solution would have become negative). The expulsion of the adsorption layer from the contact zone proves thermodynamically disadvantageous and should be accompanied by emerging of repulsion forces between the surfaces, i.e. positive disjoining pressure.

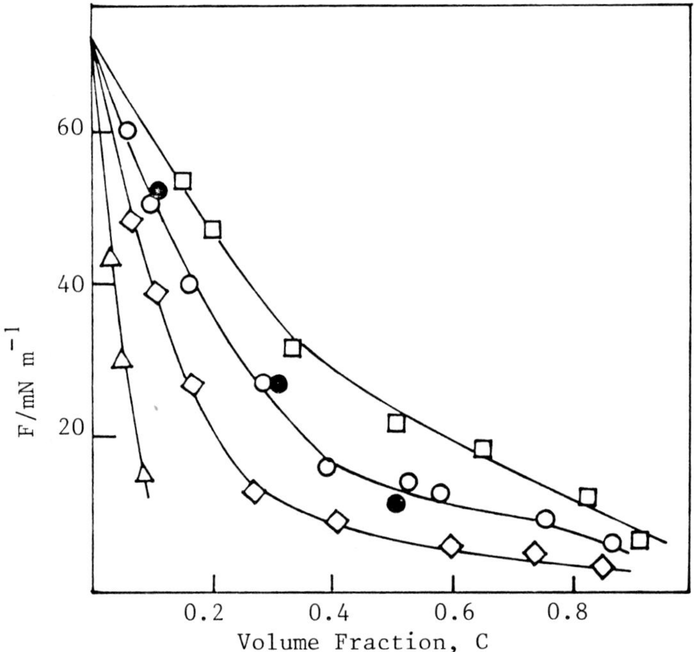

Fig.1 The effect of methyl □, *ethyl* ○, *n-propyl* ◇ *and n-butyl* △ *alcohols on the energy F of adhesion of methylated glass surfaces in water. Contour symbols show experimental values of* F_{SCS}, ●, *ethanol - the value of* $F_{SLS} - 2\ \Delta^C \sigma_{SL}$.

Fig.1 presents results of our work (with R.K. Yusupov, E.A. Amelina and V.A. Pchelin) |7| concerning the effect of alcohols on the energy of adhesion of hydrophobic particles (methylated glass spheres, $\bar{R} \sim 2$ mm) in water. The decrease in the energy of adhesion with increasing concentration of the alcohol shows a characteristic dependence on the hydrocarbon chain length of the alcohol, i.e. a qualitative correlation with the Traube rule. Likewise, the concentration dependences of the energies of adhesion for methylated glass surfaces in aqueous solutions of micelle-forming surfactants (our work with E.A. Amelina |8,10|) have a special salient point (see Fig.2) typical of surface tension isotherms of micelle-forming species.

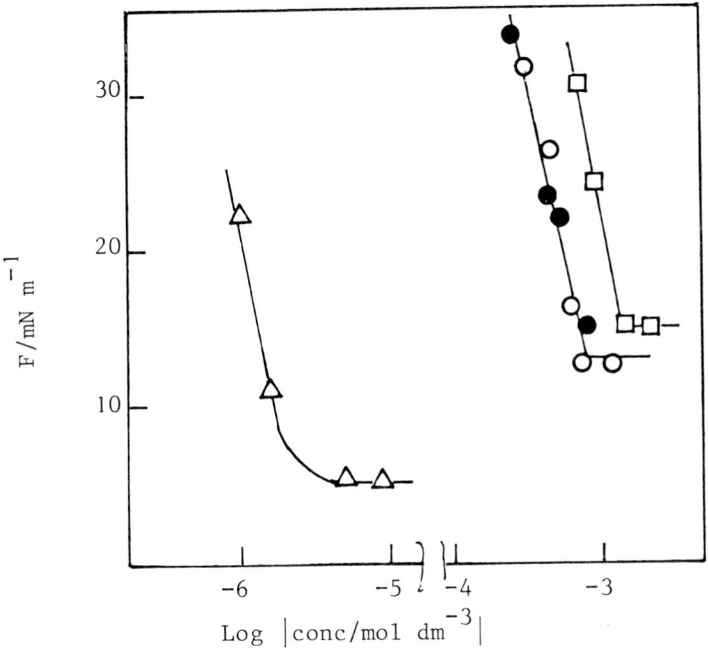

Fig.2 The effect of micelle-forming surfactants on the energy of adhesion of two methylated glass surfaces in water: □ *sodium dodecylsulfate,* ○ *cetylpyridinium bromide,* △ *polyoxyethyleneoctylphenol (OP-10). Contour symbols stand for experimental values of* F_{SCS}. ●, $F_{SLS} - 2\,\Delta^{C}\sigma_{SL}$.

It was demonstrated |7,8| that the effect of adsorption of such low molecular surfactants on the energies of adhesion of hydrophobic surfaces in aqueous media is reversible. Comparison of concentration dependences of the adhesion energy F_{SCS} with the relevant curves for $F_{SLS} - 2\,\Delta^C\,\sigma_{SL}$ (curve 2, Fig.1 and curve 2, Fig.2) shows that the decrease in the adhesion energy with increasing concentration of the surfactant can indeed be quantitatively accounted for by the reduction of interfacial tension, i.e. increase of surface pressure at the particle/medium interface. In this case the two-dimensional pressures $^C\sigma_{SL}$ were calculated from contact angle measurements according to the Young equation as the difference between the free energies of wetting by the pure solvent $\Delta^L\,\sigma_{SA} = \sigma_{SA} - \sigma_{SL}$ and the surfactant solution

$$\Delta^C\,\sigma_{SA} = \sigma_{SA} - \sigma_{SC}$$

$$\Delta^C\,\sigma_{SL} = \Delta^C\,\sigma_{SA} - \Delta^L\,\sigma_{SA}.$$

This can be done provided adsorption of the surfactant on the non-polar surface bordering on air is negligible. Fair agreement between F_{SCS} and $F_{SLS} - 2\,\Delta^C\,\sigma_{SL}$ was also observed when two-dimensional pressures were calculated according to the Gibbs equation on the basis of direct adsorption measurements for methylated glass in our work |8| and for carbon surfaces by F. van Voorst Vader |12|.

The energies of adhesion of two methylated glass surfaces even at the surfactant concentration above CMC retain finite values (5 - 10 mN m^{-1}). This is due to the fact that the doubled value of two-dimensional pressure of adsorption layers of micelle-forming species on methylated glass remains, over the entire range of concentrations, lower than the energy of adhesion of methylated glass surfaces in pure water. Significantly, surface active molecules whatever their concentration in the solution may be absent in the contact zone so that $F_{SCS} = F_{SLS} - 2\,\Delta^C\,\sigma_{SL} > 0$, as is in fact confirmed experimentally.

In contrast to the behaviour of two methylated surfaces, the energy of adhesion of acetylcellulose to methylated glass in surfactant solutions drops virtually to zero with increasing concentration (Fig.3, |10|). The two-dimensional pressure $\Delta^C\,\sigma_{S_2L}$ of adsorption layers on acetylcellulose does not exceed $\sim$ 5 mN m^{-1} even at high concentrations of the surfactant. Nevertheless, the sum $(\Delta^C\,\sigma_{S_1L} + \Delta^C\,\sigma_{S_2L})$ in the acetylcellulose-methylated glass system may be quite

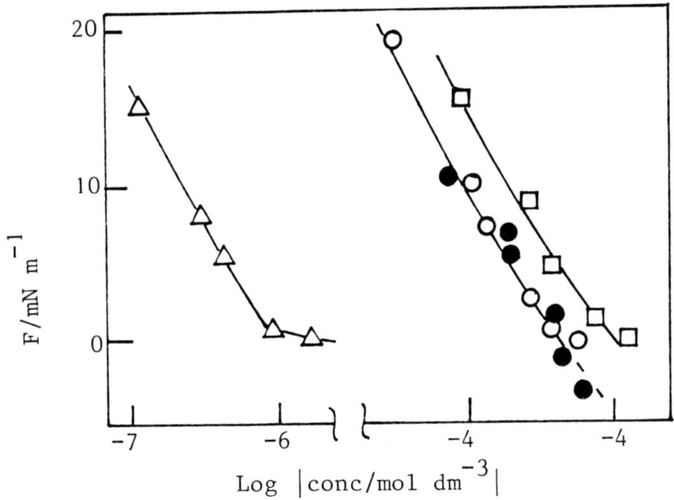

Fig. 3 Effect of micelle-forming surfactants on the energy of adhesion of acetylcellulose to methylated glass in water: □ *, sodium dodecylsulfate,* ○ *cetylpyridinium bromide,* △ *, polyoxyethyleneoctylphenol (OP-10). Contour symbols stant for experimental values of* $F_{S_1CS_2}$.
● , $F_{S_1LS_2} - (\Delta^C \sigma_{S_1L} + \Delta^C \sigma_{S_2L})$.

high due to the high values of two-dimensional pressure $\Delta^C \sigma_{S_1L}$ on methylated glass. Moreover, the energy of adhesion for this system in pure water is far lower than for two methylated surfaces. Thus the value of $(\Delta^C \sigma_{S_1L} + \Delta^C \sigma_{S_2L})$ for the acetylcellulose - methylated glass systems at high concentrations proves greater than $F_{S_1LS_2}$. At low At low concentrations

$$(\Delta^C \sigma_{S_1L} + \Delta^C \sigma_{S_2L}) < F_{S_1LS_2},$$

$$F_{S_1CS_2} = F_{S_1LS_2} - (\Delta^C \sigma_{S_1L} + \Delta^C \sigma_{S_2L}) > 0$$

and the surface active molecules are expelled from the contact zone. However, starting from a certain concentration of the surfactant the $F_{S_1LS_2} - (\Delta^C \sigma_{S_1L} + \Delta^C \sigma_{S_2L})$ curve crosses the axis of abscissae to come to the negative region

(s. curve 2 in Fig.3). At this point the surface active molecules penetrate spontaneously into the gap between the surfaces thus making the contact fail and $F_{S_1CS_2}$ drop virtually to zero. This result indicates, in particular, that oil dirt particles may be removed from the surface of moderately hydrophilic textile fibres spontaneously under the effect of surfactants adsorption without spending mechanical work.

In conclusion let us point out that in the case of small enough (colloid) solid particles their spontaneous detachment from solid surfaces or spontaneous dispersion of coagulates (peptization) may occur even if the primary minimum of F(H) has a rather great finite depth. This possibility is due to the fact that the work spent on breaking the contacts between particles is compensated for with an entropy gain as the particles are engaged in Brownian motion (spontaneous dispersion |15,16|). The authors and Amelina showed |17| that in moderately concentrated sols contacts break spontaneously so that the sol retains thermodynamical stability with regard to coagulation until the free energy of the contact W does not exceed $\sim$ 10 kT. The energy of the bond between particles in a contact may be presented as a product $W = NH_*$ where N is the adhesion force (the maximum value of the attraction force) and H_* the effective radius of attraction which, for short-range forces in a direct contact, is equal to a few Å. In view of (2) the expression $W = NH_*$ for particles of radius R may be written as

$$W = \Pi RFH_*$$

With $\sim$ 100 Å particles the condition $E \sim 10$ kT corresponds to $F \sim 10$ mN m^{-1} if we assume that $H_* \sim 2$ Å. Peptization of methylated aerosil sols with increasing concentration of alcohols in water was observed in |18|. The concentrations at which the process started were volume fractions of 0.85, 0.7 and 0.6 for methyl, ethyl and propyl alcohols respectively. As is seen from Fig.1, the values of F at these concentrations are indeed close to 10 mN m^{-1}. Similarly, peptization was observed |17| as F was decreased to $\lesssim$ 10 mN m^{-1} by adding an ionic surfactant to an aqueous dispersion of methylated aerosil.

This series of investigations involving direct measurements of adhesion forces in individual contacts between particles was started in connection with the development of a physico-chemical theory of strength of disperse structures |17|. It is a pleasure to mention that the well-known article by Professor D.H. Everett |20| (along with that by E. Mangegold |21|) was among those which prompted our work.

REFERENCES

1. Derjaguin, B.V. and Landau, L.D. (1941). *Acta Physicochim.URSS* 14, 633-662; Derjaguin, B.V. (1979). *Uspekhi khimii* 48, 675-721.
2. Verwey, E.J.W. and Overbeek, J.Th.G. (1948). "Theory of Stability of Lyophobic Colloids". Elsevier, Amsterdam.
3. Rehbinder, P.A. and Pospelova, K.A. (1950). Foreword to the Russian translation of Clayton, W. "Emulsions, their Theory and Technical Applications", pp.11-74. Inostrannaya Literature, Moscow.
4. Rehbinder, P.A. (1978). "Selected Works". Nauka, Moscow.
5. Napper, D.H. (1968). *Trans.Faraday Soc.*, 64, 1701.
6. Rehbinder, P.A. (1933). *In* "Physico-Chemical Studies of Technical Suspensions" (Ed. P.A. Rehbinder), pp. 7-59, Gostekhizdat, Moscow-Leningrad.
7. Yaminsky, V.V., Yusupov, R.K., Amelina, E.A., Pchelin, V.A. and Shchukin, E.D. (1975). *Kolloidny Zhurnal* 37, 918-925.
8. Yaminsky, V.V., Amelina, E.A. and Shchukin, E.D. (1976). *Kolloidny Zhurnal* 37, 312-320.
9. Shchukin, E.D. and Amelina, E.A. (1979). *Adv.Colloid Interface Sci.*, 11, 235.
10. Amelina, E.A., Yaminsky, V.V., Syunyaeva, R.Z. and Shchukin, E.E. (1982). *Kolloidny Zhurnal* 44, No.4 (in print).
11. Shchukin, E.D., Amelina, E.A. and Yaminsky, V.V. (1981). *Colloids and Surfaces* 2, 221-242.
12. Van Voorst Vader, F. and Dekker, H. (1974). *J.Adhesion* 7, 73-89.
13. Derjaguin, B.V. (1934). *Kolloid Z.* 69, 155.
14. Muller, V.M., Yushchenko, V.S. and Derjaguin, B.V. (1980). *J.Colloid Interface Sci.*, 77, 91-101.
15. Shchukin, E.D. and Rehbinder, P.A. (1958). *Kolloidny Zhurnal* 20, 645.
16. Pertsov, A.V. (1967). Cand. thesis. Moscow University.
17. Shchukin, E.D., Amelina, E.A. and Yaminsky, V.V. (1981). *Dokl.Akad.Nauk SSSR* 258, 419-423.
18. Benitez, R., Contreras, S. and Goldfarb, J. (1971). *J.Colloid and Interface Sci.*, 36, 146-150.
19. Shchukin, E.D. (1965). *Kinetika i kataliz* 6, 641-650; Shchukin, E.D., Margolis, L.Ya. and Rehbinder, P.A. (1964). *Dokl.Acad.Nauk SSSR* 154, 695-698.

20. Everett, D.H. (1958). *In* "The Structure and Properties of Porous Materials" p.95, London.
21. Manegold, E. (1941). *Kolloid Z.* 96, 186.

ADSORPTION OF CATIONIC SURFACTANTS ON ZEOLITE A

M.J. Schwuger, W. von Rybinski, P. Krings

HENKEL Laboratories, Düsseldorf

1. INTRODUCTION

The practical meaning of the adsorption of cationic surfactants on solid surfaces is connected with their broad application as fabric softeners. The action of the cationic surfactants as softening agents is based on the adsorption on textiles during the rinse cycle. In the subsequent wash cycle the adsorbed cationic surfactants are desorbed from the textile fiber and are released in the wash bath. A worldwide change in the formulation of laundry detergents can be observed at present, phosphates being partially substituted by zeolite A (1-4). As a consequence, cationic surfactants and zeolite A meet during the washing process and, after that, in sewage water. For this reason, the knowledge of interactions between zeolite A and cationic surfactants is of great interest.

This paper, therefore, deals with the influence of various physical chemical parameters on the adsorption in the simple model system zeolite A/cationic surfactant.

2. EXPERIMENTAL CONDITIONS

2.1 Adsorbent

Synthetic zeolite A (SASIL; registered trademark of Henkel KGaA, Düsseldorf, Germany) was used as adsorbent. The particle size was < 10 µm. The maximum of the particle size distribution amounted to 4.2 µm. The BET-surface, measured at 77 K by nitrogen adsorption, amounted to 2.6 m^2/g.

2.2 Adsorbate

Ditallowalkyl dimethyl ammonium chloride (Präpagen WK; trade name of Hoechst AG, Frankfurt, Germany) was used as adsorbate (acronym DAC). The product contains 95 % of a mixture of quaternary ammonium compounds with n-alkyl chains of 16 and 18 carbon atoms. As a comparison, some measurements were carried out with pure hexadecyl trimethyl ammonium chloride (acronym HTC).

2.3 Determination of the Amounts Adsorbed

The concentration of the cationic surfactant before and after the adsorption was determined by two-phase titration.

3. RESULTS AND DISCUSSION

Zeolite A consists of a negatively charged aluminium silicate structure, the excess charge of which is compensated for by sodium ions. In the presence of alkaline earth metal ions, the sodium counterions are largely exchanged (5). Due to the negative excess charge of the zeolite structure and the high mobility of the sodium ions, a specific adsorption of cationic surfactants is possible in a highly alkaline medium. The adsorption is probably characterized by strong electrostatic forces of interaction. The state of charge of the zeolite surface, which is influenced by the pH value and the presence of additional ions, is probably of crucial importance to the mechanism of adsorption of cationic surfactants. As the surface of zeolite A is uniform and chemically and morphologically well-defined, the interactions between the solid surface and the surfactant molecules are likely to be explained by simple adsorption models.

Fig. 1 depicts the adsorption isotherm of DAC on zeolite A. The isotherm shows a very steep slope at the start which indicates a strongly preferred adsorption of the cationic surfactant on the surface of the zeolite. The adsorption isotherm is characterized by a marked plateau and appears to be independent of the ratio of the concentration of the zeolite to that of the cationic surfactant in the range investigated. Within the extent of this flat portion of the isotherm plot, the equilibrium value is almost immediately attained. In the range beyond 2 minutes no dependence of the adsorption on the time was observed.

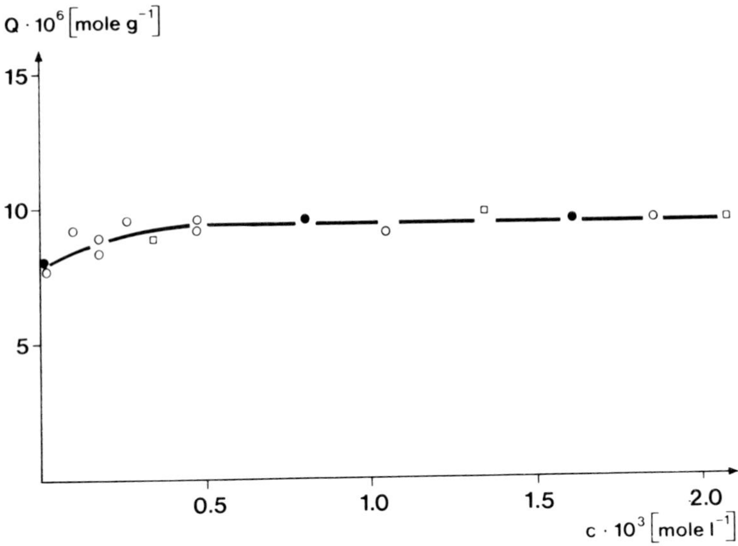

Fig. 1. *Adsorption isotherm of DAC on zeolite A for various concentrations of zeolite A*

□ - □ $c_s = 20\ g\ l^{-1}$, ● - ● $c_s = 40\ g\ l^{-1}$,

o - o $c_s = 60\ g\ l^{-1}$; T= 296 K, pH= 11

Due to the shape of the isotherm and the kinetics of adsorption, ion exchange or ion pair formation may be assumed as a mechanism for the adsorption (6). In the first case positive sodium ions, that compensate for the negative charge of the aluminium silicate structure, are exchanged by quaternary ammonium ions according to the following reaction:

$$\left[\begin{matrix} \diagdown Si \diagup \\ \diagup \quad \diagdown O \\ \diagdown Al \diagup \end{matrix}\right]^{\ominus} Na^{\oplus} + \left[\begin{matrix} R & & CH_3 \\ & N & \\ R & & CH_3 \end{matrix}\right]^{\oplus} \rightleftharpoons \left[\begin{matrix} \diagdown Si \diagup \\ \diagup \quad \diagdown O \\ \diagdown Al \diagup \end{matrix}\right]^{\ominus} \left[\begin{matrix} R & & CH_3 \\ & N & \\ R & & CH_3 \end{matrix}\right]^{\oplus} + Na^{\oplus} \qquad (1)$$

In the second case (ion pair formation), adsorption would occur on those sites of the lattice of the zeolite surface that are not occupied by sodium counterions. By virtue of the high concentration of counterions, a simultaneous process of ion pair formation and ion exchange has to be expected in the special case of the surface of zeolite A. Adsorption can only take place at the external surface as the quaternary ammonium ion cannot penetrate into the pores of zeolite A (pore area 0.16 nm^2), due to the assumed sectional area of its head group of approximately 0.43 nm^2. It can be concluded from the steep slope of the adsorption isotherm of Fig. 1 that the ion exchange equilibrium, as defined by equation 1, is strongly shifted to adsorption of the cationic surfactants.

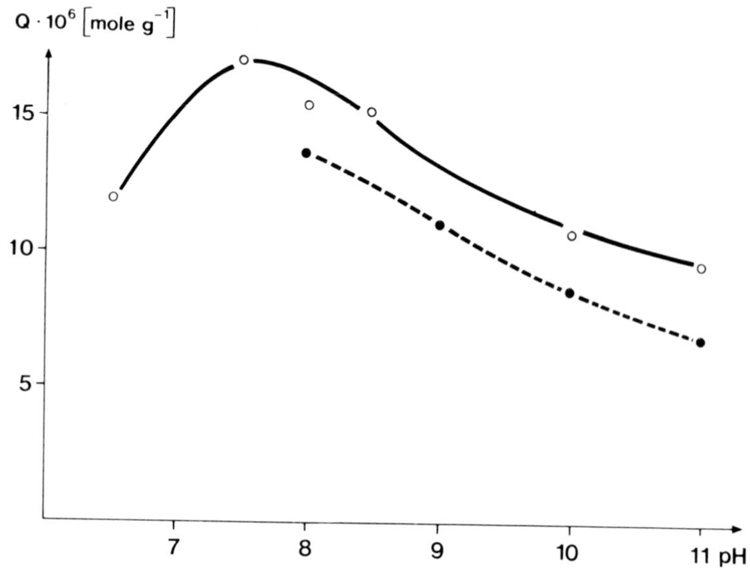

Fig. 2. *Plateau values of the isotherms of DAC (○) and HTC (●) on zeolite A as a function of pH*

$c_o = 1 \cdot 10^{-3}$ mol l^{-1}, $c_s = 20$ g l^{-1}; T= 296 K

As the state of charge of the negatively charged zeolite surface has a large influence on the adsorption of the positively charged ammonium ions, a marked dependence on the pH exists. Fig. 2 shows that, with equal shape of the isotherms, the plateau values decrease with increasing pH. Since zeolite A, as a metastable substance, is already destroyed in the acid pH range, adsorption at less than pH 7 is no more characteristic with respect to interactions between zeolite and surfactants. Reduced adsorption of surfactants at higher pH values, however, is specific for the surface of zeolite A. It does not depend on the number of n-alkyl chains of the quaternary alkyl ammonium coumpounds. The adsorbed amounts in the plateau range of the isotherms, however, are lower for HTC than for DAC. This difference can be explained by the different spaces reqüired by the monoalkyl trimethyl compounds in comparison with the dialkyl dimethyl compounds. This can be concluded from measurements of the thickness of the layer and the adsorption of cationic surfactants on layer silicates (7, 8). Because of the negatively charged zeolite surface the monoalkyl trimethyl compound is supposed to arrange in such a way, that the positively charged quaternary ammonium ions have a distance as close as possible to the zeolite surface. This causes also an arrangement of the three methyl groups as near as possible to the zeolite surface. In comparison for the tallowalkyl dimethyl compound only two methyl groups are oriented to the zeolite surface. Therefore the cross sectional area of the monoalkyl trimethyl compound (~ 0.46 nm^2) is larger than that of the dialkyl dimethyl compound (~ 0.43 nm^2).

Measurements of the electrophoretic mobility of zeolite A show that the negative charge is enhanced with increasing pH value (Fig. 3). When the results of adsorption are compared with those of electrophoretic mobility (Fig. 2 and 3), an opposite change surprisingly can be observed. Thus these results differ from many known (9, 10, 11) that indicate an increase of the adsorbed amounts of cationic surfactants with increasing alkalinity. This characteristic difference is probably due to special interactions between zeolite surface and the cationic surfactant. At the external surface of the cell of zeolite A there are, in the ideal structure, 8 aluminium atoms, of which 6 are tetrahedrically arranged in the zeolite framework. These cause a negative excess charge that is almost independent of the pH and is compensated for by sodium ions. Two aluminium atoms at the external surface have only 3 bonds in the zeolite structure and, hence, can take up a hydroxyl group at alkaline pH-values. With increasing pH, this leads to an increase of the negative charge and to bonding of additional sodium ions.

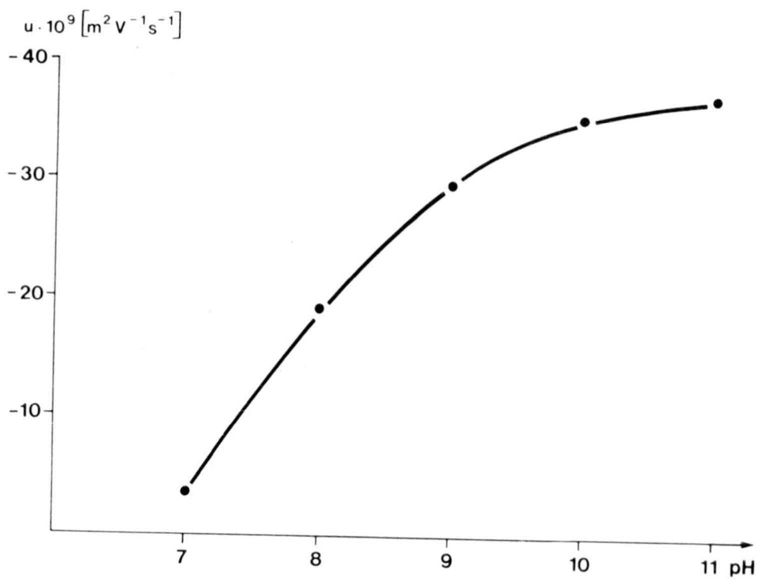

Fig. 3. *Influence of pH upon the electrophoretic mobility of zeolite A*

$c_s = 2\ g\ l^{-1}$

The sodium ions can be exchanged by other positively charged ions. When only the pH-independent exchange is considered, an approximate required space of 0.25 nm^2 for exchanged cation results from the unit cell constant of the zeolite (1.23 nm). As the required surface of the ditallow alkyl dimethyl ammonium chloride molecule with perpendicular arrangement in a closed packed monolayer is about 0.43 nm^2, only 60 % approximately of the places capable of exchange can be covered by the surfactant. Thus, a part of the negative excess charge at the external surface of the zeolite can be compensated for by sodium ions, even after adsorption equilibrium is attained. From the BET surface and the assumed required space of the surfactant molecule (0.43 nm^2), a value of about $10 \cdot 10^{-6}$ mol/g for the amount adsorbed results with monolayer adsorption. This value is reached at pH 10 in the plateau range of the isotherms (Fig. 2). At lower pH-values, the sodium ions at the surface are more weakly bonded due to a lower negative charge of the zeolite structure (Fig. 3). Hence, particularly in the pH-range from 7 to 8, additional surfactant molecules can be adsorbed in an arrangement of the alkyl chains directed to the zeolite surface via the weaker van der Waals interactions between the alkyl chains. This is a possible explanation of the apparently enhanced adsorption at lower pH-values.

Extraction tests with water at 373 K showed that the additional adsorbed amount is desorbed at pH-values lower than 10 under these conditions, whereas desorption tests at pH 10 did not result in a decrease of adsorption. This is considered a further indication of the existence of two different types of bonding in the plateau range of the isotherms.

A parallelism to the pH-dependence of adsorption of surfactants on zeolites is encountered with the intercalation of surfactants in layer silicates (12).

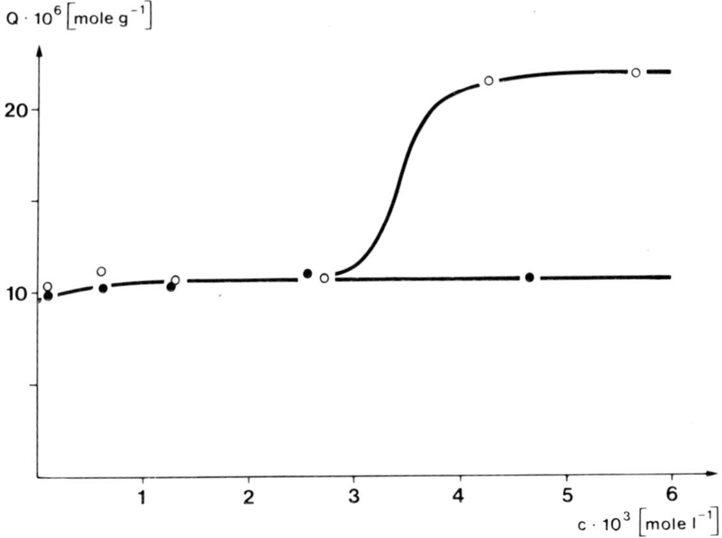

Fig. 4. *Adsorption isotherms of DAC on zeolite A (●) and zeolite A with 50 % exchange of sodium by calcium ions (o)*

c_s= 20 g l^{-1}; T= 296 K, pH= 10

As in the washing process a partial exchange of sodium ions by calcium or magnesium ions always takes place, the influence of the concentration of calcium in the zeolite on the adsorption of cationic surfactants is particulary interesting. Fig. 4 depicts the adsorption of DAC at pH 10 on zeolite A in which 50 % of the sodium ions were exchanged by calcium ions, in comparison with pure sodium zeolite A. In the range of low equilibrium concentrations, no difference in adsorption between both types of zeolites exists. The existence of a close-packed monolayer may be concluded from the BET-surface and the required place assessed for the DAC molecules.

At higher equilibrium concentrations, a second adsorption layer is formed in the case of the calcium containing zeolite type. This is shown by the second step of the adsorption isotherm, the height of which inspite of less exact data at high surfactant concentrations corresponds with the first plateau value. Whereas in the first layer the hydrophilic ammonium group is in a direction towards the zeolite surface, the polar groups are directed towards the aqueous phase in the second layer. Hence, the second layer is connected to the first adsorption layer via weak hydrophobic interactions. In case of the pure sodium type of zeolite A, the second step in the adsorption isotherm is indicated at even higher concentrations at which the measuring accuracy of the method used decreases strongly. As sodium zeolite and calcium zeolite differ only little in their state of charge of the surface, the better formation of a double-layer adsorption is only explained by the influence of the calcium ions. Due to the equilibrium of ion exchange, calcium ions exist in the solution that may give rise to enhanced adsorption based on an electrolyte effect.

According to known laws of adsorption, an addition of an electrolyte (e.g., NaCl) leads to enhanced adsorption on the zeolite surface in the range of monolayers (Fig. 5). This results in a decreased electrostatic repulsion of the positively charged groups of the cationic surfactant and hence in a denser coverage of the surface. According to the formulated mechanism of ion exchange, a possible shift of the equilibrium by the sodium ions, that is expected to give rise to a decrease of adsorption, is compensated for by the electrolyte effect. Under these conditions, the beginning of the formation of the double layer is shifted to lower equilibrium concentrations in case of the sodium type of zeolite A as well.

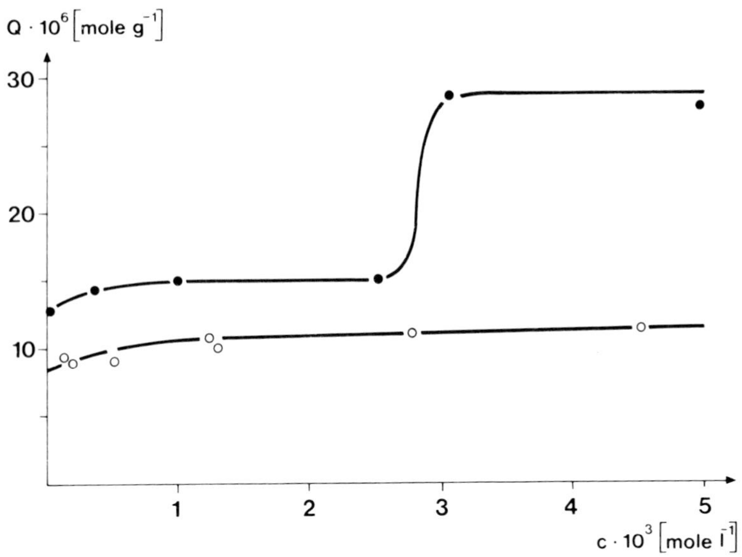

Fig. 5. *Influence of electrolyte addition upon the adsorption of DAC on zeolite A*

● - ● c_{NaCl}= 0,1 mol l^{-1}, o - o c_{NaCl}= 0;

c_s= 20 g l^{-1}; T= 296 K, pH= 10

As a consequence of adsorption of the cationic surfactants, the charge of the originally negatively charged zeolite surface reverses its sign. Fig. 6 plots the adsorbed amount with the electrophoretical mobility vs. the equilibrium concentration of the surfactant. As is shown by this comparison, a reversal of the charge occurs already at very low equilibrium concentrations and at a coverage of the surface of less than a monolayer. The point of reversal is attained at a coverage of approximately 1/3 of a monolayer as has also been determined with the adsorption of cationic collectors on surfaces of minerals (13). In the range of the flat portion of the isotherm plot, the state of charge of the zeolite surface no more changes either. Fig. 6 also shows that in the range of lower surfactant concentrations the adsorption can be described by the Langmuir equation. The use of zeolite A in laundry detergents hence has consequences with regard to the elimination of cationic surfactants in the washing process and in sewage waters (14).

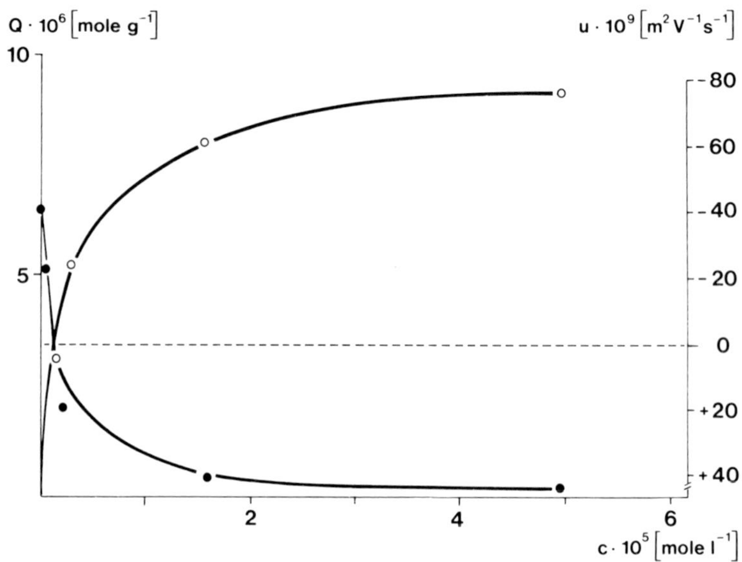

Fig. 6. *Comparison of the adsorption isotherm of DAC on zeolite A (o) and the electrophoretic mobility (•)*
$c_s = 2$ g l^{-1}; T= 296 K, pH= 10

As is shown by Fig. 7, the strong adsorption causes an almost complete elimination of cationic surfactants from aqueous solutions already with small surfactants concentrations. Hence, the removal of cationic surfactant from water and sewage water is supported by the use of zeolite-based laundry detergents. The possible influence of other types of surfactants and ingredients of laundry detergents on the adsorption of cationic surfactants on zeolite A will be described in a further publication.

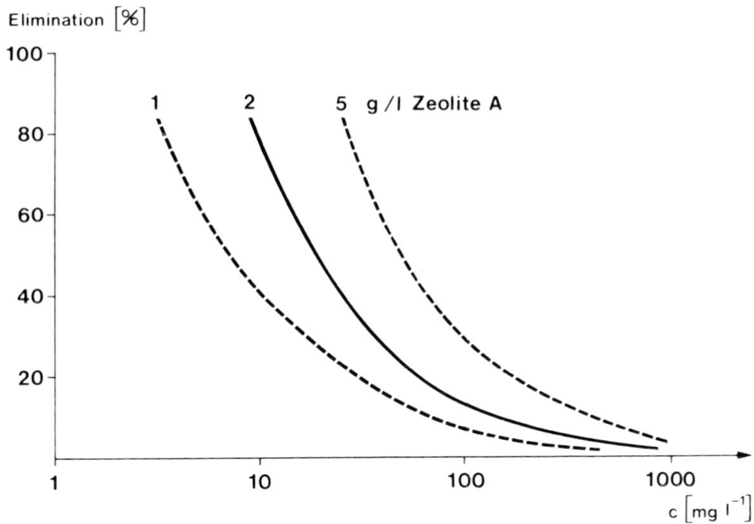

Fig. 7. *Elimination of DAC from aqoeous solution by zeolite A*
T= 296 K, pH= 10

—— *from adsorption measurement*
--- *calculated*

References

1. Schwuger, M.J.; Smolka, H.G.; Rostek, H.M. (1974), *DOS* 24 12 837, Henkel
2. Schwuger, M.J.; Smolka, H.G. (1976), *Colloid Polymer Sci.* 254, 1062
3. Berth, P. (1978), *Amer. Oil Chemists' Soc.* 55, 52
4. Krings, P.; Verbeek, H. (1981), *Tenside Detergents* 18, 260
5. Schwuger, M.J.; Smolka, H.G. (1978), *Colloid Polymer Sci.* 256, 1014
6. Rosen, M.J. (1975), *J. Amer. Oil Chemists' Soc.* 52, 431
7. Lagaly, G.; Witter, R. (1982), *Ber. Bunsenges. Phys. Chem.* 86, 14
8. Weiß, A. (1958), *Kolloid-Zeitschrift* 158, 22
9. Giles, C.H.; D'Silva, A.P.; Easton, J.A. (1974), *J. Colloid Interface Sci.* 47, 766

10. Connor, P.; Ottewill, R.H. (1971), *J. Colloid Interface Sci.* 37, 642
11. Ginn, M.E. (1970), in "Cationic Surfactants", Surfactant Science Series Vol. 4, ed. E. Jungermann, New York
12. Lagaly, G., Private Communication
13. Jaycock, M.J.; Ottewill, R.H. (1963), *Bull. Inst. Mining Met.* 677, 497
14. Bücking, H.W.; Lötzsch, K.; Täuber, G. (1979), *Tenside Detergents* 16, 2

ON HEMIMICELLE FORMATION AT OXIDE/WATER INTERFACES

S. Chander, D. W. Fuerstenau, and D. Stigter

Department of Materials Science
and Mineral Engineering
University of California
Berkeley, CA 94720, U.S.A.

INTRODUCTION

Physical adsorption of surfactant ions at solid-solution interfaces may involve hemimicelle formation through association between hydrocarbon chains of the adsorbed molecules (1-5). At present, little is really known about the phenomena of hemimicelle formation primarily due to the lack of adequate experimental data to evaluate independently the "hydrophobic effect" in the adsorbed surface layer. In this paper, the existing data on physical adsorption of surfactant ions is reviewed for the adsorption of alkyl and alkylbenzene sulfonates on alumina. A new approach is given to describe the thermodynamics of hemimicelle formation in the next section. This approach allows determination of the number of surfactant ions in a hemimicelle from the adsorption isotherms. In a subsequent section, the previous adsorption data are reanalyzed in terms of the proposed theory to calculate the size of hemimicelles.

ISOTHERMS FOR PHYSICAL ADSORPTION OF SURFACTANT IONS

Several different approaches have been taken to describe thermodynamically isotherms for the adsorption of ionic surfactants from aqueous solutions onto oxides (1-3). Here, a more general approach will be presented. We begin with the condition for electrochemical equilibrium between the surfactant ions in the adsorbed state and the bulk solution. At equilibrium,

$$\bar{\mu}^s = \bar{\mu}$$

The electrochemical potential in solution is divided into a concentration-dependent term and a reference (standard state) term

$$\bar{\mu} = \bar{\mu}^{\circ} + RT \ln a \quad (2)$$

In adsorption from dilute solutions, the activity may be replaced by concentration so that

$$\bar{\mu} = \bar{\mu}^{\circ} + RT \ln c \quad (3)$$

The separation of the electrochemical potential in the adsorbed state into a concentration-dependent term and a concentration-independent term leads to different results as discussed in the following paragraphs.

Absence of Interaction Between Adsorbed Molecules on a Homogeneous Surface.

The electrochemical potential in the adsorbed state may be written as

$$\bar{\mu}^{s} = \bar{\mu}^{\circ,s} + RT \ln f\,(c^{s}) \quad (4)$$

where the first term on the right side of Equation 4 is the concentration-independent term and the second is the concentration-dependent term. A similar approach is used by Dalahay (6) to describe adsorption on electrodes. Two specific cases can be considered depending upon the form of the concentration dependent term.

At small adsorption densities, Henry's law may be assumed for the adsorbed state and Equation 4 may be written as

$$\bar{\mu}^{s} = \bar{\mu}^{\circ,s} + RT \ln c^{s} \quad (5)$$

The widely used Stern-Grahame equation can be obtained by assuming $c^{s} = \Gamma_{\delta}/2r$, and $\Delta G^{\circ}_{ad} = (\bar{\mu}^{\circ,s} - \mu)/RT$. Then, from Equations 1, 3, and 5

$$\Gamma_{\delta} = 2rc \exp\,[-\Delta G^{\circ}_{ad}/RT] \quad (6)$$

In this derivation, ΔG°_{ad} is a constant, independent of concentration or adsorption density. For adsorption of ions on a mercury electrode, Stern defined

$$\Delta G^{\circ}_{ad} = z_{i}F\psi_{\delta} \quad (7)$$

Therefore

$$\Gamma_\delta = 2rc \exp [- z_i F \psi_\delta / RT] \tag{8}$$

This equation has been used to describe the adsorption of surfactant ions in the Stern plane even for conditions when ΔG°_{ad} varies with Γ. Modification of Equation 8 for such a condition is discussed later.

For intermediate adsorption densities, $f(c^s)$ in Equation 4 may be written as

$$f(c^s) = \frac{c^s}{c^s_{max} - c^s} = \frac{\theta}{1 - \theta} \tag{9}$$

following a Langmuir-type approach. Equation 4 then becomes

$$\bar{\mu}^s = \bar{\mu}^{\circ,s} + RT \ln \frac{c^s}{c_{max} - c^s} \tag{10}$$

From equations 1, 3, and 10, we then can write

$$\frac{\Gamma_\delta}{\Gamma_{max} - \Gamma_\delta} = 2rc \exp [- \Delta G^\circ_{ad}/RT] \tag{11}$$

This equation takes into account the change in available area of the interface as adsorption proceeds. Equation 11 reduces to the Stern-Grahame equation at small adsorption densities.

Interaction Between Adsorbed Molecules on a Homogeneous Surface.

Different approaches have been used to account for interactions between adsorbed molecules. Fuerstenau and Modi (1) introduced an additional work term to account for association of hydrocarbon chains of the adsorbed molecules.

$$\Delta G^\circ_{ad} = z_i F \psi_\delta + \phi \tag{12}$$

These investigators considered ϕ to be a function of adsorption density; it is zero until the hydrocarbon chains associate and increases gradually with increase of adsorption density. (A similar approach is used to account for a heterogeneous surface as described in the next section). Somasundaran and Fuerstenau (2) considered

$$\phi = n\phi' \tag{13}$$

where ϕ' is the van der Waals energy of interaction per CH_2 group between adjacent chains of adsorbed surfactant molecules and n is the number of associating CH_2 groups. These investigators considered ϕ' as a constant and treated n as a variable dependent upon the adsorption density. Bockris and Reddy (7) used a similar approach to incorporate the effect of electrostatic repulsion between adsorbed ions at the mercury electrode and derived an expression for the free energy of adsorption which is a function of the adsorption density. For the adsorption of various surfactants on oxides, the association of hydrocarbon chains to form hemimicelles occurs at adsorption densities where hydrocarbon chains cannot overlap if adsorption is considered to be homogeneous. Defining the surface phase as a layer of thickness equivalent to the length of the surfactant molecule, we can obtain a surface concentration. The surface concentration where hemimicelles just begin to form, can be calculated by dividing Γ_{HM} by the thickness of the adsorbed layer. At low surface potential, this surface concentration is about the same as the critical micelle concentration which suggests that both the hemimicelle formation at the interface and the micelle formation in the bulk solution involve similar interactions, namely, hydrophobic interactions. The effect of pH or surface potential on c^s_{HM} is discussed later in this paper.

As discussed in the preceeding paragraphs, the effect of hemimicelle formation was incorporated in the adsorption equation by adding an adsorption-dependent term to the free energy of adsorption. We now take an alternate approach. Consider the aggregation of m adsorbed monomers to form a hemimicelle. At equilibrium,

$$\text{m (Adsorbed Monomers)} = \text{(Hemimicelle)} \tag{14}$$

$$m\,\bar{\mu}^s = \bar{\mu}^s_{HM} \tag{15}$$

Again, separating the electrochemical potential into a concentration-dependent and a concentration-independent term,

$$\bar{\mu}^s_{HM} = \bar{\mu}^{\circ,s}_{HM} + RT \ln c^s_{HM} \tag{16}$$

Then from Equations 1, 3, and 16

$$c^s_{HM} = \frac{1}{m}\left(\frac{\Gamma_\delta}{2r} - c^s\right) = c^m \exp\left[-\Delta G^\circ_{ad}/RT\right] \tag{17}$$

(The ΔG°_{ad} in Eq. 17 is different than that in Eq. 6.)

When hemimicelles predominate, c^s, the surface concentration of surfactant molecules adsorbed as individual ions is small. Equation 17 may then be written as

$$\Gamma_\delta = 2r\ m\ c^m \exp\ [\ -\ \Delta G^\circ_{ad}/RT\ ] \qquad (18)$$

A plot of adsorption density-vs.-surfactant concentration should give a straight line of slope m. The effect of c^s in Equation 17 would be to make the transition from Region I, in which surfactants adsorb as monomers, to Region II in which they adsorb as hemimicelles, more gradual. Since the experimental data, presented in the next section, shows a sharp transition from Region I to Region II, the value of c^s in Equation 17 appears to be negligibly small.

If m is considered independent of concentration, that is, the size of the hemimicelle remains the same as concentration increases, a plot of log Γ_δ-vs.-log c should yield a straight line of slope m, the aggregation number which is a measure of the size of the hemimicelle.

Absence of Interaction Between Adsorbed Molecules on a Heterogeneous Surface.

Although copious literature exists for the adsorption of gases on heterogeneous surfaces, very little is known about the physical adsorption of surfactant ions at heterogeneous surfaces of oxide minerals. For a heterogeneous surface it is appropriate to consider the free energy of adsorption as a function of adsorption density or surface coverage. Different forms of adsorption isotherms have been derived depending upon the nature of the function considered for variation of the free energy of adsorption with coverage. In most cases the derivations of adsorption isotherms are extensions of the models for adsorption of gases on solids. These models ignore the role of the solvent in the adsorption process. In chemisorption processes, the solvent effects are relatively small and the gas-solid adsorption models may be used successfully. In physical adsorption of surfactants, however, the role of the solvent molecules may be important and requires additional investigation. Further, variation of the adsorption energy with coverage is no proof of adsorption at a heterogeneous surface since similar isotherms can be derived when there is interaction between adsorbed molecules. Because of these reasons and lack of adequate experimental data, specific models for physical adsorption of surfactant ions onto heterogeneous surfaces are not considered in this paper.

EXPERIMENTAL ISOTHERMS

In this section we review a number of isotherms for selected physically adsorbing systems. For the adsorption of sodium dodecyl sulfonate on alumina three distinct regions are observed when the isotherm is plotted on a log-log scale, as shown in Figure 1. In Region I the ions adsorb individually. If all adsorbed ions are present in the Stern layer, and the surface is homogeneous the slope of log Γ-vs.-log c plot should be unity. A constant zeta potential in Region I, shown in Fig. 1, is indicative of constant ψ_δ and therefore constant free energy of adsorption. In Region II, the adsorbed surfactant ions associate to form hemimicelles and the slope of the log Γ-vs.-log c plot gives the aggregation number, that is, the number of ions in a hemimicelle. As the adsorption density increases the number of hemimicelles increases. Figure 1 also shows that the zeta potential decreases fairly sharply with increase in concentration until it reverses sign. This marks the on-set of Region III in

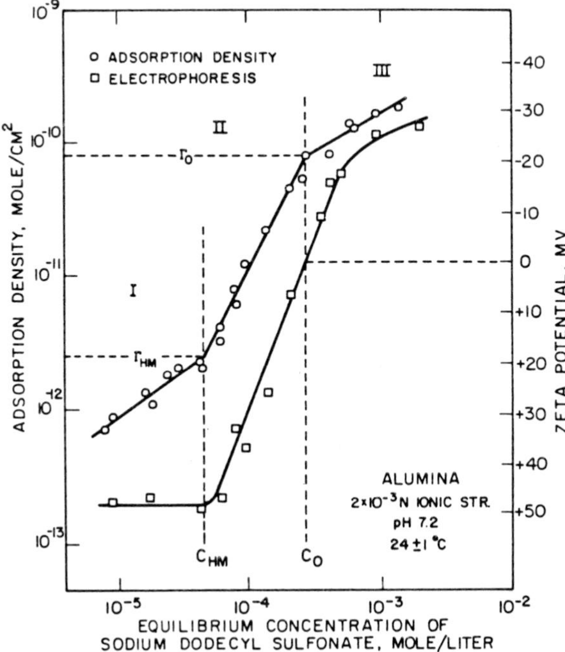

Fig. 1. The adsorption density on, and the electrophoretic mobility of alumina as a function of sodium dodecyl sulfonate concentration at pH 7.2 and ionic strength of 2X10^{-3}M NaCl (2).

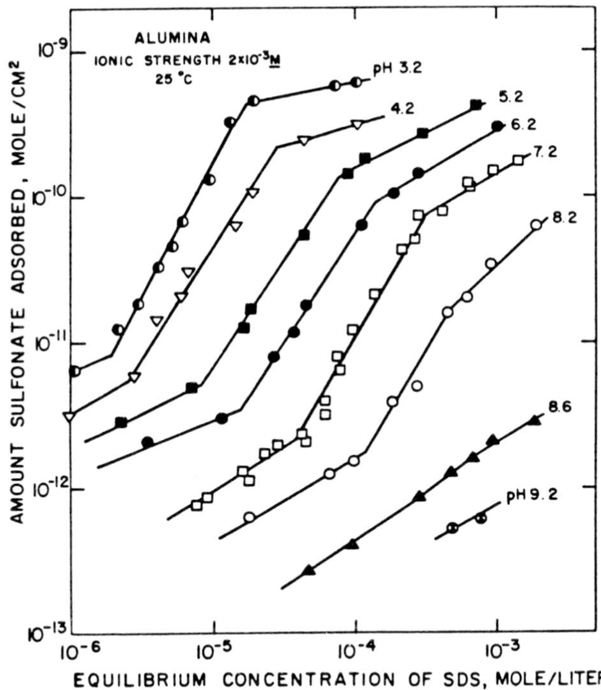

Fig. 2. The isotherms for adsorption of sodium dodecyl sulfonate (SDS) on alumina at ionic strength of $2X10^{-3}M$ NaCl for several pH's (5).

which adsorption occurs primarily through hydrocarbon chain association. The decrease in the slope of the isotherm in this region is probably due to coulombic repulsion which now opposes adsorption. The correspondence between adsorption density and zeta potential suggests that the electrostatic contribution to the free energy of adsorption is determined by the potential in the Stern plane, ψ_δ and not by the surface potential, ψ_o. The slope of the log Γ-vs.-log c plots in Figure 2 gives an aggregation number of 2 for the adsorption of dodecyl sulfonate on alumina. The size of the hemimicelle is independent of pH as seen in Figure 2 which shows that the slope in Region 2 is essentially independent of pH. An aggregation number of 2 implies that dimers are formed on the surface and that the hydrophobic association of chains contributes to the apparent increase in the free energy of adsorption. Dimerization of chains might be favored because the hydrocarbon chains can coil around each other to give maximum hydrophobic association. Figure 2 shows that the formation of hemimicelles is shifted to higher adsorption densities with decrease in pH or increase in surface potential. That is, with increasing surface potential monomers are favored over dimers.

The surface concentration of adsorbed ions where hemimicelles just begin to form, calculated by dividing Γ_{HM} by the thickness of the adsorbed layer is given as a function of pH in Table 1. At small surface potentials, the surface concentration approaches the CMC, but the aggregation of adsorbed surfactant ions to form hemimicelles is somewhat suppressed at high potentials. This can be explained in terms of the excluded volume effect of surfactant ions. At high potentials, the concentration of the counter ions increases relatively more for the smaller chloride ions than for the bulky surfactant ions. This makes it more difficult to form hemimicelles at the surface, that is, the equilibrium in Equation 14 is shifted to the left.

The effect of salt on the adsorption of sodium dodecyl sulfonate on alumina is shown in Figure 3. To explain the results, we will consider the effect of salt on the CMC. The effect of salt concentration on the CMC of surfactants may be expressed by the relation (9)

$$\log \text{CMC} = A \log C_+ + B \tag{19}$$

(it is customary to take C_+ equal to the total cation concentration for an anionic surfactant, that is, contribution to cations from the surfactant and the added salt). The value of A ranges from -0.34 to -0.67 for different surfactants (9). In this paper, A for sodium dodecyl sulfonate is taken to be the same as that for sodium dodecyl sulfate

TABLE 1

The Effect of pH on Adsorption Density and Surface Concentration Where Hemimicelles Just Begin to Form

pH	Γ_{HM} Mole/cm^2	C^s_{HM} Moles/liter
3.2	6.0×10^{-12}	3.3×10^{-2}
4.2	5.5×10^{-12}	3.0×10^{-2}
5.2	4.2×10^{-12}	2.3×10^{-2}
6.2	3.4×10^{-12}	1.8×10^{-2}
7.2	2.2×10^{-12}	1.2×10^{-2}
8.2	1.6×10^{-12}	0.8×10^{-2}

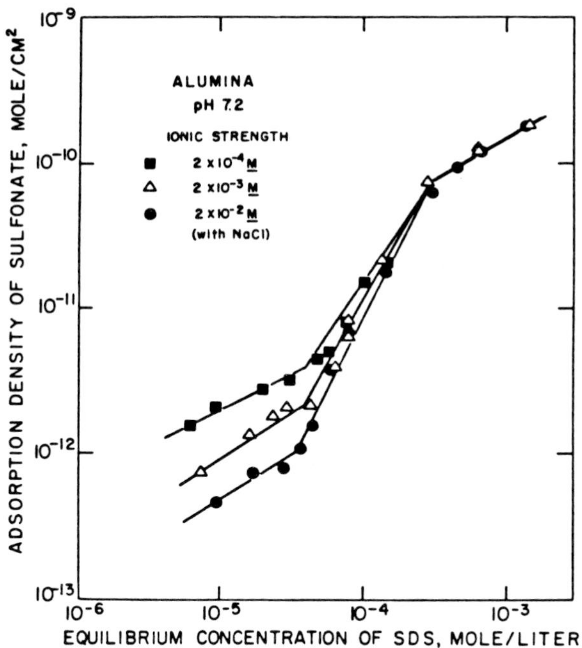

Fig. 3. The isotherms for adsorption of sodium dodecyl sulfonate (SDS) on alumina at pH 7.2 for several ionic strengths (8).

which is -0.458. The estimated CMC's are given in Table 2 for the three salt concentrations. The table also lists the adsorption densities for various electrolyte concentrations where hemimicelles just begin to form. Γ_{HM}^{calc} in Table 2 is the adsorption density calculated assuming that hemimicelles are formed when the surface concentration approaches the

TABLE 2

The Effect of NaCl Concentration on CMC and Adsorption Density for Formation of Hemimicelles.

NaCl Conc. Mole/liter	CMC Mole/liter	Γ_{HM} Mole/cm^2	Γ^{calc} Mole/cm^2
0	10^{-2}	-	1.8×10^{-12}
2×10^{-4}	9.9×10^{-3}	3.86×10^{-12}	1.8×10^{-12}
2×10^{-3}	9.2×10^{-3}	2.2×10^{-12}	1.6×10^{-12}
2×10^{-2}	6.0×10^{-3}	1.0×10^{-12}	1.1×10^{-12}

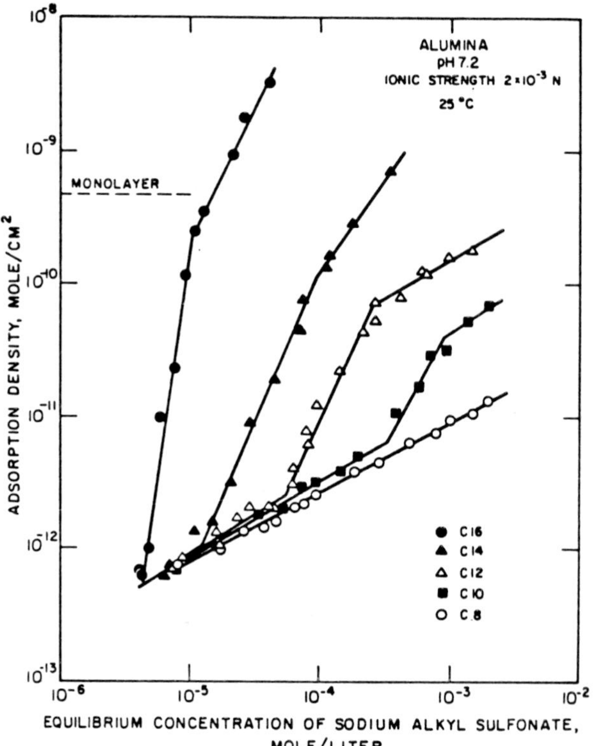

Fig. 4. The isotherms for adsorption of sodium alkyl sulfonate on alumina at pH 7.2 and ionic strength of $2X10^{-3}$ M NaCl for several hydrocarbon chain lengths (3).

CMC. The consistently larger values of Γ_{HM} compared to Γ_{HM}^{calc} are attributed to the anticipated effect of pH on CMC. Thus, we conclude that the effect of electrolyte concentration on adsorption isotherms is through the effect of salt on the aggregation of adsorbed surfactant ions to form hemimicelles. Thus, this effect is similar to the effect of salt on the formation of micelles in bulk solution.

The aggregation number depends upon the chain length, as seen from the slope of the log Γ-vs.-log C plots in Figure 4. The figure shows that there is no hemimicelle formation for the short-chain surfactant containing only 8 carbon atoms. The aggregation number is 2 for surfactant ions containing 10-14 carbon atoms and it is 7 when the number of carbon atoms in the hydrocarbon chain is 16. An aggregation number of 7 would suggest a hemimicelle containing a central chain surrounded by six chains in a hexagonal close packing.

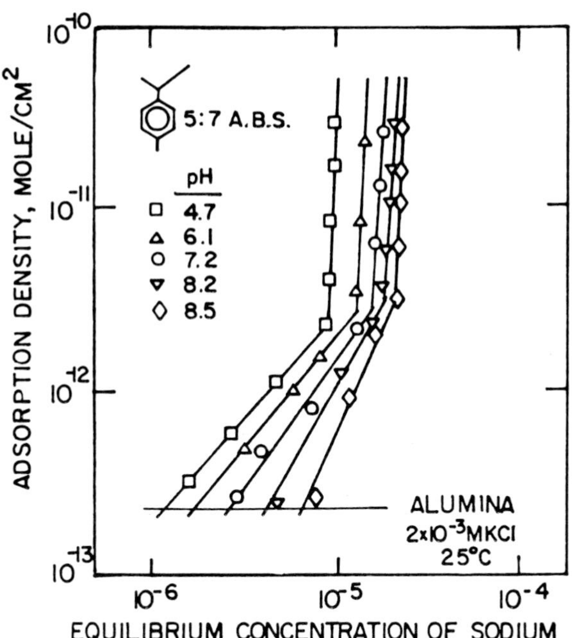

Fig. 5. The isotherms for adsorption of sodium dodecyl benzene sulfonate (5:7 isomer) on alumina at ionic strength of $2X10^{-3}$M KCl, for several pH's (4).

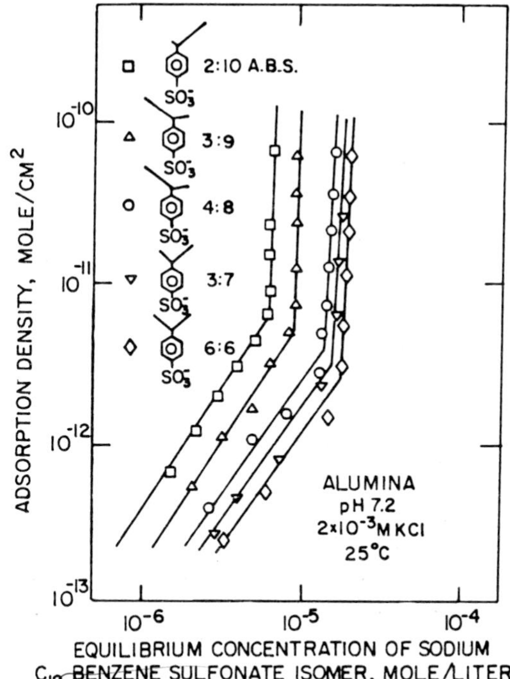

Fig. 6. The isotherms for adsorption of sodium dodecyl benzene sulfonate isomers on alumina at pH 7.2 and ionic strength of $2X10^{-3}$M KCl (4).

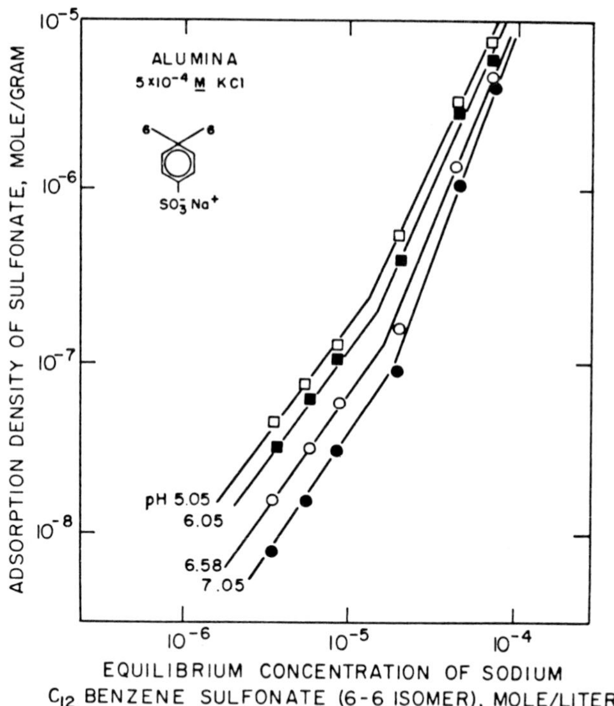

Fig. 7. The isotherms for adsorption of sodium dodecyl sulfonate on alumina (6:6 isomer) at ionic strength of $5X10^{-4}$ M KCl, for several pH's (8).

Such an arrangement is likely to give the maximum contribution to the free energy of adsorption through exclusion of hydrocarbon chains from water.

The isotherms for the adsorption of dodecylbenzene sulfonate on alumina are presented in Figures 5, 6 and 7. The slopes below the CHMC are greater by a factor of about 2 compared to the corresponding alkyl sulfonates. This is considered as indicative of dimerization in solution which can occur below the critical micelle concentration (10). The slopes below the CHMC are the same for different isomers which suggests that dimerization may occur primarily through the stacking of benzene rings. The dimerization is pH dependent however, as can be seen from the change in slope with pH of the isotherms below CHMC. Such dimerization in solution reduces the free energy gain on further association to form hemimicelles at the interface, hence, the CHMC increases when pH changes from 4.7 to 8.5. The large slope of the log Γ-vs.-log C plots in Figures 5 and 6, at concentrations above the CHMC suggests that relatively large hemimicelles are formed. The presence of the benzene ring increases the size of the hemimicelle considerably which is expected as the benzene ring is known to increase the effective length of the hydrocarbon chain. Each hemimicelle probably contains more than 20 surfactant ions. The lowest CHMC (for the 6:6 isomer) might be due to better packing of the surfactant ions in the hemimicelle, as in lipid membranes. The amount of indifferent electrolyte has a marked effect on the size of the hemimicelles as can be seen by comparing Region II slopes in Figures 6 and 7. Figure 6 is for a salt concentration of 2×10^{-3} moles/liter, whereas Figure 7 is for 5×10^{-4} mole/liter KCl.

For both dodecyl sulfonate and dodecylbenzene sulfonate (6:6 isomer) the size of the hemimicelle is increased with increase in salt concentration, as can be seen in Figure 3 and by comparing Figures 6 and 7. These results also show that the effect of salt on hemimicelle formation is similar to the salt effect on micelle formation. Thus, it is concluded that the "hydrophobic effect" which leads to the formation of micelles is also predominant in the case of hemimicelles, but the effect, however, has to be modified to include the effect of the charged interface.

SUMMARY

The thermodynamics of physical adsorption of surfactant ions is presented in this paper. The existing approaches to obtain analytical expressions for adsorption systems involving the association of adsorbed ions to form hemimicelles are reviewed and a new approach is given to analyze hemimicelle phenomena. The formation of hemimicelles is taken as a surface analog of micelle formation in bulk solution. Hemimicelle formation is dependent on the electrical nature of the interface and the salt concentration. The proposed model allows calculation of the size of the hemimicelle from the available adsorption data. The model is also used to explain the effect of pH, ionic strength, the number of carbon atoms in the hydrocarbon chain, and the effects of benzene rings in the hydrocarbon chain. Correspondence between the hemimicelle formation and micelles in bulk solution is taken as indicative of the "hydrophobic effect" in both cases. Hemimicelles are more difficult to form at lower pH's or high surface potentials. This is attributed to the exclusion volume effect of bulky surfactant ions relative to the smaller inorganic counter ions.

ACKNOWLEDGEMENT

Partial support of the National Science Foundation for this research is acknowledged.

REFERENCES

1. Fuerstenau, D. W. and Modi, H.J.(1959). *J. Electrochemical Soc.*, 106, 336-341.
2. Somasundaran, P. and Fuerstenau, D. W. (1966). *J. Physical Chem.*, 70, 90-96.
3. Wakamatsu, T. and Fuerstenau, D. W. (1968). in "Adsorption from Aqueous Solution," Adv. Chemistry Series, No. 79, ACS, New York, 161-172.
4. Dick, S. G., Fuerstenau, D. W., and Healy, T. W. (1971). *J. Colloid and Interface Sci.*, 37, 595-602.
5. Fuerstenau, D. W. and Wakamatsu, T. (1975). *Faraday Discussions of the Chemical Society*, 59, 157-168.
6. Delahay, P., (1965). "Double Layer and Electrode Kinetics". Interscience, New York.
7. Bockris, J. O'M. and Reddy, A. K. N. (1970). "Modern Electrochemistry". Vol. 2. Plenum, New York.
8. Fuerstenau, D. W. and Healy, T. W. (1967). "The Adsorption of Alkyl and Alkylaryl Sulfonates on Polar Solids", report to Federal Water Pollution Control Administration University of California, Berkeley.
9. Harkins, W. D. (1952). "The Physical Chemistry of Surface Films". Reinhold, New York.
10. Mukerjee, P. (1967). *Advances in Colloid and Interface Science*, 1, 241-276.

LIST OF SYMBOLS

a	activity
A, B	constants
c	concentration
CHMC	critical hemimicelle concentration
CMC	critical micelle concentration
m	aggregation number in a hemimicelle
n	number of carbon atoms in the hydrocarbon chain
r	one-half the adsorbed layer thickness
z_i	valence including sign
μ	chemical potential
$\overline{\mu}$	electrochemical potential
ΔG°_{ad}	standard free energy of adorption
ϕ	free energy of adsorption due to hydrocarbon chain association
ϕ'	free energy of hydrocarbon chain association per CH_2 group
Γ, Γ_{δ}	adsorption density at the surface and in Stern layer
ψ_{δ}	Stern layer potential
θ	fractional coverage

Superscripts and Subscripts: o: standard state, s: surface layer, calc: calculated, ad: adsorbed layer, HM: hemimicelle, +: salt cation, max: maximum.

ADSORPTION FROM SOLUTION ONTO HETEROGENEOUS SURFACES. EVALUATION OF TWO NUMERICAL METHODS TO DETERMINE THE ADSORPTION FREE ENERGY DISTRIBUTION

J. Papenhuijzen and L.K. Koopal

Laboratory for Physical and Colloid Chemistry
Agricultural University, De Dreijen 6
6703 BC Wageningen
The Netherlands

1. INTRODUCTION

Most solid surfaces are to some extent energetically heterogeneous. Therefore, any general theory of adsorption on ordinary surfaces should take account of heterogeneity. A natural extension of adsorption theories for homogeneous (homotattic) surfaces was originally suggested by Langmuir (1), popularized by Ross and Olivier (2) for gas adsorption and worked out for adsorption from solution by Rudziński et al. (3). In this model the surface is assumed to consist of homogeneous patches, each characterized by a certain adsorption free energy. Interactions between molecules adsorbed on different patches can be neglected if the patches are large in comparison to individual adsorption sites.

For strongly heterogeneous surfaces an alternative model has been proposed (4,5). Now the energetically homogeneous surface area to be distinguished is of the order of an individual adsorption site and the surface is characterized by a random arrangement of these sites. In this case interactions between adsorbed molecules are a function of the total surface coverage instead of being considered per patch.

In both situations the general expression for the overall adsorption isotherm, $\Theta_i(x_i)$, of an individual component i is the same. In the case of adsorption from a binary mixture the adsorption of component 2, $\Theta(x)$, can be written as (5)

$$\Theta(x) = \int_{\varepsilon_{min}}^{\varepsilon_{max}} \theta(\varepsilon,x,C)\ f(\varepsilon)d\varepsilon \qquad [1]$$

where $\theta(\varepsilon,x,C)$ is the local isotherm of component 2 on sites of equal energy, ε is the change in areal surface excess free energy upon displacement of component 1 by component 2 on the surface in units of RT. The distribution of ε is assumed to be continuous and given by $f(\varepsilon)$. The difference between patchwise and random distribution of the equal-energy sites is expressed through the local isotherm function as will be discussed below.

The experimentally measurable quantity in the case of adsorption from a binary mixture is $\Gamma_2^{(n)}$, the surface excess of component 2. The overall surface coverage Θ is related to $\Gamma_2^{(n)}$ in the following way (6)

$$\frac{n^o \Delta x}{m} = A_s \, \Gamma_2^{(n)} = n^\sigma \; (\Theta - x) \qquad [2]$$

where Δx $(= x^o - x)$ represents the difference in mole fraction of component 2 in the solution before and after contacting with adsorbent. n^o is the total number of moles present in the system. The solid is characterized by its mass m and specific surface area A_s. The amount of substance which can be accomodated on this solid per unit mass is given by n^σ. If n^σ is known Θ can be obtained from $\Gamma_2^{(n)}$. For a monolayer of equal-size molecules

$$\Theta = a_o \Gamma_2^{(n)} + x \qquad [3]$$

a_o is the molar cross-sectional area in the adsorbed layer. Hence, with some assumptions $\Theta(x)$ may be obtained from experimental data.

The two functions, $\theta(\varepsilon,x,C)$ and $f(\varepsilon)$, which determine $\Theta(x)$, are not readily accessible. In order to obtain either one of these functions the other one has to be known or chosen. In fact this choice has to be made for most cases of adsorption on solid surfaces. Common practice is then to neglect the heterogeneity, which is equivalent to assuming that the overall isotherm coincides with the local isotherm. However, in many cases the surface heterogeneity may be more important than lateral interactions, for example. In those cases it is more appropriate to make an assumption regarding the local isotherm function θ and to calculate the distribution function $f(\varepsilon)$ from Θ by inversion of eq. [1]. In favourable cases this can be done analytically (7). In general however $f(\varepsilon)$ has to be found from [1] by numerical inversion. To this end several algorithms have been proposed (8), among them three sophisticated, numerical methods known as HILDA (9), CAEDMON (10,11) and Regularization Procedure (RP) (12). In practice still little use is made of these methods. This may be because an assumption has to be made for θ or because of the complicated nature of the calculations. Moreover, some controversy exists as to the manner of solving for f(ε) from [1] (13). In the literature a few comparisons of results obtained with the different methods have been reported (12,14), but no rigorous conclusions have been drawn regarding the reliability and restrictions of the methods.

In the present contribution CAEDMON and RP are compared on the basis of rigorous tests. As we were not aware of any accurate adsorption isotherm data on surfaces with well-known heterogeneity, we have chosen to generate the required overall adsorption isotherms numerically. To this end known local isotherms and free energy distribution functions have been inserted in [1]. The f(ε) was recalculated from Θ(x) data thus obtained and compared with the true distribution function. Before presenting the results some further attention will be given to the local isotherm function and to the numerical procedures.

2. METHODS

(a) *Local isotherm*

Adsorption from a binary solution on a *homogeneous* surface is essentially an exchange process between the adsorbed phase, σ, and the bulk liquid, l. In the case of molecules of equal size the equilibrium state is governed by the constant K (6):

$$K = \frac{\gamma_1^{l}\gamma_2^{\sigma}}{\gamma_2^{l}\gamma_1^{\sigma}} \cdot \frac{1-x}{x} \cdot \frac{\theta}{1-\theta} \qquad [4]$$

where the γ's are activity coefficients. K is related to the difference in the areal surface excess free energies, $(\sigma_2^* - \sigma_1^*)$, of the interface between solid and pure components 2 and 1, respectively:

$$K = \exp[-(\sigma_2^* - \sigma_1^*)a_o/RT] \equiv \exp \varepsilon \qquad [5]$$

with R the gas constant and T the absolute temperature. The ratio $-(\sigma_2^* - \sigma_1^*)a_o/RT$ is equivalent to ε. For a regular solution from which the molecules adsorb in a regular monolayer, expressions for γ_i^l and γ_i^σ have been given (15). After substitution of these in [4], the following expression for θ is found:

$$\theta(\varepsilon,x,C) = \frac{xKC}{1+x(KC-1)} \qquad [6]$$

where C, a function of x and θ*, expresses the molecular interactions both in l and in σ:

$$C(x,\theta^*) = \exp[q^1(1-2x) - q^\sigma(1-2\theta^*)]$$

with q^1 and q^σ dimensionless interaction parameters. In the case of a patchwise topology of surface sites $\theta^* = \theta$; for a random configuration of equal-energy sites $\theta^* = \Theta$. When both the solution and the adsorbed phase are ideal, then $C = 1$, and with $x \ll 1$ eq.[6] reduces to the Langmuir equation. In the case of a regular solution and an ideal adsorbed phase $q^\sigma = 0$ and C is a function of x only. Hence, for an ideal adsorbed phase eq.[1] does not depend on the surface topology.

Eq.[6] does not apply for the situation where both components differ in size. In that case recourse has to be made to other theoretical models. The adsorption of r-mers from solution has been given considerable attention (6,16), and in principle such models can be used to describe θ, the local isotherm. However, solving $f(\varepsilon)$ from eq.[1] may then become rather cumbersome.

(b) *Numerical determination of $f(\varepsilon)$*

The integral adsorption equation [1] can be classified as a Fredholm equation of the first kind. This type of equation is known for its notorious instability with respect to solution of $f(\varepsilon)$ when $\Theta(x)$ and $\theta(\varepsilon,x,C)$ are known. This can be illustrated as follows. For any integrable kernel θ it can be proven that:

$$g(x) = \lim_{k\to\infty} \int_{\varepsilon_{min}}^{\varepsilon_{max}} \theta(\varepsilon,x,C) \sin k\varepsilon \, d\varepsilon = 0 \qquad [7]$$

Therefore, the following is true:

$$\Theta(x) = \Theta(x) + g(x)$$

or:

$$\int_{\varepsilon_{min}}^{\varepsilon_{max}} \theta(\varepsilon,x,C)f(\varepsilon)\, d\varepsilon = \lim_{k\to\infty} \int_{\varepsilon_{min}}^{\varepsilon_{max}} \theta(\varepsilon,x,C)\{f(\varepsilon) + \sin k\varepsilon\}d\varepsilon$$

From this it becomes clear that an infinitesimally small error $g(x)$ in $\Theta(x)$ due to, for example, rounding errors in numerical calculations, may give rise to non-negligible spurious oscillations $\sin k\varepsilon$ in $f(\varepsilon)$ (see also (8)). Such problems arise not only as a result of rounding errors, but also because of the way the integral in eq.[1] is numerically approximated by a discrete summation. With a suitable quadrature formula (e.g. Simpson's rule) eq.[1] may be written as:

$$\Theta(x) = \sum_{j=1}^{m} w_j \, \theta(\varepsilon_j, x, C) f(\varepsilon_j) \qquad [8]$$

where w_j is the quadrature coefficient and (m-1) equals the number of subintervals into which the considered interval $(\varepsilon_{min}, \varepsilon_{max})$ is divided. For a given set of n discrete data points $(x_i, \Theta(x_i))$ eq.[8] constitutes a set of n linear equations which may be expressed in matrix notation as:

$$\bar{\Theta} = \underline{A} \, \bar{f} \qquad [9]$$

with $\bar{\Theta} = [\Theta(x_i)]$, $\underline{A} = [a_{ij}] = [w_j \theta(\varepsilon_j, x_i, C_i)]$ an (n×m) matrix and $\bar{f} = [f(\varepsilon_j)]$. The ill-posed character of eq.[1] shows up in eq.[9] as a near-singularity of matrix $\underline{A}$. As a consequence, solving [9] is only possible when m, the number of knot points on the $(\varepsilon_{min}, \varepsilon_{max})$ interval is not too large. On the contrary, when successive values of ε_j differ by small amounts, then the linear dependency amongst successive columns of matrix $\underline{A}$ becomes more pronounced. As a result, irregularities such as spurious oscillations will override the true solution $\bar{f}$. Hence, it is clear that due care has to be taken in solving $f(\varepsilon)$ from eq.[1].

(c) *CAEDMON*

In this method initially the overall adsorption isotherm is calculated on the basis of an arbitrarily chosen distribution $f(\varepsilon)$. Then the chosen distribution is systematically adjusted in such a way that the deviation between the calculated and the experimental isotherm is minimized (10). The procedure incorporates a sophisticated non-negative least squares algorithm (11) in order to ascertain that correct solutions for $f(\varepsilon)$ be obtained. With this method (i) only non-negative solutions of $f(\varepsilon)$ and of Θ are allowed, and (ii) it is claimed that spurious oscillations of $f(\varepsilon)$, caused by the uncertainty in the calculated and experimental isotherm, can be limited.

(d) *Regularization Procedure (RP)*

The regularization method (12,17) attacks the ill-posed nature of [9] fundamentally. To this end, eq.[9] is modified to:

$$\bar{\Theta} + \bar{e} = \underline{A} \, \bar{f} \qquad [10]$$

with $\underline{A}$ and $\bar{f}$ as defined before, and $\bar{e} = [e(x)]$ an arbitrary error in Θ that is not further defined. The vector $\bar{e}$ explicitly accounts for the effects of (i) numerical errors arising from the approximation of the continuous integral eq.[1] by a discrete summation, (ii) rounding errors that occur in the calculations, and (iii) experimental error in the measured isotherm. Because in general e(x) is not known, an additional condition is required which each solution $\bar{f}$ should obey. Such a condition may be found by assuming that $\bar{f}$ is reasonably smooth. Mostly, the restriction is used that:

$$\int_{\varepsilon_{min}}^{\varepsilon_{max}} \{f''(\varepsilon)\}d\varepsilon$$

is at its minimum.

According to Twomey (17), eq.[10] and the above condition can be combined into one linear matrix equation:

$$\bar{f} = (\underline{A}^*\underline{A} + \gamma\underline{H})^{-1} A^*\bar{\Theta} \qquad [11]$$

where $\underline{A}^*$, the transpose of $\underline{A}$, is an (m×n) matrix, γ is an adjustable non-negative parameter related to the norm of $\bar{e}$, $\underline{H}$ is a constant (m×m) matrix representing the smoothness condition. Eq.[11] constitutes a set of m equations in m unknowns in which the term $\gamma\underline{H}$ serves as a non-singular correction to the near-singular matrix $\underline{A}^*\underline{A}$. Any conventional procedure to solve a set of linear equations may now be applied to find $\bar{f}$, provided γ is taken large enough.

3. RESULTS

CAEDMON and RP were extensively tested on simulated adsorption isotherm data corresponding with log-normal and bigaussian distribution functions. For most of the presented results a normalized log-normal distribution function was used:

$$f(\varepsilon) = \frac{\sqrt{\alpha/\pi}}{\varepsilon_0 \exp(1/4\,\alpha)} \exp[-\alpha\{\ln(\varepsilon/\varepsilon_0)\}^2] \qquad [12]$$

in which ε_0 determines the position of the free energy maximum and α is a measure of the width of the distribution. The overall isotherm was calculated by inserting in eq.[1] eq.[12] with $\varepsilon_0 = 1.3$ and $\alpha = 2.5$, and eq.[6] with C = 1 and K given by eq.[5]. For mole fraction ratios ranging from 0.01 to 5 the obtained surface coverages are 7% to 95%.

The energy distribution functions shown in figs. 1, 2, 3 and 4 were all calculated from these isotherm data. Results for other isotherms, each corresponding to a different distribution function, are presented in fig. 5.

The quality of recalculated distribution functions is determined to a large extent by the following parameters: (a) the smoothing parameter, γ (with RP, see eq.[11]); (b) the number of knot points, m, on the integration interval (see eq.[8]), and (c) the boundaries (ε_{min}, ε_{max}) of the integration interval.

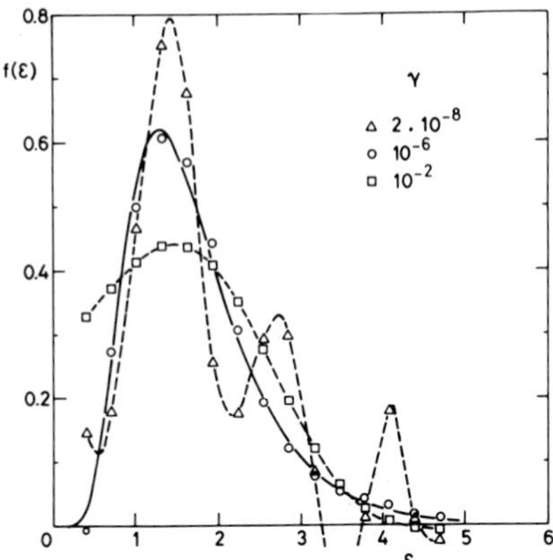

Fig. 1. Effect of γ (see text) on the distribution function as calculated with RP. The reference function is indicated by a solid line; broken lines and/or symbols represent results obtained for various values of γ.

Fig. 2. Effect of m (see text) on the distribution function as calculated with CAEDMON; —— reference, --- calculated.

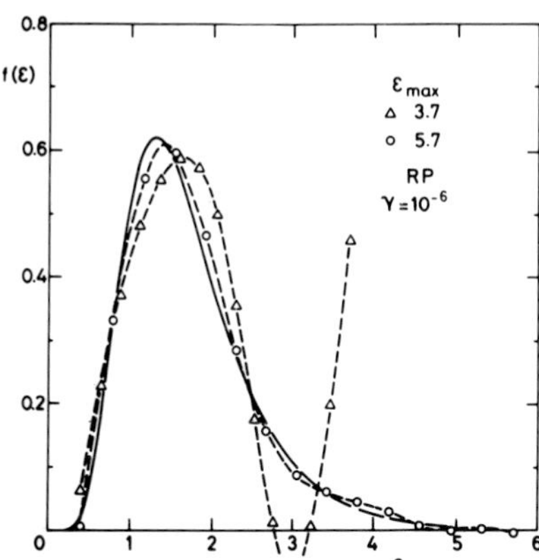

Fig. 3. Effect of ε_{max} (see text) on the distribution function as calculated with RP.
—— reference; --- calculated.

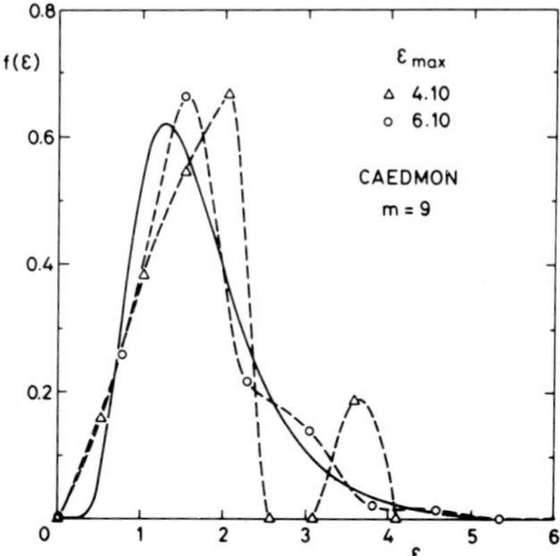

Fig. 4. Effect of ε_{max} (see text) on the distribution function as calculated with CAEDMON.
—— reference; --- calculated.

(a) *Smoothing parameter*

In RP γ should be adjusted to damp out spurious oscillations in the calculated distribution function. Its effect is shown in fig. 1 and may be explained in the following way. For $\gamma = 2.0 \times 10^{-8}$ the smoothing criterion in eq.[11], represented by matrix $\underline{\underline{H}}$, is underweighted. In the limit $\gamma = 0$, eq.[11] reduces to the original matrix equation giving free way to unrealistic oscillations. The correct distribution function is obtained with $\gamma = 10^{-6}$. For $\gamma = 10^{-2}$ the smoothing in eq.[11] is overweighted. In the limiting case of $\gamma \to \infty$ the only solution of eq.[11] is $\bar{f} = 0$, and $\bar{e}$ reaches a maximum (see eq.[10]). In conclusion, γ may only be set just as large as is necessary to damp out the unwanted oscillations in the solution of $\bar{f}$. Larger values of γ allow for a too large error $\bar{e}$ in $\Theta(x)$.

When in the overall isotherm an experimental error is present, one may expect larger values of γ to give the correct distribution function, because in general a linear relationship exists between γ and the norm of $\bar{e}$ (17). However, strict rules for the selection of γ can only be given in special cases, viz. when the accumulated error in eq.[10] is assumed to be Gaussian (18). In general the optimal γ has to be chosen by comparing results of a series of calculations performed with different values of γ.

(b) *The number of knot points, m, on* $(\varepsilon_{min}, \varepsilon_{max})$

As discussed before, the instability of the discretized integral adsorption equation [8] may become apparent as the number of knot points on the energy interval is increased in order to obtain the distribution function in more detail. A reliable numerical method should, however, yield results that do not depend upon m. This requirement was indeed found to be met by the RP method: the same distribution function was found for m = 7, 15, or 19. The maximum number of knot points is, of course, limited by the number of data points, which was 20 in the present situation. For a larger number of data points m could also be taken larger. With CAEDMON the results were disappointing, as is illustrated in fig. 2. With m = 9 the reference distribution is well reproduced, but serious problems arise when m is increased. The only way to suppress the oscillations is to decrease the number of knot points again to 9. From eq.[8] it can be seen that m determines the numerical error in the calculated $\Theta(x)$ as the discrete summation becomes more accurate for large m. However, m has no relation to rounding and experimental errors, and in that sense m is not comparable to γ in eq.[11]. Still, it is clear that in CAEDMON m is to be used as an adjustable parameter to suppress spurious oscillations. This may impose serious limitations on the number of points of $\bar{f}$ to be calculated. When m becomes too small, CAEDMON loses its potential to calculate $f(\varepsilon)$. Some more results will be presented below, in order to assess how serious this shortcoming of CAEDMON may be.

(c) *Free energy range* ($\varepsilon_{min}, \varepsilon_{max}$)

In numerical calculations the theoretical limits to the adsorption free energy range, viz. $\varepsilon_{min} = 0$, $\varepsilon_{max} = \infty$, usually are impractical. Values of ε_{min} and ε_{max} have to be selected in a narrower range, in such a way that physically realistic results are obtained by solving eq.[9]. The importance of a correctly chosen ε-range is illustrated for the RP in fig. 3 and for CAEDMON in fig. 4, where distribution functions are plotted that were calculated with different values of ε_{max}. In the case of RP optimal results are obtained when $\varepsilon_{max} = 4.70$ and with CAEDMON when $\varepsilon_{max} = 5.10$. Decreasing ε_{max} by 1 RT gives rise to serious distortions in the results with both procedures. In the case of RP (see fig. 3) the upward deviation for larger ε can be envisaged as a compensation effect for adsorption sites that were ignored in the calculation ($\varepsilon > 3.70$). The distribution found when using CAEDMON (fig. 4) shows a similar result, but in this case additional distortions occur that can only be attributed to insufficiently controlled instability.

Increasing ε_{max} by 1 RT also results in deformed distribution functions, both with RP and with CAEDMON. Problems seem to be similar in both cases. The deformations occurring when ε_{max} is too large may be understood from eq.[8], because $\theta(\varepsilon_j,x,C) \rightarrow 1$ for large ε_j. Thus, when ε_j is sufficiently large, for any isotherm point $(x_i, \theta(x_i))$ the summation may be split into two parts:

$$\theta(x_i) = \sum_{j=1}^{l} w_j\, \theta(\varepsilon_j,x_i,C_i)\, f(\varepsilon_j) + \sum_{j=l+1}^{m} w_j\, f(\varepsilon_j)$$

where ε_l is the optimal estimate for ε_{max}. When eqs.[8] and [9] are approximated in this way, matrix $\underline{A}$ will be singular as the columns (l+1) to m are linearly dependent, even for finite $f(\varepsilon_j)_{j>l}$. This explains the artifacts in the obtained distribution function. In a similar way the calculated $f(\varepsilon)$ depends upon ε_{min}. Now most distortions in $f(\varepsilon)$ occur for small values of ε.

A universal method to determine optimal estimates for ε_{min} and ε_{max} does not exist, because an acceptable degree of linear dependency amongst columns of $\underline{A}$ can not easily be specified and will, moreover, depend upon the numerical method applied. For example, we found as best estimates for RP that

$(\varepsilon_{min},\varepsilon_{max})$ = (0.40 , 4.70) and for CAEDMON $(\varepsilon_{min},\varepsilon_{max})$ = (0.004 , 5.10). However, the local isotherm $\theta(\varepsilon,x,C)$ which defines the relationship between adsorption free energy ε and mole fraction x can be used to find a first approximation of $(\varepsilon_{min},\varepsilon_{max})$. Ross and Morrison (10) proposed that the inflexion point of $\theta(\varepsilon,x,C)$ on the x-axis be employed to this end:

$$\left[\frac{d^2\theta(\varepsilon_{min},x,C)}{dx^2}\right]_{x=x_{max}} = \left[\frac{d^2\theta(\varepsilon_{max},x,C)}{dx^2}\right]_{x=x_{min}} = 0$$

In practice the most appropriate free energy range $(\varepsilon_{min},\varepsilon_{max})$ is found by trial and error, starting with an initial estimate and keeping in mind that (i) spurious oscillations should not be present, and (ii) in realistic systems the following condition usually holds: $f(\varepsilon_{min}) \simeq f(\varepsilon_{max}) \simeq 0$. However, complications arise because the smoothing parameter γ (RP) or the number of knot points m (CAEDMON), besides ε_{min} and ε_{max}, should be adjusted in order to find the most realistic distribution function.

Knowing in general the effect of the parameters γ, m, and $(\varepsilon_{min},\varepsilon_{max})$ on the calculated distribution function, we have tested RP and CAEDMON with several other energy distribution functions. Results obtained for two bi-Gaussian and a very skewed log-normal distribution function are shown in fig. 5.

Bi-gaussian functions can generally be given in the normalized form by:

$$f(\varepsilon) = \frac{1}{\sqrt{\pi}}\left[\frac{\alpha_1}{\sqrt{\beta_1}} + \frac{\alpha_2}{\sqrt{\beta_2}}\right]^{-1} \times$$

$$\times \left[\alpha_1 \exp\{-\beta_1(\varepsilon-\varepsilon_{0,1})^2\} + \alpha_2 \exp\{-\beta_2(\varepsilon-\varepsilon_{0,2})^2\}\right] \qquad [13]$$

with $\varepsilon_{0,i}$ the position of the i-th maximum, α_i the relative weight of peak i, and β_i a measure of its width.

For the distribution function shown in fig.5a, $\varepsilon_{0,1}$ = 1.10, $\varepsilon_{0,2}$ = 3.00, α_1/α_2 = 1.00, β_1 = 1.50, and β_2 = 4.00. In fig.5b a function with narrow peaks is shown: $\varepsilon_{0,1}$ = 0.80, $\varepsilon_{0,2}$ = 3.80, α_1/α_2 = 1.00, β_1 = 6.00 and β_2 = 3.00. Respectively 40 and 20 overall isotherm points were calculated as described above. The skewed log-normal distribution function of fig. 5c is given by eq.[12] using ε_0 = 0.05 and α = 0.15. In this case 0 isotherm points were calculated. Using these isotherm data distribution functions were recalculated with both RP and CAEDMON. Results as obtained after optimalization of γ, m and $(\varepsilon_{min},\varepsilon_{max})$ are shown in fig. 5. The solid lines represent the true distribution functions, the broken lines marked with symbols indicate the results obtained with RP and CAEDMON.

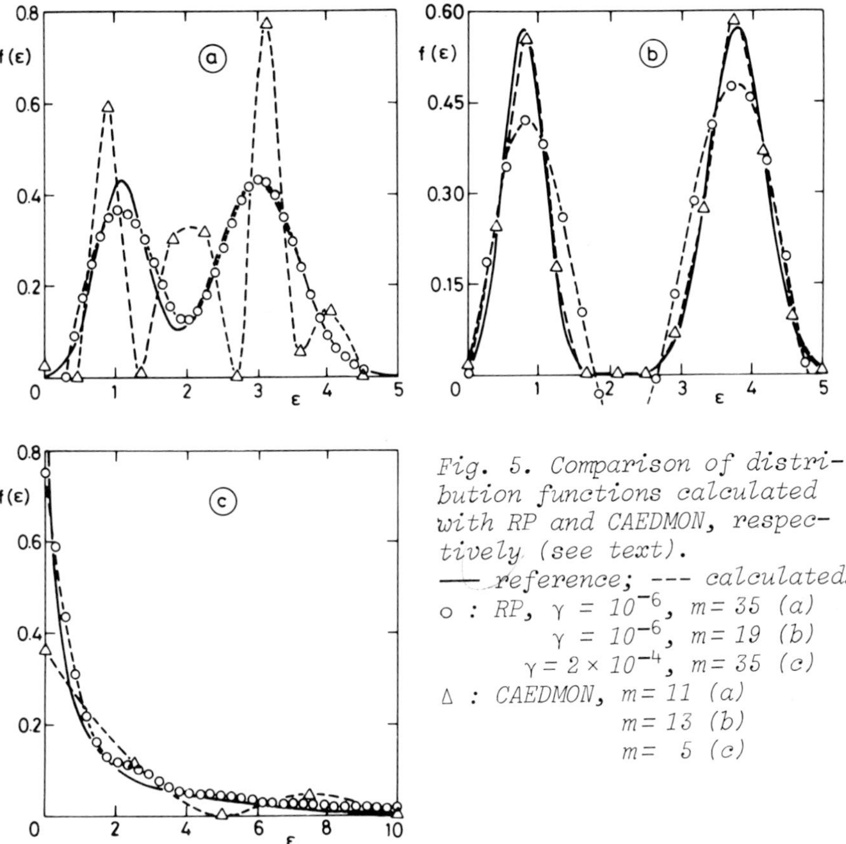

Fig. 5. Comparison of distribution functions calculated with RP and CAEDMON, respectively (see text).
—— reference; --- calculated.
o : RP, $\gamma = 10^{-6}$, m= 35 (a)
$\gamma = 10^{-6}$, m= 19 (b)
$\gamma = 2 \times 10^{-4}$, m= 35 (c)
Δ : CAEDMON, m= 11 (a)
m= 13 (b)
m= 5 (c)

It can be seen that RP reproduces the distribution functions reasonably well. A problem in the case of the distribution shown in fig. 5b is that f(ε) becomes negative in between both peaks. CAEDMON gives a better result in this case. However, for the other two distribution functions (figs. 5a and 5c) either large oscillations occur or m has to be set so small that no valuable information is obtained. Further calculations show that for most broad distribution functions (ε range 0-10 and 0-20) both methods work well.

4. DISCUSSION

In addition to the results presented here, we refer to Zolands and Myers (8) who tested a third numerical procedure called HILDA. From their results it becomes clear that HILDA, too, may yield unrealistic oscillations especially when some error is present in the data. Hence, the conclusion may be drawn that, provided the overall isotherm and the local isotherm are known, the free energy distribution $f(\varepsilon)$ can be determined most successfully with the RP. The correct $f(\varepsilon)$ is then obtained after optimalization of γ and $(\varepsilon_{min},\varepsilon_{max})$.When the distribution is broad or peaky as in Fig.5b, CAEDMON may also be used. In that case m, the number of points of $f(\varepsilon)$ to be calculated, should be small enough so that no false oscillations occur in $f(\varepsilon)$. For HILDA insufficient test data are available to reach a firm conclusion. When possible, $f(\varepsilon)$ should be calculated not only with RP but also with CAEDMON (or HILDA) in order to facilitate the proper choice of the relevant parameters.

A prerequisite for the analysis to be feasible is the availability of accurate isotherm data over the entire range of surface coverages. It is clear that no information can be attained on surface sites that are not covered by component 2 at the highest measured mole fraction of 2.

With regard to $f(\varepsilon)$ two important questions remain. (1) What kind of information is contained in $f(\varepsilon)$ in the case of adsorption from solution? (2) Is the separation between $f(\varepsilon)$ and $\theta(\varepsilon,x,C)$ feasible in practice? To answer the first question it should be realized that adsorption from solution is an exchange process. The free energy ε contains contributions from both the adsorption of component 2 and of the desorption of component 1. Owing to this, and in contrast with gas adsorption, the non-configurational entropy contribution contained in ε can *not* be neglected, as both components may exhibit different changes in translation, rotation, vibration, and (de)solvation upon de- or adsorption. Recently, Everett (19) has pointed out that even for ordinary apolar mixtures this entropy contribution may be the dominating term of ε. In aqueous solutions, where hydrophobic interactions and charge effects occur, domination of ε by entropic contributions is very likely (see e.g. (20)). As long as these contributions are independent of the surface heterogeneity, only the position of the distribution curve along the energy axis is affected and not the shape of the curve. In principle, however, the motions and (de)solvation of adsorbed molecules may depend on the type of surface site or patch. In that case also the shape of the distribution curve depends on the entropy changes upon adsorption. Hence, in general, with adsorption from solution ε can not be considered as an enthalpic measure of the surface sites only; it also contains enthalpic and entropic contributions depending upon the adsorbate system. Only in rare cases, such as for example the benzene-cyclohexane system where the entropic contribution to ε is small (19), $f(\varepsilon)$ can be considered as a measure of the heterogeneity of the adsorbent as such.

Our second question is concerned with the separation of $f(\varepsilon)$ and $\theta(\varepsilon,x,C)$. For strongly heterogeneous surfaces, in our view it seems more appropriate to make an assumption for θ and to calculate $f(\varepsilon)$ than to do the reverse. Inherent in such treatment is that all errors in the assumed θ ultimately accrue in $f(\varepsilon)$. Serious errors may arise from the neglect of (i) differences in size between the molecules (21), and (ii) lateral interactions (8). In a few cases the lateral interaction parameter q^{σ} for a specific adsorbate system may be estimated from adsorption studies on homogeneous surfaces. However, for not too large values of Θ it might be expected that on a heterogeneous surface lateral interactions are less important than on homogeneous surfaces, especially in the case of a random topology of the equal-energy sites. On such a surface adsorption on high energy sites lying far apart will be favoured over adsorption on neighbouring sites characterized by low adsorption energy, unless nearest neighbour interaction outweighs the difference. In view of this, the best strategy would probably be to use a relatively simple local isotherm and to accept that the calculated distribution function is somewhat dependent on this choice.

For comparative purposes, heterogeneity analysis from measured adsorption isotherms may be a powerful tool; for example (i) a study of the differences in $f(\varepsilon)$ obtained for different adsorbate systems adsorbed on one adsorbent could give valuable information on the adsorbent, and (ii) surface treatments such as activation or graphitization could be followed by analysing adsorption isotherms of a selected adsorbate system and monitoring the relative changes in $f(\varepsilon)$.

ACKNOWLEDGEMENTS

Thanks are due to Dr. R.S. Sacher for sending the Fortran subroutine CAEDMON on magnetic tape, and to Mr. K. Vos who performed some of the computations.

REFERENCES

1. Langmuir, I. (1918). *J.Amer.Soc.* 40, 1361.
2. Ross, S. and Olivier, J.P. (1964). *In* "On Physical Adsorption", Interscience Publ. New York.
3. Rudziński, W., Oŝcik, J. and Dabrowski, A. (1973). *Chem. Phys.Lett.* 20, 5.
4. Hill, T.L. (1949). *J.Chem.Phys.* 17, 762.
5. Borowko, M. and Jaroniec, M. (1980). *Revue Roumaine Chimie* 25, 475.
6. Everett, D.H. (1973). *In* "Colloid Science" (Ed. D.H. Everett), Specialist Periodical Reports, Chem.Soc. London, Vol. 1, Ch. 2.

7. Cerofolini, G.F. (1974). *Thin Solid Films* 23, 129.
8. Zolandz, R.R. and Myers, A.L. (1979). *Progr.Filtr.Sep.* 1, 1.
9. House, W.A. and Jaycock, M.J. (1978). *J.Colloid Polymer Sci.* 256, 52.
10. Ross, S. and Morrison, I.D. (1975). *Surface Sci.* 52, 103.
11. Sacher, R.S. and Morrison, I.D. (1979). *J.Colloid Interface Sci.* 70, 153.
12. House, W.A. (1978). *J.Colloid Interface Sci.* 67, 166.
13. Morrison, I.D. and Ross, S. (1977). *Surface Sci.* 62, 331; Dormant, L.M. and Adamson, W.A. (1977). *Surface Sci.* 62, 337.
14. House, W.A. and Jaycock, M.J. (1977). *J.Chem.Soc. Faraday Trans.I* 73, 942; House, W.A. (1978). *J.Chem.Soc. Faraday Trans.I* 74, 1045.
15. Everett, D.H. (1965). *Trans.Faraday Soc.* 61, 2478; Ash, S.G., Everett, D.H. and Findenegg, G.H. (1968). *Trans. Faraday Soc.* 64, 2639.
16. Scheutjens, J.M.H.M. and Fleer, G.J. (1982). *Advan.Colloid Interface Sci.* (accepted).
17. Twomey, S. (1963). *J.Ass.Comp.Mach.* 10, 97.
18. Merz, P.H. (1980). *J.Comp.Phys.* 38, 64.
19. Everett, D.H. (1981). *J.Phys.Chem.* 85, 3263.
20. Conway, B.E. (1977). *Advan.Colloid Interface Sci.* 8, 91.
21. Dhar, H.P., Conway, B.E. and Joshi, K.M. (1973). *Electrochim.Acta* 18, 789.

INTEREST AND REQUIREMENTS OF LIQUID-FLOW MICROCALORIMETRY IN THE STUDY OF ADSORPTION FROM SOLUTION IN THE SCOPE OF TERTIARY OIL RECOVERY

R. Denoyel, F. Rouquerol and J. Rouquerol

Centre de Thermodynamique et de Microcalorimétrie du CNRS
26, rue du 141ème R.I.A., 13003 Marseille - France

1. INTRODUCTION

As one knows, adsorption is among the phenomena which are limiting to-day the economical interest of the chemical oil recovery processes. Adsorption by the porous rocks of the oil field results indeed in a change in composition of the flooding medium and therefore in a change in efficiency. Our aim is to study the surfactant/rock interaction in a rather straightforward manner, by combining the information available from "adsorption isotherms" with that obtainable from direct enthalpy measurements.

Our first approach has been to use a Tian-Calvet microcalorimeter in one-batch experiments in which the surfactant solution was added to a fine suspension of ground rock (1). The method is sensitive (allowing one to measure enthalpies of displacement usually ranging from only 10 to 50 mJ m^{-2}) and generally applicable to any powder provided its grain size is small enough (i.e. smaller than 10 µm) to keep the necessary stirring power at a satisfactorily low value. Moreover, the long-term stability of the system allowed one to follow slow structuring or rearrangements of the already adsorbed layer over periods of up to 8 h when necessary.

Nevertheless, we felt it necessary to develop a complementary microcalorimetric approach which would allow - at the expense of a possible restriction in the choice of the systems - one to study the *desorption* phenomenon and also to study more quickly *the influence of the numerous parameters able to control the adsorption* of a solute : pH of the solution, ionic strength, presence of other solutes giving rise to competitive adsorption, specific features of the molecule studied (aliphatic chain length, hydrophilic-lipophilic balance, molecular mass, linear or "branched" structure, various chemical functions, ...). To meet the above goals, we were naturally led to choose a *liquid-flow* set-up where a bed of powdered adsorbent (but here, coarse enough to avoid any clogging) is permanently flowed across by a solution with which it eventually reaches an adsorption equilibrium.

Time-saving brought by such a liquid-flow microcalorimeter was stressed at various times and especially by :

- A.J. Groszek, who developed an extremely convenient equipment (2), actually closer to a DTA system (with the same punctual temperature detection) than to a calorimeter and therefore better suited for the detection of thermal effects than for their quantitative measurement (3). It was recently used by Bocquenet and Siffert (4) in the scope of oil recovery.

- R. Berg et al. (5) and Findenegg et al. (6) who make use - for adsorption experiments - of P. Monk and I. Wadso's microcalorimeter (7).

- T. Morimoto and H. Naono (8) whose apparatus may be considered as a development of the Groszek's one.

On the basis of the above works we found it useful to develop a versatile microcalorimeter which could be operated with a wide range of specific surface areas (since, in our case, the ground rocks often have specific surface areas smaller than 1 m^2g^{-1}) and of sample masses (because of sampling problems occurring in some cases). After describing the apparatus and its way of operation we will make the thermodynamic analysis of the experiment. Results obtained with a selected "reference" system will be given and discussed, together with the possibilities of the method.

2. EXPERIMENTAL METHODS AND MATERIALS

1. *Description of the adsorption microcalorimeter*

The equipment developed is presented on Fig. 1. It is both of the liquid-flow and heat-flow type. It basically consists of a twin cylinder aluminum block I. Each cylinder encases an easily removable assembly, itself mainly consisting of the heat flow-meter and exchanger J, of the adsorption cell K and of the lid L. The solution to be adsorbed is introduced through the inlet port C, flows trough the adsorbent E (housed between two 40 μm-bore stainless steel sieves which may be replaced, if necessary, by teflon 5 or 10 μm filters)

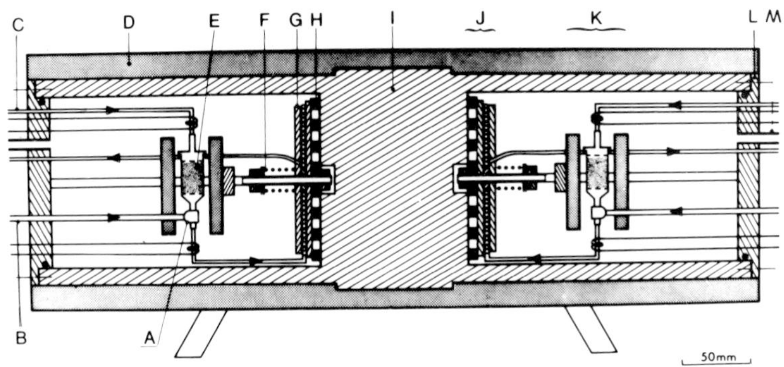

Fig.1 Twin liquid-flow adsorption and mixing microcalorimeter

and carries out the heat evolved by the adsorption phenomenon to the heat exchanger (2.8 m long, 1.4 mm bore teflon tubing wound between an insulating disk G on one side and an aluminum one on the other) itself in close contact with the heat flow-meter H. The latter is made of semi-conductive elements (Cambion, type 801) and the whole assembly is pressed by a spring F against the aluminum block I acting as a heat sink. The solution then leaves the calorimeter through the exit port.

The solution is in contact with the "teflon" tubing, with the "Kel-F" adsorption cell, with the adsorbent and, when used, with the stainless steel sieves. An extra inlet port B allows one to by-pass the adsorption cell in order to determine easily the enthalpies of dilution or mixing which are needed, as we shall see, for a proper presentation - and interpretation - of the adsorption data. The "Kel-F" mixing cell A includes a turbulence chamber with two 0.5 mm bore inlet and outlet channels. This device plays a similar part to that played by the 3 disks with 0.3 mm holes formerly used by Wadso (7).

The block is surrounded with an insulating shield D and the whole system is immersed into a high performance water bath (modified Tronac 1005 thermostat ; checked short term temperature fluctuations : 2.10^{-4} K).

To ensure a satisfactory dryness of the electrical heat flow-meter the microcalorimeter can be evacuated through a stainless-steel flexible tubing attached to port M.

The flow-meters are connected in the usual "differential" way. The electrical signal is amplified (Sefram amplifier "Amplispot") and then recorded, so that the final recording range is usually from 2.5 to 50 μV full scale.

Various Joule effect resistors located before and after the adsorption cell allows one to calibrate the microcalorimeter.

The adsorption liquids are usually injected with the help of two peristaltic pumps (Gilson "Minipuls" 2). The flow rates chosen usually lie around 1 $cm^3.min^{-1}$.

2. *Carrying out an experiment*

The sample is first introduced into the adsorption cell (between the sieves), the calorimetric chambers are evacuated and then filled with dry air and the whole system is immersed into the water thermostat where a satisfactory thermal equilibrium is reached in about 10 hours.

The microcalorimeter is connected with the peristaltic pumps (through an 8 m long teflon tubing exchanger coiled around the insulating cylinder D and therefore immersed into the water thermostat) and with the amplifier and recorder.

The assembly is then ready for a set of experiments. The order of the experiments is of course chosen by taking into account the reversible (or irreversible) character of the various phenomena, so as to make the best use of the sample and of the time-consuming first temperature equilibration.

In a usual experiment, the sample is pre-equilibrated with a flow of pure solvent (water in the present case) and successively brought in contact with solutions of various compositions. *The composition of the solution coming out* from the calorimeter (i.e. after adsorption or desorption has taken place) is *continuously monitored* by means of a differential refractometer (Waters Associates, model R-403) so that the amount adsorbed is known at any time and that the final adsorption equilibrium may be checked both by microcalorimetry and refractometry.

3. *Thermodynamic analysis of the experiment*

The amount of heat Q measured during the adsorption phenomenon must be translated into the corresponding changes in the state functions of the adsorption system. For that purpose, let us consider a system enclosing, in its initial state, (i) the adsorption cell (filled with adsorbent and pure solvent) and (ii) the amount of two-component solution necessary to reach the adsorption equilibrium under constant liquid flow conditions. The changes in concentration (and then the thermal effects either due to adsorption or to dilution phenomena) are supposed to take place only in the adsorption cell and to be measured continuously.

The initial and final state of the system are defined in Table I with the following symbols :

- Subscript 1 stands for the solvent, subscript 2 for the solute ;
- $h_2(c_e)$ is the apparent molar enthalpy of solute 2 in solution at the equilibrium concentration c_e (we take the case of a *dilute* solution with the asumption that the molar enthalpy of the solvent is the same as in its pure state, h_1^*);
- h_2^σ and h_1^σ are (following Everett's notations (9)) the partial enthalpies of solute 2 and solvent 1 in the adsorbed state.

Table I

	initial state		final state	
	in cell	out of cell	in cell	out of cell
amount of solute { adsorbed / in solution	0 0	0 $n_{2,i}$ $(c=c_e)$	$n_{2,f}^\sigma$ $n_{2,\alpha}$ $(c=c_e)$	0 $n_{2,f}=\int_i^f dn$ $(0\leqslant c\leqslant c_e)$
amount of water { adsorbed / desorbed	$n_{1,i}^\sigma$ 0	0 0	$n_{1,f}^\sigma$ 0	0 Δn_1^σ
system enthalpy	$H_i=n_{2,i}h_2(c_e)+n_{1,i}^\sigma h_{1,i}^\sigma$		$H_f=n_{2,\alpha}h_2(c_e)$ $+\int_i^f h_2(c)dn$ $+n_{2,f}^\sigma h_{2,f}+n_{1,f}^\sigma h_{1,f}^\sigma$ $+\Delta n_1^\sigma h_1^*$	

From the expression of the initial and final enthalpies of the system (H_i and H_f) listed in the bottom of table I and taking into account (also from table I) that

$$n_{2,i} = n_{2,\alpha} + n_{2,f}^\sigma + \int_i^f dn$$

one gets :

$$\Delta H = H_f - H_i = n^{\sigma}_{2,f}\,(h^{\sigma}_{2,f} - h_2(c_e)) - n^{\sigma}_{1,f}(h^{\sigma}_{1,i} - h^{*}_{1,f})$$

$$+ \Delta n^{\sigma}_1\,(h^{*}_1 - h^{\sigma}_{1,i}) + \int_i^f (h_2(c) - h_2(c_e))\,dn$$

This overall enthalpy change is the one measured calorimetrically and it may also be written :

$$\Delta H \text{ measured} = \Delta_{dis} H + \Delta_{dil} H \text{ with}$$

|1| $$\Delta_{dis} H = n^{\sigma}_{2,f}\,(h^{\sigma}_{2,f} - h_2(c_e)) - n^{\sigma}_{1,f}\,(h^{\sigma}_{1,i} - h^{*}_{1,f}) + \Delta n^{\sigma}_1\,(h_1 - h^{\sigma}_{1,i})$$

and

$$\Delta_{dil} H = \int_i^f |h_2(c) - h_2(c_e)|\,dn$$

The latter term is calculated from a set of calorimetric measurements (where the solution is diluted from the starting concentration c_e to a number of lower concentrations) and from the continuous concentration recording obtained for instance from a differential refractometer during the adsorption experiment. Introducing this concentration c and the liquid flow rate f we get :

$$\Delta_{dil} H = \int_0^{\infty} |h_2(c) - h_2(c_e)|\;f.c.dt$$

Relationship |1| giving $\Delta_{dis}H$ is close to that used by Liphard et al. (10), with an extra term (the 3rd one on the right side) taking into account the partial enthalpy change of the adsorbed solvent.

4. System studied

The surfactant chosen is the commercial octylphenol-oxyethylene "Triton X-100" marketed by Rohm and Haas with the following expanded formula :

$$C_8H_{17}-C_6H_4-(CH_2-O-CH_2)_{9-10}OH$$

Its stated purity (in oxyethylene) is 99 % although it is not monodisperse in chain length (hence an approximate number of oxyethylene groups 9-10 in the formula). The main interest of this class of surfactants comes from their good solubility in water, even in the presence of divalent cations such as Ca^{++} or Mg^{++} often found on natural rocks.

Instead of a natural quartz or sandstone, the adsorbent chosen was a very pure chromatographic silica gel ("Spherosil XOC 005, marketed by Péchiney-Ugine-Kuhlmann) with a specific surface area of 14 m^2g^{-1} and a grain size larger than 100 µm. This grain size is very convenient for a liquid flow experiment both because it prevents any clogging and because it keeps at a satisfactorily low value the heat effect due to the liquid flow through the sample bed. This heat effect usually varies in an a-priori unknown manner during the course of the adsorption phenomenon, so that it is advisable to keep it small.

3. EXPERIMENTAL RESULTS

Fig. 2 gives the surface excess isotherm of Triton X-100 adsorbed at 25 °C, from a water solution, onto Spherosil. This isotherm was determined by the *static method*, where a number of capped test tubes, containing similar amounts of adsorbent but various concentrations of solution, were rocked overnight in a thermostat and then centrifuged. The residual concentration in the supernatant liquid was finally determined by differential refractometry. We checked that similar results were obtained by the *dynamic method* (which is a kind of frontal chromatography method) during a calorimetric experiment. The latter method was

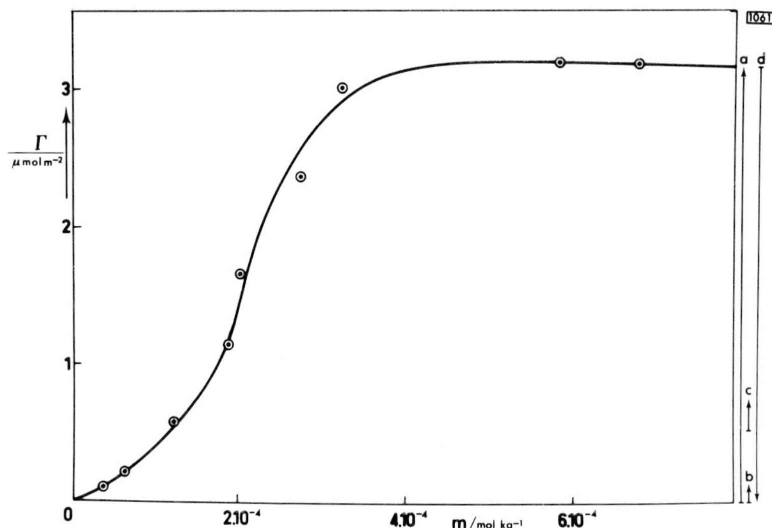

Fig. 2. Surface excess concentration of TX-100 on Spherosil at 25 °C. On right side : adsorption (or desorption) paths corresponding to Fig. 3 experiments.

also used to check the *reversibility* of the adsorption phenomenon. During a set of 3 successive adsorption-desorption cycles, on the same sample, the amounts of surfactant involved were found to be equal within 1 %.

On the right side of Fig. 2 the 4 arrows a, b, c, d indicate the adsorption (or desorption) paths corresponding to the 4 experiments reported in Fig. 3.

Fig. 3 is a direct photography of the recordings, where C denotes the microcalorimetric curve and R the refractometric one.

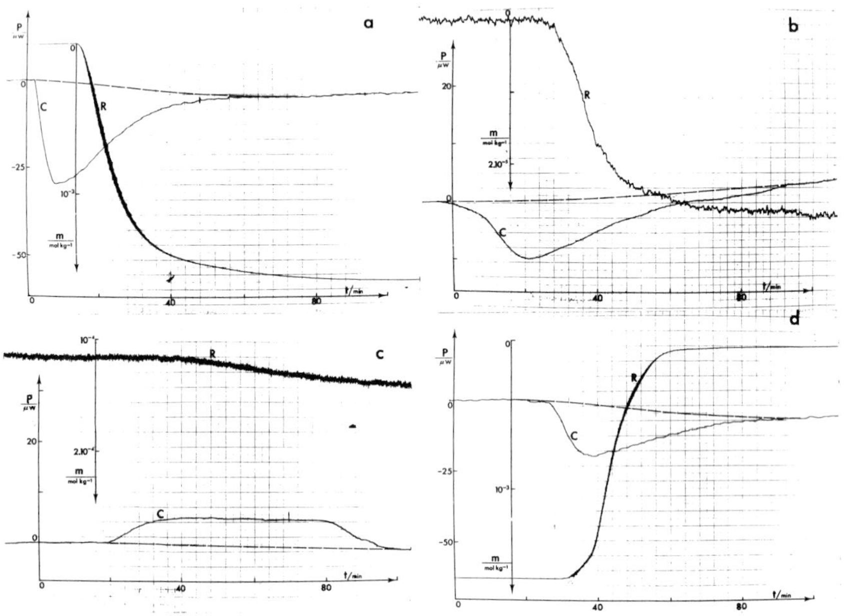

Fig. 3. Calorimetric (C) and Refractometric (R) recordings obtained for TX-100/Spherosil adsorption or desorption (see corresponding paths on right side of Fig. 2).

4. DISCUSSION

Experiment 3a corresponds to a direct *adsorption* up to saturation (cf. Fig. 2) with the following conditions : total surface area of Spherosil in the adsorption cell : 2.7 m^2 ; final (equilibrium) concentration of Triton X-100 in the water solution : 1.6 10^{-3} $mol.l^{-1}$; total amount of heat measured in 1h 30 : 40 mJ. As normally expected this adsorption phenomenon is *exo*thermal. Nevertheless, if one looks at the reverse experiment 3d, which is a *desorption* from saturation (cf. Fig. 2), he is amazed to see that the recorded enthalpy change is *again* *exothermal*. On the other hand, adsorption from a solution around the critical micellar concentration (CMC) (experiment not reported on Fig.3) seems to be athermal (within the experimental uncertainties). And if the very small adsorption step 3b is carried out up to a final equilibrium concentration of only 3.5 10^{-5} mol.1^{-1} (cf arrow b on Fig. 2), we detect again the expected exothermal effect already seen in Fig. 3a for a much wider coverage. Now, if another small adsorption step is carried out somewhat further, between two concentrations of 1.16 10^{-4} and 1.4 10^{-4} mol.1^{-1}, respectively (experiment 3c, see arrow c on Fig. 2), we detect an *endothermal* effect. These apparently contradictory data may be interpreted as follows :

- at low coverage (cf. experiment 3b) the main interaction is likely to be between the adsorbate and adsorbent, so that the displacement enthalpy is of the usual sign for a physisorption phenomenon.

- at higher coverages (cf. experiment 3c) the so called "lateral" interactions become predominant between the adsorbed surfactant molecules. This view is supported by the fact that such interactions, which are predominant in the formation of micelles, are also known to give rise, in the latter case, to an endothermal effect (11). Moreover, these lateral interactions could be thought to give rise to a "cooperative effect" during the course of adsorption, therefore leading, after Giles'analysis (12), to an S-shaped isotherm. This is actually the case, as it can be seen in Fig. 2. Finally, such an interaction is more easily understood if the adsorbed molecules stand more or less normal to the surface, so that their "tails" may interact ; this is consistent with the limited area of 52 $Å^2$ per molecule which may be derived from the amount adsorbed at saturation (i.e. from the plateau of the isotherm).

- in the case of adsorption from a solution close to (and slightly under) the CMC, there is a compensating effect between both phenomena above.

- when the adsorption is carried out from a solution above the CMC then one measures mainly the enthalpy of demicellization. In our experiment 3a the measured enthalpy change is of 5 kJ per mole of surfactant, whereas the mere demicellization enthalpy (or "dilution enthalpy") corresponding to the amount adsorbed would be of 5.3 kJ mol^{-1} (from (11)). The adsorption phenomenon or, better said, the displacement phenomenon (water by a surfactant) is, here, in itself, nearly athermal when integrated over a coverage of about 1.

The important part played by the "lateral" interactions in the adsorption layer was stressed, these last years, by Cases (13), Somasundaran (14) and by Fuerstenau (15). Competition between two types of interactions (one exothermal, the other endothermal) could also be an explanation for the crossing-over of adsorption isotherms observed by the latter between 25 and 50 °C.

At this stage of the work, liquid flow microcalorimetry appears as an interesting tool, able to bring novel and safe information, especially if associated with a refractometric or UV detector which is actually required to derive appropriate enthalpy data. Its versatility comes from the possibility of carrying out, successively, with a given sample, adsorption, desorption and dilution experiments, and also from the fact that the experimental chamber is large enough (≃ 1000 ml) to accomodate a wide range of adsorption cells.

REFERENCES

1. Rouquerol, J. and Partyka, S. (1981). *J.Chem.Tech.Biotchnol* 31, 584-92.
2. Groszek, A.J. (1979). *Proc.Roy.Soc.London* A 314, 473-98.
3. Allen, T. and Patel, R.M. (1970). *J.Appl.Chem.*, 20, 167-71.
4. Bocquenet, Y. and Siffert, B. (1980). *J.Chimie Physique* 77, 295-302.
5. Berg, R.L., Noll, L.A. and Good W.D. (1977). *Proceedings 3rd ERDA Symposium*, pp. 10-1/B 10-8, Tulsa, Okla.
6. Kern, H.E., Piechocki, A., Brauer, U. and Findenegg, G.H. (1978). *Progr.Colloid and Polymer Sci.*, 65, 118-24.
7. Mark, P. and Wadsö, I. (1968). *Acta Chemica Scandinavia* 22, 1842-52.
8. Morimoto, T. and Naono, H. (1972). *Bull.Chem.Soc.of Japan* 45, 700-5.
9. Everett, D.H. (1972). *Pure and Applied Chemistry* 31, 579-638.
10. Liphard, M., Glanz, P., Pilarski, G. and Findenegg, G.H. (1980). *Progr.Colloid and Polymer Sci.*, 67, 131-40.
11. Paredes, S., Tribout, M., Ferreira, J. and Leonis, J. (1976). *Colloid and Polymer Sci.*, 254, 637-42.
12. Giles, C.H., D'Silva, A.P., Easton, I.A. (1974). *J.Colloid Interface Sci.*, 47, 766.
13. Cases, J.M., Canet, D., Doerler, N. and Poirier, J.E. (1982). *In* "Adsorption at the gas/solid and liquid/solid interfaces", (Eds J. Rouquerol and K.S.W. Sing), Elsevier, Amsterdam.
14. Hanna, H.S. and Somasundaran, P. (1977). "Physico-chemical aspects of adsorption at solid/liquid interface, Part I" *In* Improved oil recovery by surfactant and polymer flooding, 205-51, Academic Press.
15. Wakamatsu, T. and Fuerstenau, D.W. (1968). *Advance.Chem.Ser.*, 79, 161-171.

INTERACTION OF HEAVY METALS WITH CHITIN AND CHITOSAN. II. CADMIUM AND ZINC.

Billie Jo Suder and J. P. Wightman

Chemistry Department
Virginia Polytechnic Institute and State University
Blacksburg, Virginia 24061, U.S.A.

I. INTRODUCTION

The natural chelating marine polymer chitin (poly-N-D-glucosamine) and its deacetylated derivative chitosan may be useful for removing heavy metal ion wastes from discharge water. Many researchers have explored the feasibility of this approach. Muzzarelli has reviewed much of this work and, in addition, has compiled an excellent and comprehensive summary of the characteristics and properties of chitin and related materials(1). Hauer conducted experiments involving Kytex H chitosan with a number of heavy metals including Cd, Cr, Cu, Pb and Zn(2). Masri and Randall have explored the possibility of heavy metal ion removal from actual waste materials from manufacturing plants(3). A recent report by Kurita and co-workers(4) has demonstrated that Cu(II) and Hg(II) are adsorbed by chitin and its congeners varying in amino group content. Eiden, Jewell and Wightman have recently reported on the uptake of Cr(III) and Pb(II) by chitin and chitosan (5). This study indicated that uptake of the metal ion on chitosan was much greater than that on chitin and suggested that the metal uptake mechanism involves nucleation and growth of nodules on the polymer surface.

In order to further study nodular formation and attempt to establish the uptake mechanism, we report here a study of the uptake of Cd(II) and Zn(II) by chitin and chitosan. During this investigation, the effects of concentration and pH on the uptake isotherms were also studied. SEM/EDAX and ESCA were used to characterize chitin and chitosan before and after equilibration with the metal ions.

2. MATERIALS AND EXPERIMENTAL METHODS

Materials

Chitin was obtained from the Velsicol Chemical Corporation in the form of large flakes. It was ground in a blender and sieved to consistent particle sizes before use. Chitosan (Velsicol Chemical Corporation) was received in a finely divided state and was sieved to constant particle size before using. All other reagent grades used were from Fisher.

Procedure

Stock aqueous solutions of Cd(II) and Zn(II) were prepared and standardized against commercial standards. These stock solutions were then diluted to give standards of the appropriate concentrations. Fifty ml aliquots of each of these standard solutions were placed in stoppered 125 ml Erlenmeyer flasks and equilibrated with 0.100 g of chitin or chitosan for approximately 24 hours. Unless otherwise noted, the samples were shaken in an automatic shaker at room temperature during the equilibration period. The equilibrated supernate was decanted or filtered from the polymer samples before analysis. It was necessary to filter the liquid from the polymer when very small particle sizes were used, otherwise consistent analyses could not be obtained. Metal ions were analyzed using either a Varian 175 or a Varian Techtron 1100 atomic absorption spectrophotometer. After determining the change in the metal ion molar concentration, ΔC, the quantity $V\Delta C/w$ was calculated where V is the solution volume (50.0 ml) and w is the weight (0.1 g) of the polymer used. This quantity corresponds to the moles of metal ion taken up per gram of polymer. Sodium hydroxide (0.01M) or nitric acid (0.01M) was used to correct the pH of runs that were made at constant pH.

Samples of chitin, chitosan and these two polymers after equilibration with all the solutions studied in this work were examined using an AMR 900 scanning electron microscope (SEM) with an International 7070A energy dispersive analysis of x-rays (EDAX) accessory. After filtering, washing with deionized water and air drying, the equilibrated polymers were mounted on sample probes with copper conductive tape and coated with an Au/Pd alloy to minimize sample charging.

Electron spectroscopy for chemical analysis (ESCA) was also used to characterize all polymer samples. A duPont 650 electron spectrometer with a magnesium x-ray source was used for this determination. The carbon 1s photoelectron peak at

284.6 eV was used to reference the binding energies for the other elements examined. The samples of polymer were mounted on double sided tape for analysis.

3. RESULTS

Uptake of Cadmium (II) Nitrate on Chitin and Chitosan

In Figure 1, the uptake isotherms for Cd(II) are given

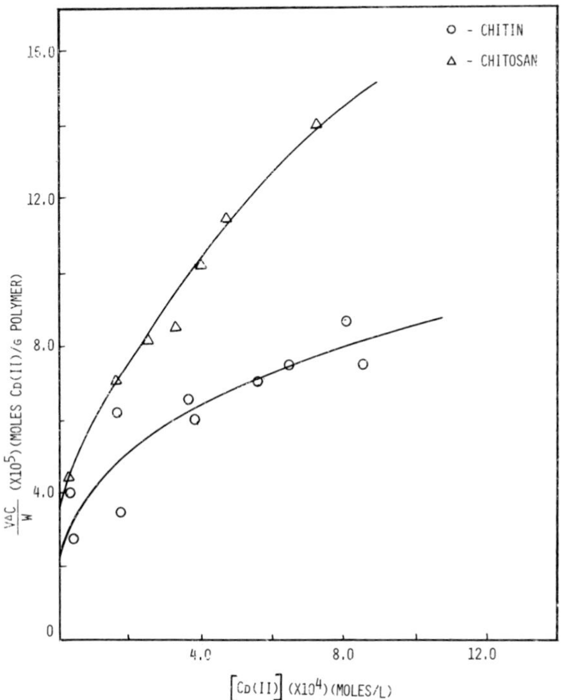

Fig. 1 Cadmium (II) uptake isotherms on chitin and chitosan

for chitin and chitosan. A significantly greater sorption of Cd(II) on chitosan compared to chitin was observed. This result is consistent with previously reported results whereby chitosan showed a greater capacity for heavy metals than chitin (1,5). During these initial experiments the pH of the solutions was determined both before and after equilibration with the polymers. It is interesting to note that the final pH values for equilibrated solutions of both metal ions were higher as the initial (or final) concentration of the metal ion became smaller. This is what would be expected since the metal ions studied are Lewis acids and the lower the concentration, the higher the pH. Fig. 2 gives the uptake isotherms for the Cd(II)/chitosan system at pH 5.0, 6.0 and 7.5. The isotherm at constant pH 7.5 tends to maximize at lower concentrations than does the uncontrolled pH isotherms

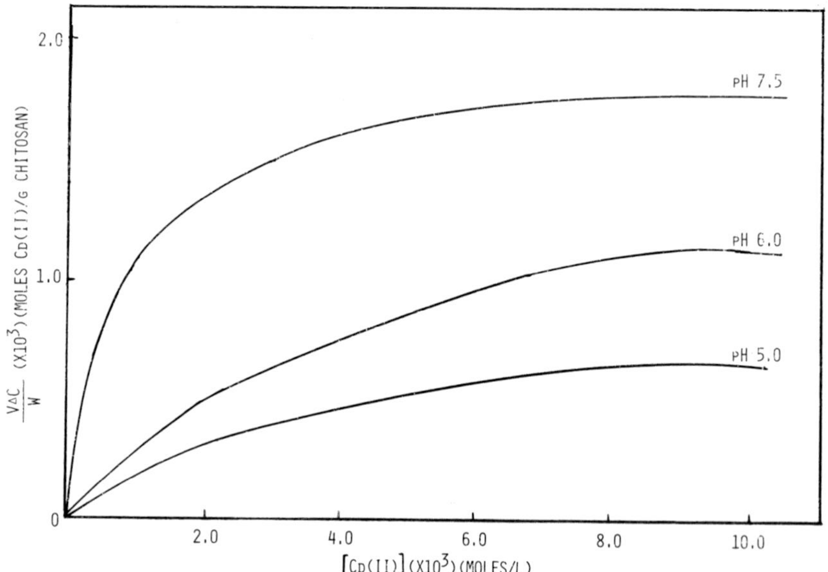

Fig. 2 Uptake of cadmium (II) on chitosan at different pH values

(See Fig. 1.). Above pH 7.5, the equilibrated solutions became cloudy indicating a precipitation of the metal hydroxide. In the vicinity of pH 7.5, then, the uptake mechanism appears to be most efficient. It is possible that in this slightly alkaline environment, all the amine groups on the chitosan are free amines, thus allowing them to more readily interact with the Cd(II) ions.

Uptake of Zinc(II) Nitrate on Chitin and Chitosan

Isotherms generated for chitin and chitosan with Zn(II) are given in Fig. 3. Again, the $V\Delta C/w$ values for chitosan were approximately double those obtained for chitin.

Scanning Electron Microscopy/Energy Dispersive Analysis of X-rays (SEM/EDAX)

SEM photomicrographs were taken of both chitin and chitosan prior to equilibration with heavy metal ion solutions. Representative SEM photomicrographs have been published (5). Calcium was detected by EDAX (energy dispersive analysis of x-rays) only in the chitosan sample. Small flake-like

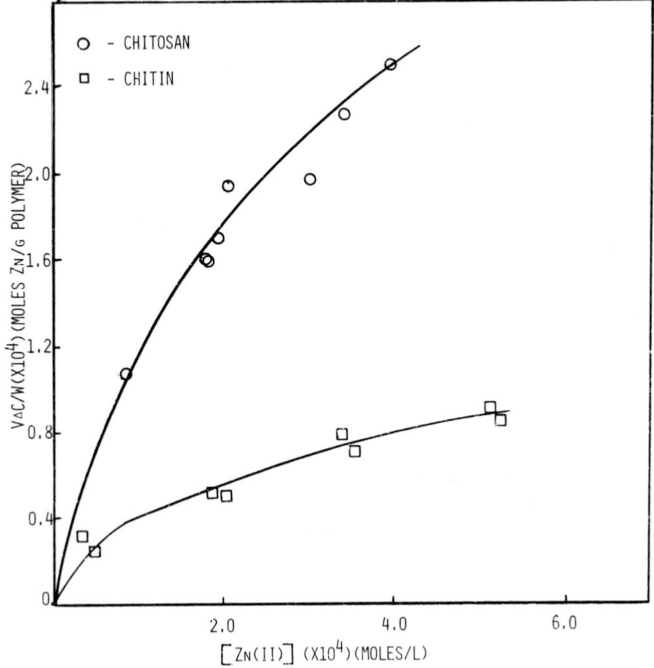

Fig.3 Zinc(II) uptake isotherms on chitin and chitosan

particles were noted on the surface of the chitosan flakes which EDAX indicated were high in calcium. In neither polymer were any heavy metal peaks of any significant concentration noted by EDAX.

After equilibration with Cd(II), the SEM photomicrographs shown in Fig. 4 were obtained. A weak Cd peak was noted by EDAX analysis. While some flakes could be noted on the polymer surface, no large nodules characteristic of the Eiden-Jewell effect (5) were observed. In the Eiden et al. studies (5), the samples were not shaken continuously during the equilibration period. This appears to suggest that agitation of the samples precludes large nodular formation. At higher magnifications, a greater number of nodules were observed but these also showed minimal Cd content by EDAX.

Very little change occurs in the character of the polymer surface after equilibration with Zn(II). A weak Zn EDAX signal was obtained for the Zn(II)/chitosan system and none at all was observed for the Zn(II)/chitin system. As was the case of Cd, at higher magnifications, small nodules were observed on chitosan which EDAX analysis indicated were low in Zn concentration.

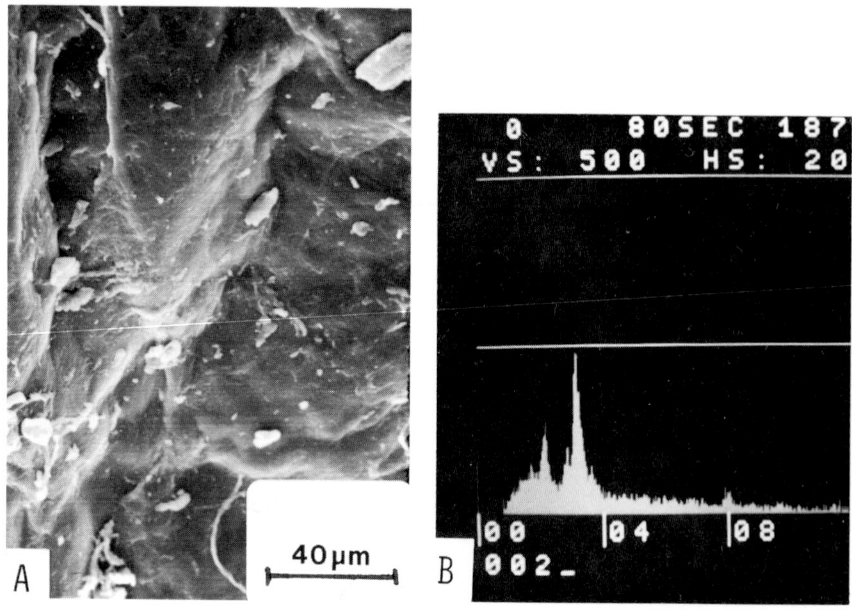

Fig.4 SEM photomicrograph (A) and EDAX spectrum (B) of chitosan after equilibration with Cd(II)

Electron Spectroscopy for Chemical Analysis (ESCA)

The ESCA results for chitosan equilibrated with various ions are listed in Table I. Cd and Zn were detected at significant levels on equilibrated chitosan samples. The binding energy of the O 1s photopeak is higher for chitosan equilibrated with Cd(II) compared to unequilibrated chitosan; whereas, a lower binding energy O 1s photopeak was found for Zn(II). Additional work is necessary to assign these observed binding energy shifts to polymer-metal interaction or to metal oxide (hydroxide) formation.

The atomic fraction ratios Cd/N and Zn/N calculated from the ESCA spectra are plotted as a function of the initial concentration of the equilibrating solution in Fig. 5. The fact that the Cd/N ratio is larger than the Zn/N ratio at all solution concentrations is consistent with the greater uptake of Cd(II) compared to Zn(II) by chitosan reported above. Further, the ratios are independent of solution concentration. The isotherms (Figs. 1 and 3) show increased Cd(II) and Zn(II) uptake by chitosan with increasing solution concentration. However, the ESCA and isotherms results are not in fact inconsistent. The ESCA technique is limited to analysis of the top 5 nm. Thus, saturation of the top

TABLE I

ESCA Parameters for Chitosan, Cd(II)/Chitosan and Zn(II)/Chitosan

Photopeak	Chitosan(5) B.E.*	A.F.**	Cd(II) B.E.	A.F.	Zn(II) B.E.	A.F.
N 1s	399.0	0.042	399.1	0.037	398.6	0.050
O 1s	532.2	0.177	533.5	0.230	531.5	0.238
Cd 3d 5			407.1	0.009		
Zn 2p 3					1020.2	0.010

*Binding energies (B.E.) in eV
**Atomic fraction (A.F.) - balance due to carbon.

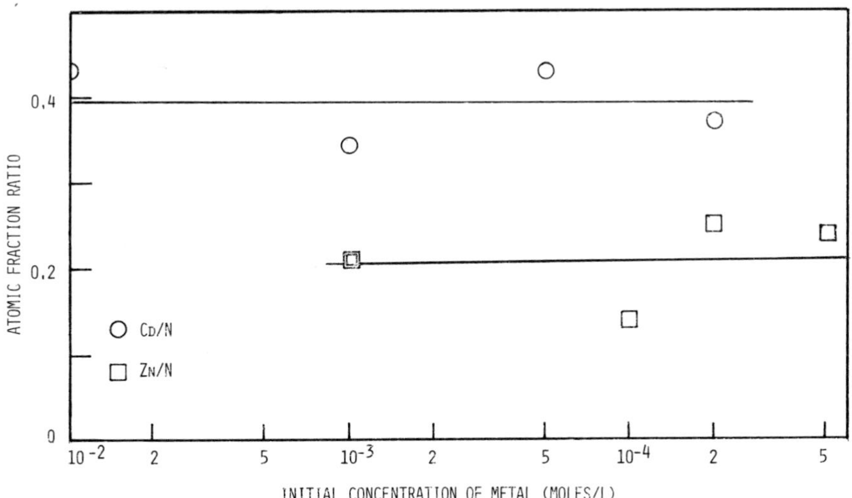

Fig. 5 ESCA atomic fraction ratios of chitosan samples equilibrated with Cd(II) and Zn(II) solutions

5 nm of chitosan by both Cd(II) and Zn(II) occurs even at the lowest solution concentrations. Further uptake of Cd and Zn could occur by continued diffusion into chitosan and nodular formation.

4. DISCUSSION

The Nature of the Uptake Process

Numerous models have been proposed for the adsorption of gases on solid surfaces (7). In contrast, there are few models to describe adsorption from solution. Everett has contributed significantly to an understanding of adsorption from non-aqueous solutions and his research is summarized in part in several recent reviews (8,9). Healy and co-workers (10,11) have addressed this problem as applied to adsorption from aqueous solution and in particular onto oxide surfaces. The use of polymers as adsorbents and in particular chitin and chitosan is an enormous complicating factor.

The shapes of the uptake isotherms shown in Figs. 1 and 3 suggest that the uptake process is a complex one. Nonetheless, plausible uptake mechanisms can be suggested based on the isotherm results coupled with the SEM/EDAX and ESCA results. Simple ion adsorption on the chitin or chitosan surface does not appear to be the dominant mechanism as is inferred from much of the literature. This conclusion is based primarily on the SEM/EDAX analysis where under moderate magnification, metal-containing aggregates are observed on the polymer. Thus, the formation of metal-containing nodules on the polymer surface has to be considered a possible uptake mechanism. Absorption presumably by diffusion of metal ions into the polymer cannot be excluded from consideration. Thus a combination of nodular formation, ion adsorption and ion absorption account for the total uptake.

The greater uptake of chitosan compared to chitin may well be due to differences in the nitrogen moiety of the two polymers. The N-acetyl D-glucosamine groups of the chitin could act as specific chemical binding sites for metal ions capable of forming complex ions but would not be expected to be as efficient as the free amine sites in chitosan. That is, the free amine groups in chitosan are much better ligands for binding to metal ions like Cd(II) and Zn(II) than the N-acetylated amine groups would be.

Effectiveness for Waste Water Purification

Fig. 6 describes the effectiveness of chitin and chitosan at removing the ions studied. The lower concentration end of

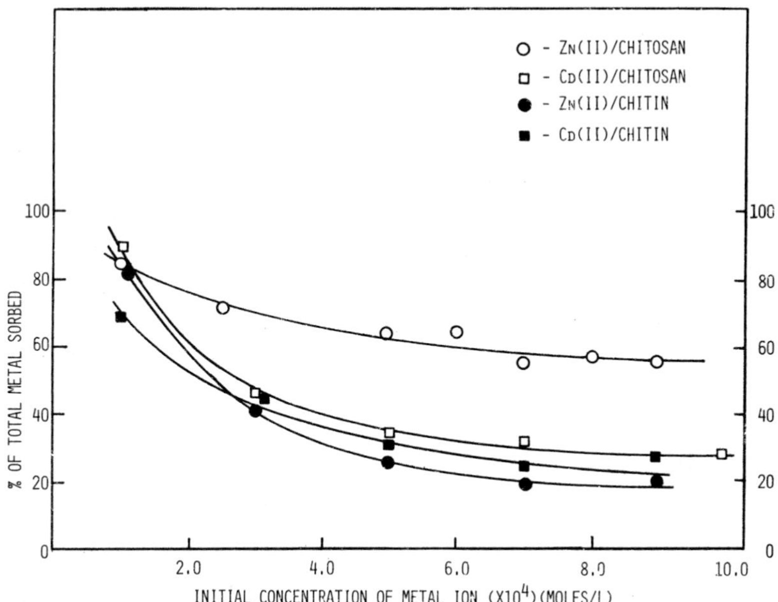

Fig. 6 Percentage of uptake of Cd(II) and Zn(II) by chitin and chitosan

these graphs from zero to 1 x 10^{-4} M, corresponds to Cd(II) and Zn(II) concentrations of zero to 11 and 6.5 ppm, respectively. In this range, chitosan appears to be quite effective at removing the ions; while chitin is less effective except at the extremely low concentrations. Since this concentration range for metal ions corresponds to that of many waste waters, chitosan would appear to hold promise as a sorbent to remove metal ions. At higher concentrations of the metal ions, the effectiveness of these polymers is considerably lower.

5. REFERENCES AND ACKNOWLEDGEMENTS

This research was done in part under the 1980 NSF-Undergraduate Research Participation Program at Virginia Polytechnic Institute and State University, Dr. J. W. Viers, Director.

This work was sponsored in part by the Office of Sea Grant, NOAA, U. S. Department of Commerce, under Grant No. NA81AA-D-00025 and the Virginia Sea Grant Program through Project No. R/UW-3. The U. S. Government is authorized to produce and distribute reprints for governmental purposes, notwithstanding any copyright that may appear hereon.

1. Muzzarelli, R. A. A. (1976). "Chitin". Pergamon Press, Oxford.
2. Hauer, H. (1978). *In* "Proc. First Intl. Conf. on Chitin/Chitosan", (Eds. R. A. A. Muzzarelli and E. R. Pariser). pp. 263-276, MIT Sea Grant Program, Cambridge.
3. Masri, M. S. and Randall, V. G., ibid, pp 277-287.
4. Kurita, K., Sannau, T., and Iwakura, Y. (1979). *J. Appld. Polym. Sci.* 23, 511.
5. Eiden, C. A., Jewell, C. J. and Wightman, J. P. (1980). *J. Appld. Polym. Sci.* 25, 1587.
6. Clark, jr., J. M. and Switzer, R. L. (1977). *In* "Experimental Biochemistry" 2nd edn, pp 106-107. W. H. Freeman, San Francisco.
7. Gregg, S. J. and Sing, K. S. W. (1967). "Adsorption, Surface Area, and Porosity", Academic Press, London.
8. Brown, C. E. and Everett, D. H. (1975). *In* "Colloid Science" (ed. D. H. Everett). Vol. 2, ch. 2. The Chemical Society, London.
9. Everett, D. H. and Podoll, R. T. (1979). *In* "Colloid Science" (ed. D. H. Everett). Vol. 3, pp. 63-149. The Chemical Society, London.
10. James, R. O. and Healy, T. H. (1972). *J. Colloid Interf. Sci.* 40 2.
11. Furlong, D. N., Yates, D. E. and Healy, T. W. (1981). "Stud. Phys. Theor. Chem." 11, 367-432.

POLYELECTROLYTE ADSORPTION FROM SALINE SOLUTIONS

J. Marra, H.A. van der Schee, G.J. Fleer,
and J. Lyklema

Laboratory for Physical and Colloid Chemistry
Agricultural University, De Dreijen 6
6703 BC Wageningen
The Netherlands.

1. INTRODUCTION

Various phenomena involving polyelectrolytes at interfaces have substantial practical interest, and hence it is not surprising that many experimental studies can be found in literature. The quality of these investigations is very divergent, but all interpretations suffer from the virtual absence of an adequate polyelectrolyte adsorption theory. The sole available theory, due to Hesselink (1), involves too severe approximations to be considered "adequate".

At the same time the theory of the adsorption of uncharged macromolecules has seen considerable progress. The difficult problem of the effect of the lateral interaction between the polymer chains was successfully attacked by Silberberg (2) and Hoeve (3), although both models made drastic simplifications in the calculation of the polymer conformations. A much more sophisticated treatment to account for all possible conformations was presented by Ash, Everett and Findenegg (4). However, due to computational problems their model could only be applied to oligomers. Extensions to longer chains were given by Roe (5) and more recently by Scheutjens and Fleer (6,7). In the latter model, the statistics of the polymer conformations are treated in sufficient detail to discriminate contributions of trains, loops, and tails. Moreover, effects of polydispersity and adsorption reversibility are now also understood (8,9).

Recently, one of us (10) has successfully extended the theories of Roe (5) and Scheutjens-Fleer (6,7) to the adsorption of charged macromolecules. Since in this theory the partition function of the system is minimized with respect to the potential and segment distributions simultaneously, without any *a priori* assumption about these distributions, it may be considered quite general. Therefore it is worthwhile to use carefully designed experiments to verify its predictions.

In this paper we present experimental results for the adsorption of polystyrene sulphonate on silica and compare these with those predicted from the above theory. Polydispersity problems were avoided as much as possible by working with polymer fractions of narrow size distributions. We studied the effect of electrolytes up to very high concentrations. One of the salient features turned out to be the continuing influence of electrolytes at concentrations far beyond those where diffuse double layers are usually considered to be fully compressed. As we shall see, this feature is also predicted by theory, invoking electrostatic arguments only.

2. EXPERIMENTAL METHODS

a. Polystyrene sulphonate (PSS)

We used commercially available products ex Pressure Chemical Comp.(Pittsburgh, USA) with narrow molecular weight distributions ($\bar{M}_w/\bar{M}_n \lesssim 1.10$). Average relative molecular weights $\bar{M}$ ranged from 6,500 to 1,060,000 , corresponding to degrees of polymerization from 35 to 5760. The concentration of PSS in solution was determined from the absorbance at 212 nm (in NaCl solutions), or 226 nm (in $MgCl_2$ solutions).

b. Silica

The adsorbent was Cab-O-Sil M-5 (Cabot Inc.) with a BET (N_2) specific surface area of ca. 200 $m^2.g^{-1}$ for moderately heated samples and ca. 165 $m^2.g^{-1}$ for silica heated for 16 h at 925 ^oC.

c. Adsorption measurement and pretreatment of the silica

In the course of the experiments a compromise had to be found between two conflicting requirements: (i) On original, i.e. hydroxylated Cab-O-Sil very little PSS adsorbs, but adsorption does take place if the silica is preheated at high temperatures. (ii) Preheated silica is rather hydrophobic and disperses poorly in aqueous solutions. In addition, if PSS is adsorbed on freshly dispersed preheated silica, it slowly desorbs due to rehydroxylation of the surface. After a series of systematic experiments to assess these effects the following procedure was adopted: (a) the silica was preheated for 16 h at 925 ^oC, (b) this silica was dispersed by ultrasonication during 30 s and left in contact with water at pH 5.3

for 20 min (the initially high rate of rehydroxylation is considerably reduced under these conditions), (c) the pH was reduced to 2 and PSS was adsorbed (at pH 2 rehydroxylation is still slower; moreover, we found evidence that the adsorption of PSS reduces the rate of rehydroxylation), (d) after equilibration (1 h for the lowest $\bar{M}$, 16 h for the highest $\bar{M}$) the adsorbed amount was measured by depletion. In this way the results were reproducible within 4%. This satisfactory repeatability indicates that there have been no major problems with incomplete dispersion of the silica, although we have to make allowance for the possibility that not the entire surface is "seen" by the polyelectrolyte. For this reason adsorbed amounts will be expressed per gram of adsorbent.

d. Viscometry

The intrinsic viscosity $[\eta]$ of PSS samples in saline solutions was measured in a Viscometer ex Fica, France, equipped with a capillary that had a flow time of ca. 230 s for water.

3. EXPERIMENTAL RESULTS

a. Viscometry

In studying the adsorption of polyelectrolytes from saline solutions, invariably the problem of interpreting the role of the low molecular weight electrolytes presents itself. One obvious effect of salt is its reduction of the electrostatic interaction between charged groups on the polyion, another is its influence on solvent quality. As yet, it is virtually impossible to discriminate between those two contributions. For uncharged polymers it is possible to define a polymer-solvent interaction parameter χ, as introduced by Flory and Huggins. One of the methods to determine this parameter is from the molecular weight dependency of $[\eta]$, using, e.g., a Flory-Fox-Schaefgen or a Burchard-Stockmayer-Fixman plot. For polyelectrolytes such an approach fails, and no suitable alternative has been formulated as yet (11). Even if experimental data can be fitted to one of the above plots (which appeared to be the case for our PSS samples) and if a kind of "effective" solvency parameter can be derived from it, this sheds no further light on the physical meaning of such a parameter. Therefore, we will make no attempt to quantify the solvent power of PSS in saline solutions.

Nevertheless, it is useful to get some impression of the behaviour of PSS in electrolyte solutions. To that end we give in table 1 the coefficients ν of the semi-empirical Mark-Houwink equation $[\eta] = KM^{\nu}$ for various electrolyte concentrations c_s.

TABLE 1

Exponents of the Mark-Houwink equation. T = 25 °C.

c_s/M	0.01	0.1	1	2	2.5	3
ν (NaCl)	1.01	0.91	0.80	0.72		0.64
ν ($MgCl_2$)	0.86	0.73	0.70	0.59	0.47	

The values of ν and their decrease with increasing c_s suggest strongly that both screening and solvency effects play a role. For uncharged linear flexible polymers ν does not usually exceed 0.8 . The higher values found here are indicative of a somewhat elongated shape for the polyion, tending to a rodlike conformation at low c_s owing to electrostatic repulsion. On the other hand, with increasing salt concentration ν becomes lower, approaching 0.5 at around 2.5 M $MgCl_2$. In NaCl solutions a concentration beyond the range for which measurements have been carried out would have been necessary to reach the point ν = 0.5 . According to Takahashi *et al.* (12) the Θ-point in NaCl is situated at c_s = 4.2 M. Therefore, it is probable that incipient phase separation occurs around 2.5 M $MgCl_2$ and 4.2 M NaCl, and some parallel effect might be expected to show up in the adsorption data. However, it is not straightforward to identify Θ-conditions with χ = 0.5 , as is the case for uncharged polymers. For polyelectrolytes the "effective" interaction is a compromise of Flory-Huggins type interaction and electrostatic repulsion, even if the latter is very short-range in concentrated solutions. Consequently, the "pure" χ-parameter (excluding the electrostatic effect) could very well differ from 0.5 under Θ-conditions.

b. Adsorption data

Fig. 1 collects typical isotherms at 0.44 M NaCl. As is usually found for the adsorption of uncharged polymers from poor solvents, the adsorbed amount Γ increases with $\bar{M}$ and this trend is paralleled by an increasing high-affinity character of the isotherm. If the effects found for uncharged polymers are also valid for polyelectrolytes in concentrated salt solutions, one would expect that the increase in Γ is due to the growth of the average size of loops and tails. Although the various fractions are fairly homodisperse, the isotherms are rather rounded, a feature which is at variance with the theory for uncharged polymers (8). Perhaps this is related to the heterogeneity of the silica surface.

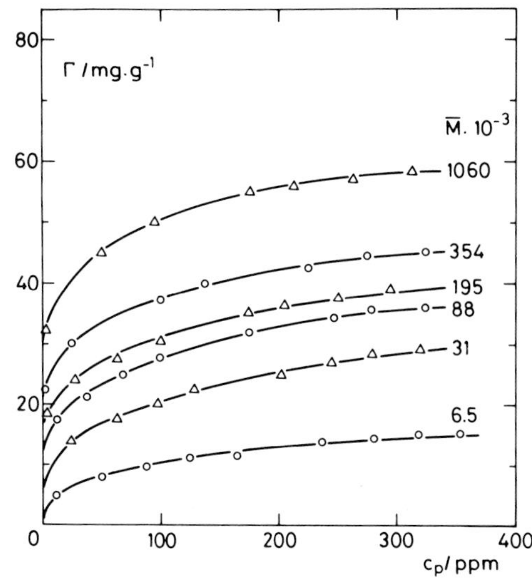

Fig. 1. Adsorption isotherms of PSS on activated Cab-O-Sil. The relative molecular weights $\bar{M}$ are indicated. pH = 2.0; [NaCl] = 0.44 M.

The effect of electrolyte concentration on the adsorbed amount at c_p = 200 ppm is shown in fig. 2. This influence is very pronounced and continues to be so far above the concentrations where the diffuse part of the double layer is fully compressed as judged by coagulation or electrokinetic studies with hydrophobic sols. Either the remainder of the double layer is still effective in modifying the interaction between charged segments, the electrolyte is influencing the solvent quality, or both effects are acting. The first point could be

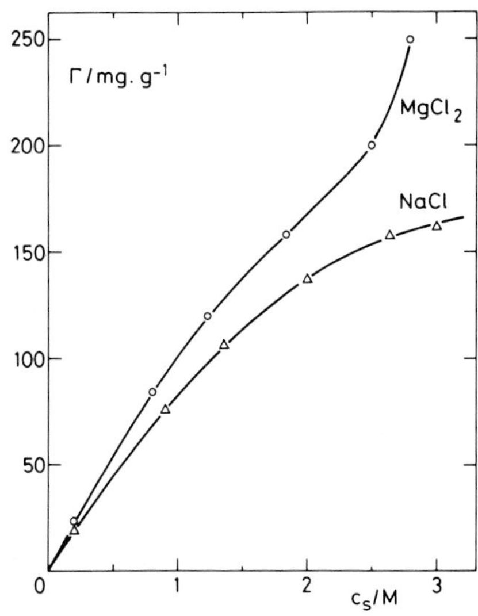

Fig. 2. Effect of electrolyte concentration on the adsorption of PSS on activated Cab-O-Sil. $\bar{M}$ = 354,000; pH = 2; c_p = 200 ppm.

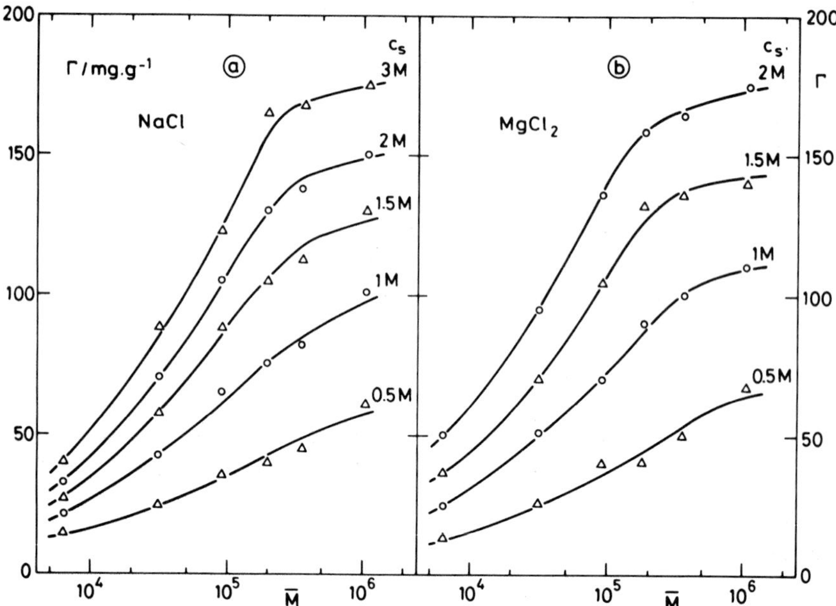

Fig. 3. Molecular weight dependence for the adsorption of PSS on activated Cab-O-Sil. The salt concentration is indicated. pH = 2; c_p = 200 ppm.

related to the three-dimensional character of the adsorbed layer, in which small ions can interpenetrate between the polyion charges, in contrast to the situation of interacting charged surfaces as for hydrophobic sols. The second point seems to show up in the strongly increasing adsorption in $MgCl_2$-solutions around the Θ-point. The latter effect was also found for other molecular weights; it appeared that the $MgCl_2$-concentration, at which Γ rises steeply, decreases slightly with increasing $\bar{M}$.

In fig. 3 the molecular weight dependence is illustrated at various concentrations of NaCl and $MgCl_2$. After an initial rise of Γ with log $\bar{M}$, the curves flatten at high $\bar{M}$, although a real plateau is absent in the experimental range studied. For uncharged polymers in Θ-solvents it has been shown, both experimentally and theoretically (9), that Γ increases linearly with log $\bar{M}$ for long chains. From the present data a similar solvent effect does not seem likely, since the flattening of the curves becomes more pronounced at high c_s where a behaviour approaching that of uncharged polymers would be expected. If the curves in fig. 3 are interpreted as if the polymer were uncharged, the stronger flattening at high c_s would suggest a decreasing χ with increasing c_s, which is contrary to the viscosimetric results. Thus, an interpretation solely in terms of solvent power, ignoring electrostatic effects, does not appear to be valid.

In addition to these observations, the following facts emerge: (i) In the absence of electrolyte there is no detectable adsorption, apparently because the repulsion between the charged groups on the polyion inhibits its accumulation near an interface. (ii) Differences between NaCl and $MgCl_2$ are noticeable, but they are smaller than would correspond with their valencies or ionic strength. Obviously, this could be related to different solvent qualities, but also an effect of the adsorption energy could play a role. As adsorption of a segment implies desorption of solvent and/or small ions, a higher affinity of Mg^{2+} ions for the surface as compared to Na^+ ions would imply a lower net adsorption energy from $MgCl_2$ solutions.

After a description of the theoretical model in the next section, we shall return to the interpretation of the experimental results in section 5.

4. POLYELECTROLYTE ADSORPTION THEORY

A logical choice to set up a theory for polyelectrolyte adsorption is to start from a suitable theory for adsorbing uncharged macromolecules and incorporate the electrical contribution to the free energy. The only existing polyelectrolyte adsorption theory (1) is based upon the model of Hoeve (3) and uses a pre-set exponential segment density distribution. Although such an exponential distribution, in which tails are neglected, is to some extent a reasonable approximation for uncharged chains, its validity for polyelectrolytes is doubtful since the electrostatic effects may be expected to influence the concentration profile drastically.

A better approach, therefore, is to use a theory which does not make an *a priori* assumption about the segment distribution. For uncharged polymers, there are two theories satisfying this condition, those of Roe (5) and Scheutjens-Fleer (6,7). In Roe's theory the canonical partition function Q_o is written in terms of the concentration profile (making an approximation that amounts to the neglect of end effects (7)), whereas Scheutjens and Fleer write it in terms of the individual conformations of chains resulting in a given overall profile. In both cases the equilibrium segment distribution is found from maximizing Q_o, taking into account the appropriate boundary conditions. Both theories have now been extended to polyelectrolyte adsorption. Anticipating a full account of this theory elsewhere (10), we now give briefly the outline.

Applying Roe's model, Q_o is known for a given concentration profile of uncharged polymer. The partition function Q_e for a polyelectrolyte with the same profile is found as $Q_o\exp(-F_{el}/kT)$, where the electrical free energy F_{el} is a function of the profile, and k and T have their usual meaning. To evaluate F_{el} for a given segment profile, the charges of the polyions are supposed to be smeared out in planes through the centres of lattice sites parallel to the adsorbing surface. The ensuing plane charge density in layer i is:

$$\sigma_i = z_p \alpha e \phi_i / a^2 \quad [1]$$

where ϕ_i is the segment volume fraction in i, e the elementary charge, α the degree of ionization, and a^2 the surface area per lattice site. We have restricted ourselves to univalent charged groups ($z_p = \pm 1$). In principle, α may depend on i, but in the present case we consider only constant α ($\alpha = 1$).

The small ions can distribute themselves freely between the plane charges due to the polyions. Their distribution n(x) is described by the Boltzmann equation:

$$n(x) = n_*\{\exp[-z\psi(x)/kT] - \exp[z\psi(x)/kT] - z_p n_{p*}\exp[z_p e\psi(x)/kT]\} \quad [2]$$

Here $\psi(x)$ is the potential at some distance x. The first term of [2] gives the contribution of added (z-z) electrolyte with bulk concentration n_*, the second term that of the counterions of the polyelectrolyte itself. The bulk concentration n_{p*} of the latter ions is related to the polyion volume fraction ϕ_* in the bulk solution:

$$n_{p*} = \alpha\phi_*/a^3 \quad [3]$$

In addition, Poisson's law applies throughout. The Poisson-Boltzmann equation can be solved numerically by second order Runge-Kutta integration with the following boundary conditions:

(a) At any plane charge σ_i there are field strength discontinuities $-\sigma_i/\varepsilon$, where $\varepsilon = \varepsilon_r\varepsilon_o$ is the dielectric permittivity which in our approach is taken invariant.

(b) At the surface the field strength is $-\sigma_o/\varepsilon$, where σ_o is the surface charge density.

(c) For high i the Debye-Hückel approximation is satisfied.

In this way the potential distribution at a given segment profile is obtained. The electrical free energy is found from a suitable charging process (13):

$$F_{el} = La^2 \int_0^1 \frac{d\lambda}{\lambda} \int_{x=0}^{\infty} \psi'(x,\lambda)\rho'(x,\lambda)dx \quad [4]$$

Here λ is a charging parameter, ρ' and ψ' are the space charge density and potential during the charging, and L is the number of lattice sites on the surface. For low potentials (i.e., high c_s) a simpler calculation is possible by using the Debye-Hückel linearization. In that case analytical expressions for $\psi(x)$ and F_{el} are available so that numerical integrations are not needed.

Now Q_e for a given segment density profile is known. The equilibrium profile is found by maximizing Q_e with respect to this profile by an iterative procedure in which the influence of profile changes on F_{el} is taken into account. In the above analysis the volume of the ions is neglected, but in an additional study it was incorporated by treating the ions as individual components in Roe's multicomponent adsorption theory. The outcome is nearly the same, apparently because of the relatively low average volume fraction of ions.

It is also possible to extend the model of Scheutjens-Fleer to polyelectrolyte adsorption. In this case the segmental weighting factors p_i that contain the configurational entropy and the interaction energy, as expressed in the polymer-solvent interaction parameter χ and the adsorption energy parameter χ_s, have to be multiplied by a Boltzmann factor, i.e. $\exp[-z_p\psi_i/kT]$. The potential ψ_i can be found in a way analo-

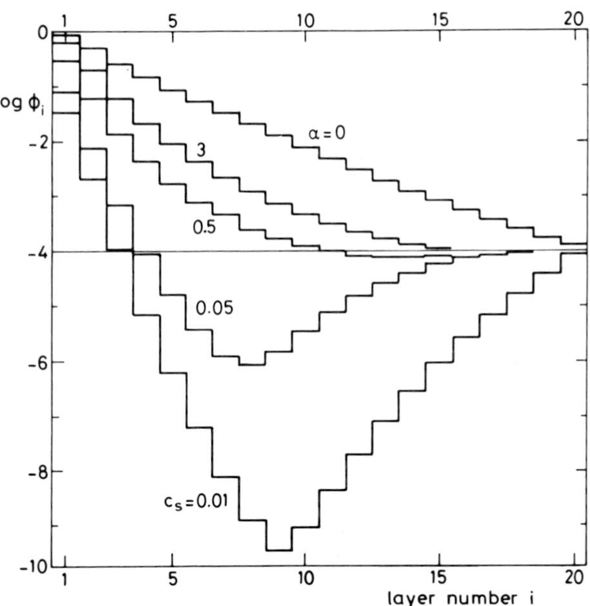

Fig. 4. Logarithmic concentration profiles for a chain of 2000 segments at $\phi_* = 10^{-4}$. The salt concentration c_s (in M) of univalent electrolyte is indicated. $\alpha = 1$, except for the upper curve. $\chi = 0.5$; $\chi_s = 2$; $\sigma_0 = 0$; $a = 0.71$ nm.

gous to that discussed above. For details we refer to (10). The numerical elaboration of the Scheutjens-Fleer model is much more complicated than that of Roe. Below we shall use the latter model in most cases, although for comparison purposes one result obtained with the extended Scheutjens-Fleer theory will be mentioned.

Fig. 4 gives a typical example of segment density profiles at various salt concentrations and for an uncharged polymer. For low c_s the curve displays a minimum which originates from the high potential which is generated by the adsorbing chains, repelling other molecules from the outer layers. At higher c_s the screening by the salt suppresses the minimum and increases the adsorbed amount, while the profile becomes gradually similar to that of uncharged polymer. However, even at $c_s = 3$ M this screening is not yet complete and adding more salt still increases the adsorbed amount. This is also illustrated by the excess surface coverage θ_{ex} (expressed in monolayers): for the five profiles in fig. 5 θ_{ex} amounts, from bottom to top, to 0.0341, 0.0913, 0.3769, 0.9284, and 2.0322 monolayers, respectively. The conclusion is that even at very high salt concentration purely electrostatic effects still greatly influence the adsorption. As said, this is probably related to the interpenetration of small ions *between* the individual polyion charges, which is not analogous to the two-dimensional screening effect of salt between two impenetrable colloid particle surfaces.

5. APPLICATION OF THEORY TO EXPERIMENTS

In order to compare experimental results with our theoretical model, a choice must be made for the parameters a, χ and χ_s. Closely related to the choice for a is the value of r, the number of segments per chain, and the conversion of the theoretical excess coverage θ_{ex} (in monolayers) to Γ (in $mg.g^{-1}$).

For the assessment of a two different approaches are possible. To account for chain flexibility, a should correspond to the size of one statistical segment, usually a few monomer units. On the other hand, the theory assumes one charge per segment. These two conflicting requirements can be met to some extent by using the concept of counterion condensation, which predicts that so many (univalent) counterions "condense" on a linear highly charged polyion that effectively only one charge per Bjerrum length remains (14). The Bjerrum length equals $e^2/4\pi\varepsilon kT$ or 0.71 nm, under ambient conditions. Assuming that also in a three-dimensional adsorbate the Bjerrum length can be used as a scaling distance, we choose $a = 0.71$ nm. This implies that parallel to the surface the plane charge density (considered to be smeared out in our model) remaining after counterion condensation corresponds to one charge on an area equal to the square of the Bjerrum length. A consistent application of the condensation concept would lead to an even larger value for a in the presence of bivalent counterions, thereby also changing the value taken for the chain flexibility. Therefore, we prefer the same value for a in NaCl and $MgCl_2$ solutions.

Since the length of a monomer unit of PSS is 0.25 nm (11), only 1/3 of the polyion groups remains effectively charged so that r corresponds to 1/3 of the degree of polymerization. An estimate for the conversion of a monolayer coverage of "Bjerrum cells" to $mg.m^{-2}$ and, thence, to $mg.g^{-1}$, can also now be made. In view of the molecular dimensions a reasonable value for a monolayer of PSS monomer units is about 1.5 $mg.m^{-2}$. A layer of fully packed lattice sites of edge 0.71 nm would accomodate three times as much. However, for steric reasons it seems unlikely that such a lattice site is completely occupied, and we estimate that $\theta_{ex} = 1$, calculated for "Bjerrum cells", corresponds to ca. 3 $mg.m^{-2}$ or about 500 $mg.g^{-1}$. In the following figures we present the experimental Γ in $mg.g^{-1}$ and the theoretical surface coverage θ_{ex} in monolayers, with the scales for both differing by a factor of 500. This presentation allows the comparison of trends without placing too

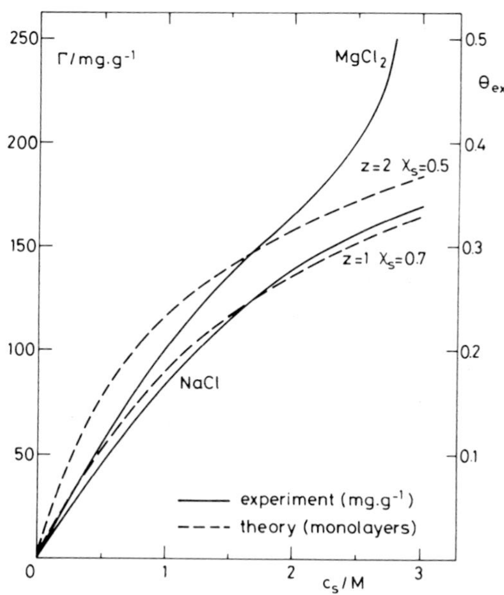

Fig. 5. Experimental and theoretical dependency of the adsorbed amount on c_s. The experimental curves are the same as those in fig. 2; the theoretical ones were calculated for $\alpha = 1$, $\sigma_0 = 0$, $\chi = 0.5$, $a = 0.71$ nm, $r = 640$, and $\phi_* = 2 \times 10^{-4}$.

much emphasis on the precise conversion of Γ to θ_{ex}. Even if this conversion factor would be wrong by, say, 50%, the main conclusions about the dependencies of Γ or θ_{ex} on variables such as c_s and $\bar{M}$ remain valid.

The last choice to be made is adopting values for χ and χ_s. For lack of any information on χ we shall use arbitrarily χ = 0.5. The parameter χ_s is treated as adjustable. It turned out that the best agreement was found for $\chi_s = 0.7$ in NaCl solutions and $\chi_s = 0.5$ in $MgCl_2$. This difference could be due to the fact that Mg^{2+} ions adsorb more strongly on SiO_2 than Na^+ (15). Since all experiments were carried out at pH 2, close to the p.z.c. of SiO_2, we used in the calculations $\sigma_o = 0$.

Fig. 5 gives a comparison of the theoretical and experimental salt concentration dependency. For NaCl solutions, purely electrostatic effects explain satisfactorily the increase of Γ with c_s, even in concentrated solutions. Changing the parameters χ and χ_s affects the level of the theoretical curves, but not their overall shape so that the general trend seems well established. For $MgCl_2$ the agreement is less perfect,

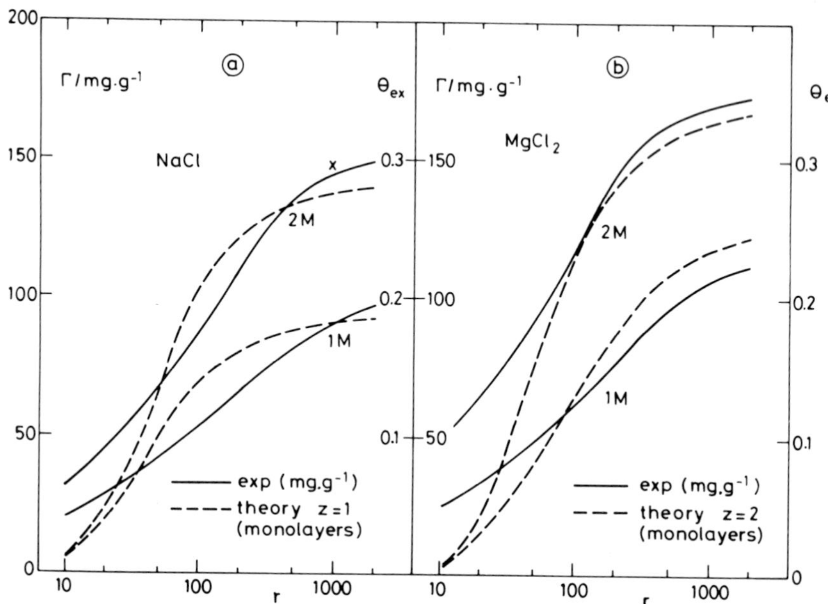

Fig. 6. Experimental and theoretical molecular weight dependency of the adsorbed amount. The experimental curves are the same as those in fig. 3; the theoretical ones were calculated for $\alpha = 1$, $\sigma_o = 0$, $\chi = 0.5$, $\chi_s = 0.7$ (NaCl) or 0.5 ($MgCl_2$), $a = 0.71$ nm, $\phi_* = 2 \times 10^{-4}$, and c_s as indicated.

especially at high c_s. In this case probably solvency effects play a role; full agreement between theory and experiment can only be obtained by using a χ parameter that is a function of c_s. Theoretical and experimental results for the molecular weight dependence of the adsorbed amount are given in fig. 6. At low chain lengths the experimental data are higher than the theoretical predictions. Possible reasons could be some polymer polydispersity and surface heterogeneity, both effects tending to increase Γ, especially at low Γ. At high r, the theoretical curves seem to flatten somewhat too strongly. This is presumably related to the fact that Roe's theory for uncharged polymer is known to underestimate the adsorbed amount due to the neglect of tails. This aspect is corroborated by one point (the cross at r = 1000 in fig. 6a) that was calculated with the extended Scheutjens-Fleer theory, which gives a slightly higher θ_{ex}. Unfortunately, at present there are not enough numerical data to further examine the molecular weight dependence of long chains by the more sophisticated Scheutjens-Fleer theory. As said, the latter model allows the calculation of trains, tails, and loops separately. For the one point indicated in fig. 6a the fraction of trains is 0.49 , that of loops 0.44 , and that of tails 0.07 , and the average tail length is 49 segments. Thus also for polyelectrolytes adsorbing from saline solution tails seem to play an important role.

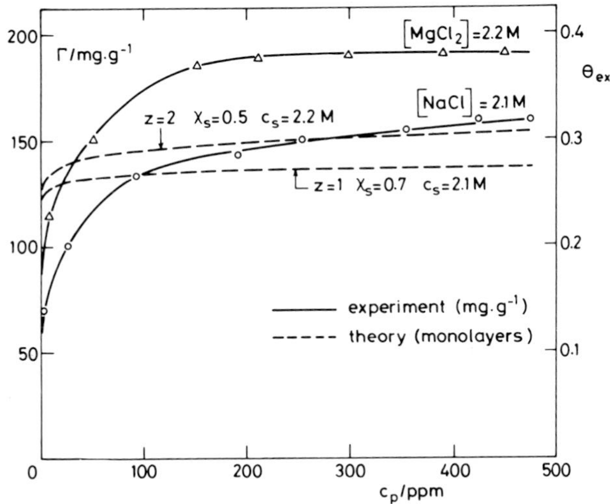

Fig. 7. Experimental and theoretical adsorption isotherms for $r = 350$. The theoretical isotherms apply to $\alpha = 1$, $\sigma_0 = 0$, $\chi = 0.5$, and $a = 0.71$ nm. The values of χ_s and c_s are indicated.

Finally, in fig. 7 we compare experimental and theoretical adsorption isotherms. The agreement is only moderate: the experimental isotherms are much more rounded than the theoretical ones. It seems unlikely that polydispersity could be solely responsible for this difference. A possible reason for this discrepancy is again the heterogeneity of the surface, as discussed above.

6. CONCLUSION

Comparison of experiments, covering a wide range of variables, with a new theory of polyelectrolyte adsorption in saline solutions has shown that a satisfactory agreement is emerging. One of the most salient features of this work is that the effect of electrolytes can be well represented by the Poisson-Boltzmann picture up to concentrations of order 1-3 M, which is far beyond the usually accepted limit of applicability.

7. REFERENCES

1. Hesselink, F.Th. (1977). *J.Colloid Interface Sci.* 60, 448-466.
2. Silberberg, A. (1968). *J.Chem.Phys.* 48, 2835-2851.
3. Hoeve, C.A.J. (1970). *J.Polym.Sci.* *C* 30, 361-367.
4. Ash, S.G., Everett, D.H. and Findenegg, G.H. (1970). *Trans.Faraday Soc.* 66, 708-722.
5. Roe, R.J. (1974). *J.Chem.Phys.* 60, 4192-4207.
6. Scheutjens, J.M.H.M. and Fleer, G.J. (1979). *J.Phys.Chem.* 83, 1619-1635.
7. Scheutjens, J.M.H.M. and Fleer, G.J. (1980). *J.Phys.Chem.* 84, 178-190.
8. Cohen Stuart, M.A., Scheutjens, J.M.H.M. and Fleer, G.J. (1980). *J.Polym.Sci., Polym.Phys.Ed.* 18, 559-573.
9. Scheutjens, J.M.H.M. and Fleer, G.J. (1981). *In* "The Effect of Polymers on Dispersion Properties" (Ed. Th.F. Tadros), pp. 145-168. Academic Press, London/New York.
10. Van der Schee, H.A. (1983). Thesis Agricultural University, Wageningen. To be published.
11. Soumpasis, D.M. and Bennemann, K.H. (1981). *Macromolecules* 14, 50-54.
12. Takahashi, A., Kato, T. and Nagasawa, M. (1967). *J.Phys. Chem.* 71, 2001-2005.
13. Verwey, E.J.W. and Overbeek, J.Th.G. (1948). "Theory of the Stability of Lyophobic Colloids" p. 58. Elsevier, Amsterdam/New York.
14. Manning, G.S. (1978). *Quart.Rev.Biophys.* 11, 179-246.
15. Tadros, Th.F. & Lyklema, J. (1969). *J.Electroan.Chem.* 22, 1-7.

INFRARED AND MICROCALORIMETRIC STUDIES OF THE CONFORMATION OF ADSORBED POLYMERS WITH CARBONYLGROUPS AT THE SILICA/CCl_4, $CHCl_3$ INTERFACE

E. Killmann, M. Korn, M. Bergmann

Institut für Technische Chemie der Technischen Universität München
Lehrstuhl für Makromolekulare Stoffe
Lichtenbergstraße 4, D 8046 Garching, BRD

1. INTRODUCTION

In order to understand polymer adsorption from solution it is important to gain insight into the manner, energetics, kinetics, adsorbed amounts, conformation of the adsorbed macromolecules, competitive adsorption and the dependence of these features on the chemistry and structure of the adsorbent surface, on the solvent, on the chemical character of the polymer, its molar mass, the molar mass distribution and on variables such as concentration, temperature, pH and the kind and content of any electrolytes present. The experimental work on polymer adsorption is mostly concerned with the determination of the amounts adsorbed, the adsorption isotherms, the average fraction of adhesive segments by quantitative spectrometry (IR, NMR, ESR), the thickness and concentration of adsorbed layers by ellipsometry and hydrodynamic techniques (viscosity, diffusion, sedimentation, electrophoresis). For complete characterization of the adsorbed macromolecular layer, especially with the purpose of comparing the experimental results with new theoretical approaches and parameters (1) it would be desirable to obtain numbers, sizes and distributions of trains, loops and tails or (because this seems unrealistic) to measure the overall segment distribution perpendicular to the surface (2). According to nearly all theories a critical adsorption enthalpy per adsorbed segment is necessary for the existence of adsorption and this enthalpy determines the conformation, i.e. the distribution of the segments as trains, loops or tails. Therefore, the determination of the energetics of polymer adsorption should be given more attention.

In this study the quantitative IR-Spectrometry of adsorbed polymers with carbonyl groups especially from CCl_4 or $CHCl_3$ suspensions on nonporous silica enables the determination of the number of adsorbed carbonyl segments from the polymer and of the number of the interacting silanol groups on the silica surface. Microcalorimetric measurements of adsorption and immersion enthalpies of the same systems complete these measurements and clarify the energetics of polymer adsorption. The adsorbed amounts, the manner of attachment, the fraction of adhered carbonyl polymer segments and of surface silanolgroups, the adsorption enthalpy and the reversibility of adsorption were determined as a function of the coverage, the molar mass and also of the surface density of adsorption sites. Additionally, for comparison, measurements of the same type have been done with several low molecular mass substances, e.g. alcohols, ethers, esters and ketones, under the same conditions.

2. MATERIALS AND EXPERIMENTAL METHODS

Pyrogenic, colloidal nonporous, amorphous Silica (Aerosil 200, Degussa Wolfgang) was used as adsorbent at 25 $\pm$ 1°C. The polarity and linkage ability of the surface SiOH groups and their mutual steric arrangement determine to a great extent the behaviour of the silica in adsorption processes. Siloxan bridges (Si-O-Si) on the surfaces are possible adsorptive sites for hydrogen bridge donors.

The pretreatment of silica, the preparation of the silica CCl_4, $CHCl_3$ suspensions, the merging of the suspensions with the adsorptive, and the tempered agitation of the system have already been reported in detail (3).

Linear polymers with C = O groups in the main or side chain e.g. Polyvinylpyrrolidone, Polyethyleneglycol and low molar mass alcohols, ethers, esters, ketones were investigated (3,4): poly(caprolactone) (PCL) with M_n = 10700, M_w = 33 000, poly(vinylacetate) (PVAc) with M_n = 9,600 and M_w = 25,500, poly(methacrylic acid n-butyl ester) with (a)M_w = 2,400 and M_w = 3,200 (PBMA 2,400), (b) M_n = 55,000 and M_w = 101,000 (PBMA 55,000), (c) M_n = 1,570,000 and M_w = 4,020,000 (PBMA, 1,570,000), poly(vinylpyrrolidone) (PVP) with (a) M_n = 300 (OVP 300) (b) M_w = 360 000 (PVP K 90) and Poly-(ethyleneglycol) PEG with (a) M_n = 600 (PEG 600) (b) M_n = 6300 (PEG 6000) (c) M_w = 35000 (PEG 40 000). The low molar mass adsorptives can be seen in Table 1.

The concentration difference of the adsorptive before, c_E, and after, c_L, adsorption, related to the concentration of the solid, c_s, (gl^{-1}) results in the amount adsorbed A,

$$A = \frac{c_E - c_L}{c_S} \text{ monomer mole } g^{-1}_{\text{Aerosil}} \qquad (1)$$

The concentration of the adsorbed polymer is given by $c_A = Ac_S = c_E - c_L$ monomer mole liter^{-1}.

Two degrees of coverage can be defined:

$$\theta = \frac{A}{A_\infty} \quad (2) \quad \text{and} \quad \theta_m = \frac{A}{A_m} \qquad (3)$$

where A_∞ is the amount adsorbed in the adsorption isotherm plateau. The monolayer capacity A_m can then be calculated using a formula from Emmett and Brunauer (5).

A second aliquot from each adsorption experiment was placed in a cuvette of thickness, d, and its infrared spectrum between 2800 and 4000 cm^{-1} was recorded after a special compensation procedure. The decrease in the concentration, ΔC_{OH}, of the unperturbed surface SiOH groups due to the adsorption and the interaction with the adsorbed segments of the adsorptives was measured by the decrease of the absorbance, E_{OH}, of the IR-band at 3695 cm^{-1} in CCl_4 resp. at 3670cm^{-1} in $CHCl_3$ according:

$$C_{OH} = \frac{E_{OH}}{\varepsilon_{OH} d} \qquad (4)$$

Where $\Delta C_{OH} = C_{OH}^{o} - C_{OH}$ the fraction p_{OH},

$$p_{OH} = \frac{\Delta C_{OH}}{C_A} \qquad (5)$$

represents the ratio of the number of SiOH groups occupied by the adsorbate to the total number of polymer segments in the adsorbed layer.

In a modified procedure a method first reported by Fontana and Thomas (6) was used for determining the fraction of adhered carbonyl polymer segments. Thereby, three different compensation procedures were used to determine the concentrations of the free, C_{CO}, and adsorbed polymer segments, C_{COH}, from the absorbances of the bands of the free (E_{CO}) and adsorbed (E_{COH}) carbonyl groups. With $C_{CO} = C_A - C_{COH}$ the fraction of adhered polymer segments is expressed by

$$p_{COH} = \frac{c_{COH}}{c_A} \qquad (6)$$

A multiple interaction quotient Q, representing the decrease of the number of unperturbed SiOH groups in relation to the number of adsorbed carbonyl groups can be determined for every coverage,

$$Q = \frac{-\Delta c_{OH}}{c_{COH}} = \frac{-\Delta c_{OH}}{p_{COH}\ c_A} = \frac{p_{OH}}{p_{COH}} \qquad (7)$$

The microcalorimetric measurements of the adsorption enthalpies were carried out in an isoperibolic LKB-Mikrocalorimeter. The standard preconditioned Aerosil suspension were equilibrated in the measuring cell and then mixed with the solution of the adsorptive by breaking a small ampule.

3. RESULTS

Low Molecular mass substances

The amount of the investigated low molar mass substances adsorbed fit the Langmuir plot in the relevant concentration range, as is shown in Fig. 1. A comparison of the saturation values of the amount adsorbed A_∞ with the monolayer capacities, calculated from the formula given by Emmet and Brunauer (5) is shown in Table 1.

The multiple interaction quotient Q defined by equ.(7) with $p_{COH} = 1$, shows values lower and higher than one.

With the exception of ethanol and ethyleneglycolmonoethylether (Fig.3), the integral adsorption enthalpies of all low mass substances are proportional to the amounts adsorbed, the differential molar adsorption enthalpies are constant; demonstrating an isoenergetic adsorption at all coverages (Fig.2).

Polymers

The adsorption isotherms of polymers PCL, PVAc, PBMA, PVP, PEG plotted as amount adsorbed, A, vs final supernatant concentration, c_L, show the typical high affinity character with a steep slope at low, and a plateau region at high concentrations (Fig. 4). The plateau values are included in Table 1 and compared with the monolayer capacities, A_m. In the calculation of A_m for the polymer a density of the adsorptive equal to that of a substance similar to the repeated unit was used.

The IR-absorbances E_{OH}, E_{OHO}, $E_{C=O}$, E_{COH} at the corresponding frequencies from CCl_4 at ν_{OH} = 3695 cm^{-1}, ν_{OHO} = 3430 cm^{-1}, $\nu_{C=O}$ = 1731 - 1741 cm^{-1}, ν_{COH} = 1705 - 1712 cm^{-1} depending on the type of polymer and from $CHCl_3$ at ν_{OH} = 3670 cm^{-1}, ν_{OHO} = 3250 cm^{-1}, $\nu_{C=O}$ = 1677 cm^{-1}, ν_{COH} = 1660 cm^{-1} for PVP were measured in order to determine the fraction of adhered segments.

TABLE 1

Degree of Polymerisation P_n, Monolayer capacity A_m, Amount adsorbed in saturation A_∞, Enthalpies $\overline{\Delta H}$ and $\Delta H_{P,A}$.

Adsorptive, number in Fig. 12	P_n	A_m	A_∞	$\overline{\Delta H}$	$\Delta H_{P,A}$
		(10^{-4} mole $g^{-1}_{Aerosil}$)		(KJ monomermole^{-1})	
1 1-Hexanol	1	8,5	5,4	25,2	49,0
2 Ethanol	1	15,2	14,1	25,0	48,8
3 Ethyleneglycol-monoethylether	2	11,8	18,4	26,6	50,4
4 2-Butanone	1	11,5	3,8	23,9	47,7
5 2-Hexanone	1	9,3	2,7	29,5	53,3
Diethylether	1	11,3	5,0	-	-
6 Di-n-propylether	1	8,6	2,2	32,3	56,1
7 Acetic acid-n-butylester	1	8,9	1,9	21,4	45,2
Propionic acid-n-butylester	1	7,6	1,6	-	-
8 CCl_4	1	10,9	-	-	-
9 Cyclohexane	1	-	-	-	26,7
10 PCL	94	8,9	18,2	12,7	36,5
11 PVAc	110	10,8	33,6	14,6	38,4
12 PBMA 2400	17	10,4	7,6	14,2	38,0
13 PBMA 55000	387	10,4	10,2	10,4	34,2
14 PBMA 1570000	11050	10,4	10,2	10,4	34,2
15 PEG 600	13	20,7	26,4	25,1	48,9
16 PEG 6000	140	20,7	40,4	16,1	40,3
PEG 40000	800	20,7	45,4	16,5	40,3
17 OVP 300	3	9,2	6,0	30,4	54,1
18 PVP K 90	3300	9,2	17,1	4,2	26,6

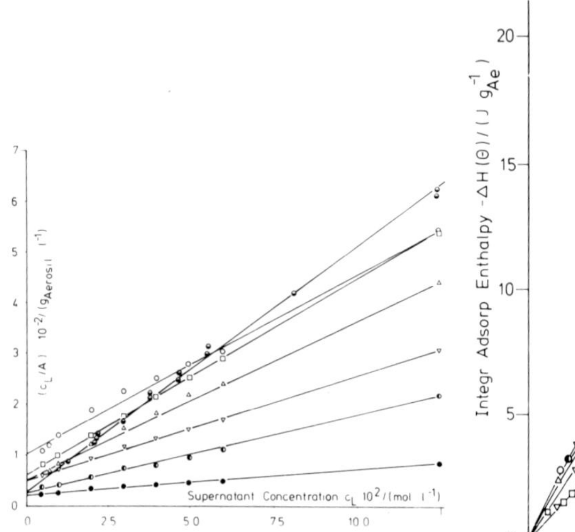

Fig. 1. Langmuir plots; CCl_4*: ○ di-n-propylether, □ acetic acid-n-butyl ester, △ 2-hexanone, ▽ butanone, ◐ hexanol, ● ethanol;* $CHCl_3$*: ◒ ethylpyrrolidone.*

Fig. 2. Integral adsorption enthalpy $\Delta H(\theta)$ *versus amount adsorbed A; symbols as fig. 1.*

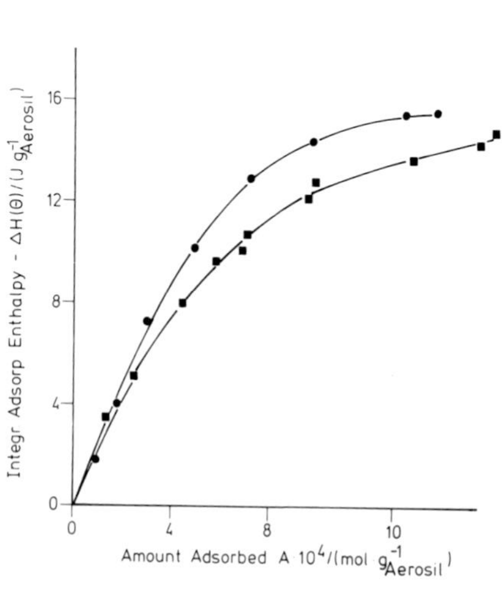

Fig. 3. Integral adsorption enthalpy $\Delta H(\theta)$ *versus amount adsorbed A;* CCl_4*: ● ethanol, ■ ethyleneglycol monoethyl ether*

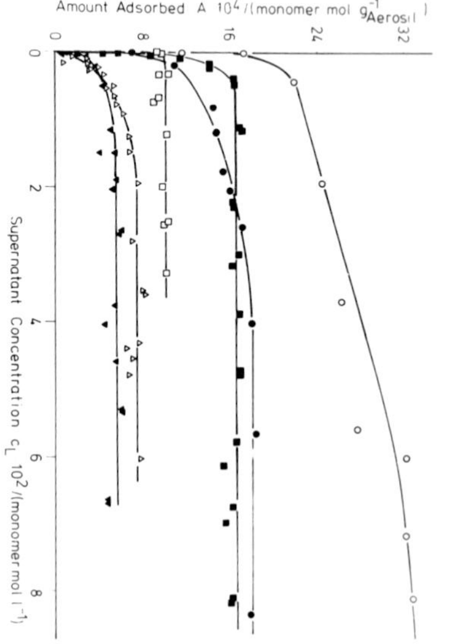

Fig. 4. Adsorption isotherms; CCl_4*: ● PCL, ○ PVAc, △ PBMA 2400, □ PBMA 55000, 1570000;* $CHCl_3$*: ▼ OVP 300, ■ PVP K 90*

The fraction of interacting SiOH-groups, p_{OH}, the fraction of adhered polymer segments, p_{COH}, and the multiple interaction quotient, Q, are plotted against coverage in Figs.5-8.

In Fig. 9 the integral adsorption enthalpies $\Delta H(\Theta)$ are plotted as a function of the amount adsorbed A for PCL, PMBA, PVAC, PEG from CCl_4 and for PVP from $CHCl_3$.

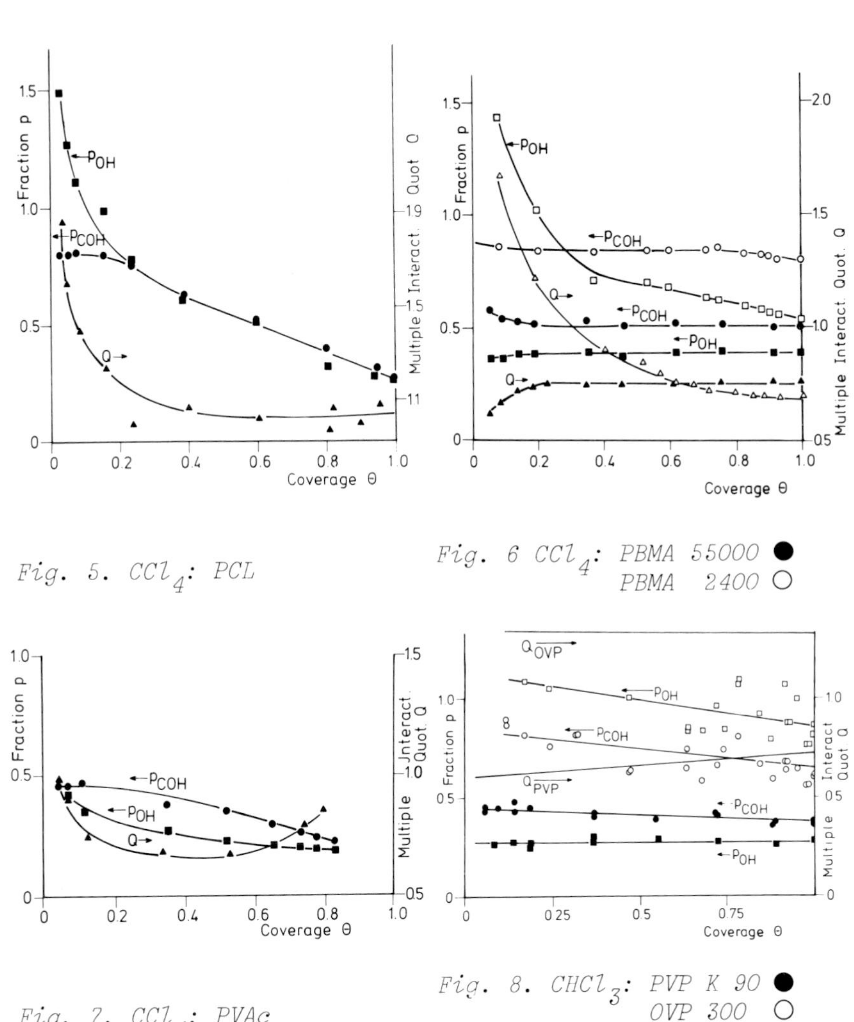

Fig. 5. CCl_4: PCL

Fig. 6 CCl_4: PBMA 55000 ●
PBMA 2400 ○

Fig. 7. CCl_4: PVAc

Fig. 8. $CHCl_3$: PVP K 90 ●
OVP 300 ○

Figs. 5 - 8. Fractions of adhered segments p_{COH} ●○, p_{OH} ■□; multiple interaction quotient Q▲△ versus coverage $\Theta = A/A_{\infty}$.

DISCUSSION

Low Molecular Mass Substances

The Langmuiric behaviour (Fig. 1) of the low molecular mass substances is evidence that the adsorption of the small molecules on the silica surfaces out of solution occurs on unoccupied, homogeneous surfaces in a monolayer by a reversible adsorption process. The reversibility is also indicated by some desorption experiments with dilution after adsorption. The comparison of the saturation values (Table I) shows, that the A_∞ values, with the exception of the alcohols, are lower than the A_m-values for the monolayer capacity and lower than the total concentration of the surface SiOH-groups. In contrast to this behaviour, Stählin (7) has reported that $A_\infty = A_m$ for the adsorption of ethers on Aerosil from the gas phase. Furthermore by adding an A_∞-value ($A_\infty = 1,64 \cdot 10^{-4}$ mol g^{-1}) for propionic acid n-butylester in the same system from Joppien (8) we can conclude that in each substance class the A_∞-values of alcohols, ketones, ethers and esters decrease with molar mass. According to Fig. (2), with exception of ethanol and ethyleneglycol monoethylether (Fig. 3) all investigated adsorptives release a constant differential adsorption enthalpy. Therefore, an energetic equivalence of all adsorption sites at low and high coverages can be assumed.

The most plausible interpretation for the incomplete occupation of all available sites and for the differences between the saturation values of the single adsorptive classes is provided by the influence of the solvation envelope of CCl_4, which hinders sterically the adsorption of further molecules. Such a solvation effect seems not to be decisive for the adsorption of polymers, since an almost complete coverage of the available SiOH-groups takes place in the plateau region. The reason for this is that the segments of the adsorbing polymer chain, which are located in the vicinity of already adsorbed segments, are forced, due to their covalent binding, to approach the surface and forced desolvation takes place.

Ketones, esters and ethers can only act as H-bridge acceptors and thus only donor groups such as SiOH groups or assymetrical strained siloxan bridges, as reproted by Morrow and Cody (9) are possible adsorption sites. In this respect the divergent behaviour of ethanol, ethylene glycol monoethylether and 1-hexanol can probably be attributed to their ability to act as hydrogen bridge donors and thus to adsorb at other surface adsorption sites in addition to SiOH groups.

It is evident from table I that the values of A_∞ for ethanol and ethyleneglycol monoethylether reach their approximate monolayer capacities A_m and exceed the available SiOH concentration. The differential adsorption enthalpies $\Delta H(\theta)$ decrease strongly for these substances at coverages where the amount adsorbed A exceeds the concentration of available SiOH-groups ($A_\infty = 7,9 \cdot 10^{-4}$ mol SiOH $g^{-1}_{Aerosil}$) (Fig. 3). The adsorption in this region takes place at other, less energetic adsorption sites.

Contrary to this behaviour, the saturation value A_∞ of hexanol is smaller than its monolayer capacity and the concentration of the total SiOH-groups and corresponds remarkably to the concentration of the unperturbed SiOH groups. In this case the differential adsorption enthalpy remains constant over the whole range of coverage (Fig. 2).

The values of the multiple interaction quotient, Q indicate that the interaction with the Silanol groups via hydrogen bridges does not take place in a simple 1 : 1 ratio (Q = 1) as has often been assumed. Coverage independent constant Q values were found for: di n - propylether Q = 1,4, acetic acid n-butyl ether Q = 1,3 and 1-hexanol Q = 0,7. The Q values decrease from Q = 1,7 at θ = 0,14 to Q = 1.1 at θ = 0,4 and then remain constant until saturation for 2-hexanone and from Q = 1,5 at θ = 0,15 to Q = 0,8 at θ = 0,5 for 2-butanone.

Values of $Q > 1$ according to equ. (7) suggest that, on average, more than one SiOH group interacts with one adsorbate group and implies, that, at least some of the unperturbed SiOH groups arrange in pairs, the members of which are unable to form mutual hydrogen bridges with themselves. One of these possible arrangements is the geminal OH-pairing suggested e.g. by Peri and Hensley (10). A second possibility is that two isolated SiOH groups or members belonging to two different geminal pairs simultaneously interact with one functional group of an adsorptive. This phenomenon of multiple interaction has already been reported by Davidov et al. (11).

Values $Q < 1$ could mean that adsorption additionaly takes place at adsorption sites other than unperturbed SiOH groups as in the case of 1-hexanol at low coverages.

Polymers

Adsorption isotherms, multiple interaction quotient With the exception of the Oligomer PBMA 2400 the adsorption isotherms of the polymers (Fig. 4) show the typical high affinity character. The plateau values A_∞ exceed the monolayer capacities and the concentration of surface SiOH-groups depending on the polymer type (Table I). As can be seen from the Figs. 5 - 8, the polymer carbonyl groups do not interact with the surface SiOH groups in the stochiometric ratio Q = 1:1. In the case of the oligomer PBMA (Fig. 6) and of PCL, which has the most distant C=O segments, (Fig. 5) the polymer carbonyl segment occupies nearly two SiOH groups at low coverage. The decrease of the Q values with increasing coverage θ can be explained in terms of increasing mutual competition of the adsorbing carbonyl groups.

All other polymer substances PBMA (Fig. 6), PVAC (Fig. 7) PVP (Fig. 8) show almost constant Q values with Q ~ 0,75. The interpretation of these values is consistent with the interpretation for the low molar mass substances given above and the Q values lower than 1 have to be attributed mainly to the additional interaction of the carbonyl groups with strained siloxane bridges (9). This is confirmed by the fact that also with Aerosil annealed at 1000 ^{o}C, which possesses practically no surface hydrogen bridged SiOH groups, Q values lower 1 have been observed.

Fraction of adhered segments From the discussion above one can conclude that the fraction of adhered segments $p_{C=O}$ is the most reliable parameter characterizing the conformation of the adsorbed chain. Therefore, an identical conformation can be assumed for the whole coverage range for all PBMA- and PVP-samples. Values of p_{COH} = 0,8 - 0,9 for PBMA 2400, of p_{COH} = 0,5 for PBMA 55,000 and 1,570,000 (Fig. 6), p_{COH} = 0,75 - 0,6 for OVP 300 and p_{COH} = 0,35 for PVP K 90 (Fig. 8). indicate flat conformations at any adsorbance. This is also confirmed by the plateau values A_{∞} which do not exceed the monolayer capacities (Table 1). The difference in the p_{COH}-values for the different molar mass PBMA and OVP, PVP is also demonstrated by the higher slopes for the oligomers than for the polymers in the plots of the adsorption enthalpy versus amount adsorbed (Fig. 9).

The decrease of the p_{COH} values for PCL from p_{COH} = 0,8 to 0,27 and for PVAC from p_{COH} = 0,45 to 0,2 indicates a change from a flat conformation to a conformation with more segments in loops and tails. In discussing the conformation of polymers adsorbed on colloidal solids, the colloidal character of the solid is often, inadmissibly, underestimated. It is obvious that no equivocal conclusion can be drawn about the conformation of the adsorbed polymer from the knowledge of the fraction of adhered segments alone. In addition to the frequently discussed segments in loops and trains a high fraction of segments may also be involved in tails protruding into solution (1) and in interparticular bridges. In certain cases some of the possible conformations can be excluded. If the amount adsorbed significantly exceeds the monolayer capacity, or if a very low fraction of adhered segments p_{COH} is indicated, then totally flat conformations of the whole chain can be excluded. If the fraction of P_{COH} reaches values near one, then a flat conformation can nearly be assumed. Additional information may be obtained by the investigation of the adsorbed layer thickness on monodispersed. spherical particles with hydrodynamic methods (e.g. diffusion measurements by photoncorrelation spectroscopy (2)), or by studies of stability/flocculation behaviour (12), which yield more information about the extent of bridging.

Adsorption Enthalpy

The microcalorimetric measurements allow the determination of the integral adsorption enthalpy $\Delta H(\theta)$ as an analytical measure of the fraction of adhered polymer segments p_{COH} and of the energetics of the adsorption interaction. This is the sum of the exothermic intrinsic binding process $n\,\Delta H_{P,A}$ and the endothermic terms for the desorption of the corresponding solvent molecules from the surface $\Delta H_{L,A}$ and for the desolvation of the solvent from the adsorbing segment $\Delta H_{P,L}$. The final term $\Delta H_{P,P}$ is associated with the

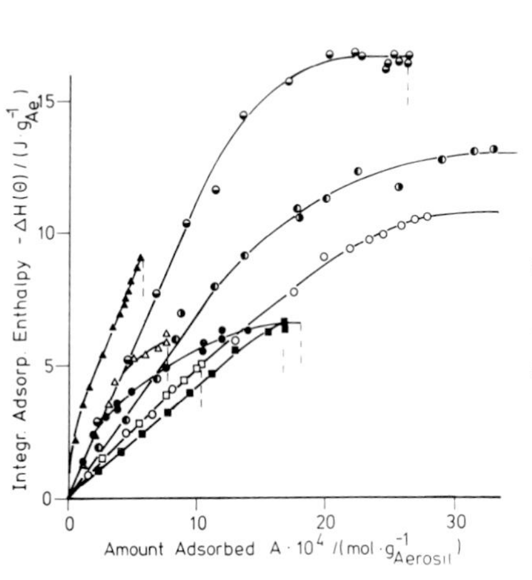

Fig. 9. Integral adsorption enthalpy $\Delta H(\theta)$ versus amount adsorbed A; CCl_4: ◒PEG 600, ◐PEG 6000, ◑PEG 40000; other symbols as in fig. 4.

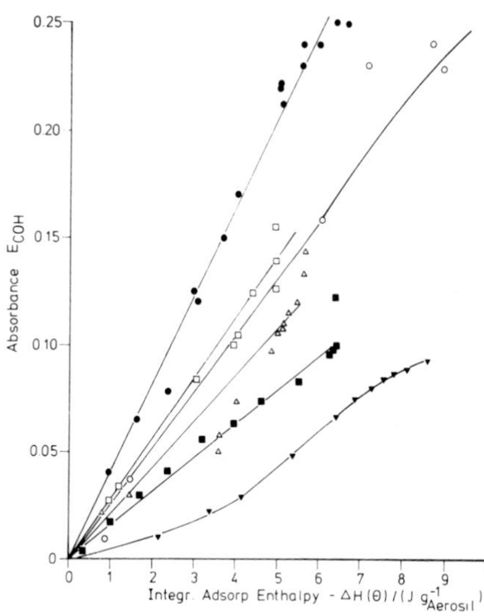

Fig. 10. Absorbance E_{COH} versus adsorption enthalpy; symbols as in fig. 4.

change of polymer interaction in the adsorbed layer compared with the solution state and can be either exo- or endothermic.

$$-\Delta H\ (\theta) = -\ n\ \Delta H_{P,A} + n\ \Delta H_{L,A} + n\ \Delta H_{P,L} + \Delta H_{PP} \qquad (8)$$

Neglect of the last two terms with $n = A \cdot p_{COH}$ equ. 9 results:

$$\Delta H_{P,A} = \frac{\Delta H(\theta)}{A \cdot p_{COH}} + \Delta H_{L,A} \quad \mathrm{kJ \cdot monomer\ mole^{-1}} \qquad (9)$$

With the assumption that during the adsorption of one segment one specifically adsorbed CCl_4 or $CHCl_3$-molecule desorbs the value of 23,8 kJ mol^{-1} in CCl_4 and 16,1 in $CHCl_3$ can be employed for $\Delta H_{L,A}$. The net binding enthalpy $\Delta H_{P,A}$ can then be calculated from equ. 9. The term $\Delta H(\theta)/n$ corresponds to the differential adsorption enthalpy $\overline{\Delta H}$. $\overline{\Delta H}$ can be evaluated in the case of the monofunctional adsorptives as the slope of the straight line of the plots $\Delta H(\theta)$ versus A (Figs. 2.3).

For the polyfunctional adsorptives $\overline{\Delta H}$ and $\Delta H_{P,A}$ can be calculated according to equ.9 with the use of the fraction of adhered segments p_{COH} from IR. The neglect of the last two terms is justified because separate dilution and solution enthalpy measurements result in small values for $\Delta H_{L,P}$ and $\Delta H_{P,P}$ in comparison to the other contributions. Values of $\overline{\Delta H}$ and $\Delta H_{P,A}$ are included in Table I.

Taking corresponding values of E_{COH} and $\Delta H(\theta)$ from the plots E_{COH} versus A and $\Delta H(\theta)$ versus A (Fig. 9), respectively then plotting them in the form of E_{COH} versus $\Delta H(\theta)$ at equally adsorbed amounts (Fig. 10) one obtains straight lines within the limit of the errors. This linearity confirms that the integral adsorption enthalpy $\Delta H(\theta)$ is an analytical measure of the number of adhered segments and that $\Delta H_{P,A}$ and $\overline{\Delta H}$ are constant molar heats within the whole region of coverage. This validates the assumption of an isoenergetic adsorption of the investigated polymers on the standard preconditioned surface.

The determined binding enthalpies $\Delta H_{P,A}$ (Table I) can be correlated with the square root of the frequency shift $(\Delta\nu_{OH})^{1/2}$ of the maximum of the absorption band of the hydrogen bridges with respect to the maximum of the band of the unperturbed SiOH groups. For the low molar mass substances a clear agreement, within measurement accuracy, can be seen with results determined for the adsorption from the gaseous phase by Hertl and Hair (13) (Fig. 11). This concordance makes it evident that this correlation applies also for the adsorption of monofunctional adsorptives from solution, that the determined adsorption enthalpies are reliable and that the evaluation procedure for getting $\Delta H_{P,A}$ according to equ. 9 as well as the assumption of one desorbed CCl_4 molecule per one adsorbed segment is correct for the low mass material. All these adsorptives occupy an equal or somewhat smaller area than the CCl_4 molecules released from the surface. Under these conditions the molar immersion enthalpy of Aerosil surface groups in CCl_4 $\Delta H_{L,A}$ can be applied to the calculation of the net binding enthalpy $\Delta H_{P,A}$ (equ.9).

In contrast to the monofunctional substances the binding enthalpies $\Delta H_{P,A}$ for the polymers are distinctly below the correlation line (Fig. 11). The most plausible explanation for this fact is that during a polymer adsorption the flat adsorbing chain displaces more than one solvent molecule per adsorbed polymer segment. The measured adsorption enthalpy is then diminished by the contribution $\Delta H_{L,A}$, which is necessary for the displacement of additional solvent molecules.

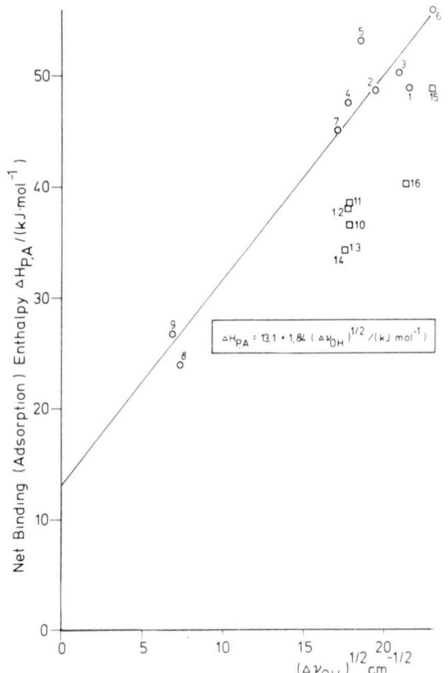

Fig. 11. Net binding (adsorption) enthalpy $\Delta H_{P,A}$ versus spectral frequency shift $\Delta\nu_{OH}$; ads. number refer to table 1.

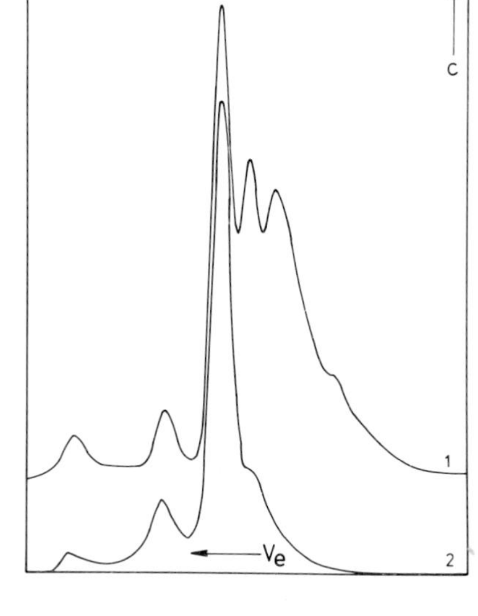

Fig. 12. GPC-Chromatograms of the initial (1) and of the supernatant (2) OVP 300 - solution after adsorption.

Reversibility and Preferential Adsorption

For the low mass substances a full reversible adsorption behaviour accompanied by a fast equilibration process has been proved by desorption measurements. In contrast to this behaviour the polymers investigated show irreversible adsorption. In particular the oligomer PBMA 2400 showed no desorbed portions even after extracting the dispersion in a soxhlet for 24 hours. Also with PVP K 90 no desorption was recognizable after treatment with $CHCl_3$ over several days.

Comparison of GPC-Chromatograms (Fig. 12), of the initial polymer solutions before adsorption and of the supernatant solution after adsorption of the oligomer OVP 300, demonstrates the variation of the molar mass distribution to lower molar mass fractions. Thus, preferential adsorption of the higher molar mass fraction is indicated.

REFERENCES

1. Scheutjens, J.M.H.M. and Fleer, G.J. (1980). *J.Phys.Chem.* 84, 178.
2. Barnett, K., Cosgrove, T., Crowley, T.L., Tadros, Th.F. and Vincent, B. (1982). *Proceedings Conf.* "Effect of Polymers on Dispersion Properties", Academic Press, London.
3. Korn, M., Killmann, E. and Eisenlauer, J. (1980). *J. Colloid Interface Sci.* 76, 7, 19
4. Bergmann, M. (1983). Ph.D. thesis, Technische Universität München.
5. Emmett, P.H. and Brunauer, S. (1937). *J.Am.Chem.Soc.* 59, 1553.
6. Fontana, B.J. and Thomas, J.R. (1961). *J.Phys.Chem.* 65, 480.
7. Stählin, W. (1976). Ph.D. thesis, Ludwig-Maximilians-Universität München.
8. Joppien, G.R. (1974). *Makromol. Chem.* 175, 1931.
9. Morrow, B.A. and Cody, J.A. (1973). *J.Phys.Chem.* 77, 1465.
10. Peri, J.B. and Hensley, A.L. (1968). *J.Phys.Chem.* 72, 2926.
11. Davidov, V.Y., Kiselev, A.V. and Lygin, V.J. (1963). *Kolloid. Zh.* 25, 152.
12. Eisenlauer, J., Killmann, E. and Korn, M. (1980). *J. Colloid. Interface Sci.* 74, 120.
13. Hertl, W. and Hair, M.L. (1968). *J. Phys. Chem.* 72, 4676.

MONTE CARLO STUDY OF THE CONFIGURATIONAL BEHAVIOUR OF ADSORBED POLYMER MOLECULES

A. Higuchi*, D. Rigby[+] and R.F.T. Stepto

Department of Polymer Science and Technology, The University of Manchester Institute of Science and Technology, Manchester, M60 1QD, UK.

*present address: *Department of Polymer Science, Tokyo Institute of Technology, O-Okayama Meguro-ku, Tokyo, Japan.*

[+]present address: *Department of Materials Science and Metallurgical Engineering, University of Cincinnati, Cincinnatti, Ohio, 45221, U.S.A.*

INTRODUCTION

The configurational behaviour of chain molecules interacting with a surface is fundamental to the understanding of the properties of adsorbed polymer layers. The behaviour is determined by the chain structure and by polymer-solvent and polymer-surface interactions. Many theoretical and computational investigations have been reported in the literature, and these have been reviewed recently by Dickinson and Lal(1). Most have focussed on the single chain and the use of computer simulations has enabled the effects of local interactions(2)(chain structure) and non-local interactions(2)(excluded volume) to be examined. The present paper is distinct from other computational investigations in that it uses an off-lattice chain model and continuous potential functions representing non-local (polymer)segment/segment and segment/surface interactions. In addition, chain structure is accounted for in a realistic fashion in that the Abe-Jernigan-Flory (AJF) rotational-isomeric-state (RIS) model of the unperturbed polymethylene chain(3) is used to define local interactions.

The use of off-lattice chains has the disadvantage that longer computations are required for convergence of the numerical results, compared with the computations required for lattice chains. Thus, only shorter chains can be studied using present (scalar) processors. In this paper, preliminary results are considered, mostly for chains up to 40 skeletal bonds in length together with a few results for chains of up to 80 skeletal bonds. Comparisons are made with earlier lattice results(4) and, in particular, the limitations of the conventional train-loop-tail model of an adsorbed chain are discussed.

CONTINUUM MODEL FOR POLYMER ADSORPTION

Chain Parameters

The parameters for the polymethylene chain according to the AJF RIS model(3) are bond length(ℓ) 0.153nm, valence-angle supplement(θ) 68°, and conformational angles(ϕ) for trans(t), gauche$_+$(g_+) and gauche$_-$(g_-) states of three skeletal bonds 0, 112.5°, and -112.5°, respectively. The energies associated with the t and $g_\pm$ states are 0 and ε_σ = 1674 J mol^{-1} (400 cal mol^{-1}), together with an extra energy ε_ω = 6276 J mol^{-1} (1500 cal mol^{-1}) for every pair of contiguous gauche states of opposite sign ($g_\pm g_\mp$). The model reproduces measured values of the unperturbed mean-square end-to-end distance ($\langle r_o^2 \rangle$) and its temperature coefficient ($d\ \ell n \langle r_o^2 \rangle / dT$) for long polymethylene chains.

Solution Parameters

The chain parameters characterise the local interactions arising from the chain structure; the solution parameters characterise the *effective* interactions between segments and solvent. For an off-lattice model a continuous potential function may be used which reflects such interactions more closely than the nearest-neighbour interactions between lattice sites most frequently used in lattice calculations (4).

The chosen potential function was the Sutherland, where

$$\Delta\varepsilon_{12} = \infty, \quad 0 \leqq r < r_o \tag{1}$$

and

$$\Delta\varepsilon_{12} = cr^{-6}, \quad r \geqq r_o . \tag{2}$$

The potential has the minimum of parameters consistent with dispersion-force attraction between non-polar groups and strong repulsive forces at small distances. It can reflect good and poor solvent conditions through the value of c, which can be positive or negative. The values of c used in the calculations were 0 and -0.65 J nm6 mol^{-1}, reflecting good (athermal) and poor solvent conditions with respect to the unperturbed (θ-)dimensions of the chains. r_o was given the value 0.27nm, as reflecting the effective 'hard-core' size of methylene groups subject to pairwise interactions(2).

The symbol $\Delta\varepsilon_{12}$ in equations(1) and (2) signifies the energy of pairs of volume elements (of segmental size) in the pure components at separation r relative to that in the solution for two segment/solvent pairs at the same separation. Thus,

$$\Delta\varepsilon_{12} = \varepsilon_{11} + \varepsilon_{22} - 2\varepsilon_{12}, \tag{3}$$

where ε_{11} and ε_{22} are the energies in pure solvent and pure polymer and ε_{12} is the energy when one of the solvent volume elements is interchanged with a segment. This representation of the segment/solvent interaction energy is a generalisation of that used in lattice models to characterise nearest-neighbour interactions(5). In the computations, the separations(r) of pairs of segments separated chemically by at least five skeletal bonds are noted and each pair is given an energy $\Delta\varepsilon_{12}(r)$, according to equations(1) and (2). The sum of these energies over all pairs gives the non-local interaction energy of a configuration.

Segment/Surface Potential Energy and Adsorption Criteria

The potential function used was a $(\text{distance})^{-3}$ one, representing the effects of dispersion-force interaction between a segment and an infinite plane. The surface was located at d = 0 and the energy parameter was $\Delta\varepsilon_s(d)$, where

$$\Delta\varepsilon_s = \infty, \quad 0 \leqq d \leqq d_o, \tag{4}$$

and

$$\Delta\varepsilon_s = c_a/d^3, \quad d \geqq d_o \,. \tag{5}$$

Negative values of c_a represent segment/surface attraction, and values equal to -40, -45, -50, -60, -75 and -100 J nm^3 mol^{-1} were used in the computations.

$\Delta\varepsilon_s$ is also a parameter measuring interactions in the presence of solvent. Thus, cf. equation(3),

$$\Delta\varepsilon_s = \varepsilon_{2S} - \varepsilon_{1S}, \tag{6}$$

where ε_{2S} and ε_{1S} are the potential energies of a segment and a solvent volume element at distance d from the surface.

The distance between the surface and a segment just touching it, d_o, was put equal to 0.27nm, as for r_o. However, as no segment can approach closer to the surface than

d_o, it is convenient to define distances

$$d_a = d - d_o \tag{7}$$

relative to the'hard-core' surface. These distances can be used to define adsorption criteria, such that segments at distances d_a less than a certain amount are counted as being adsorbed. The criteria can be chosen at will and represent a generalisation of the criterion used in lattice calculations(4) according to which a segment is adsorbed if it is on the surface.

A convenient way of defining adsorption criteria which relates to the potential function yet is independent of c_a is to use the parameter

$$y = \Delta\varepsilon_s(d)/\Delta\varepsilon_s(d_o) = d_o^3/d^3 \tag{8}$$

defining the fractional energy at a distance d on the basis of that at d_o. The significance of y is illustrated in figure 1, and the values of y used in the present work are listed. The largest value of y(=0.9) corresponds most closely to the criterion used in lattice calculations(4), and the use of a wide range of criteria allows investigation of the effects of the criterion used on the configurational properties deduced. Possible effects of changing the criterion in terms of the train-loop-tail model are illustrated in figure 2.

COMPUTATIONS AND COMPUTATIONAL METHOD

The numbers of skeletal bonds(n), and the values of the chain parameters(ℓ, θ, ϕ, ε_σ, ε_ω), solution parameters(r_o,c) and segment/surface parameters(d_o,c_a) used in the computations have been given in the preceding sections. The temperature was set at 298.2K.

A Metropolis sampling method(2,4) was used to generate an asymptotic equilibrium ensemble of configurations. An initial configuration was generated randomly with regard to its sequence of conformational angles and orientation of a reference bond to the surface. It was then placed at a random distance from the surface such that at least one segment fell within the largest adsorption criterion (value of d_a) being considered. Thus, the positions of the segments, groups of skeletal atoms, were not only defined in their own relative coordinate system (from the values of ϕ along the chain) but also in the external coordinate system, in which the surface was the plane z = 0.

The configurational energies (E_{config}) of the initial and subsequent configurations were evaluated as

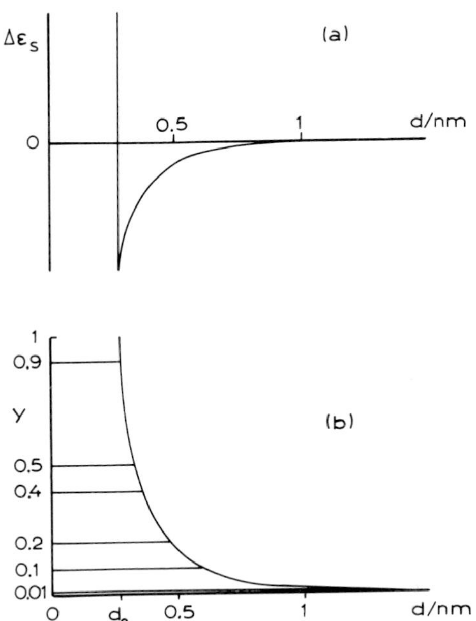

Fig. 1. Definition of adsorption criteria. (a) Segment/surface potential function according to equations (4) and (5) with arbitrary value of c_a and d_o = 0.27nm. (b) y of equation (8) plotted versus d, showing the criteria used in the present work, viz. y = 0.01, 0.1, 0.2, 0.4, 0.5 and 0.9 for which $d_a(=d-d_o)$ = 0.983, 0.312, 0.192, 0.096, 0.070 and 0.010nm, respectively. y = 1 when d = d_o.

$$E_{config} = E_{local} + E_{non\text{-}local} + E_{segment/surface}, \quad (9)$$

that is

$$E_{config} = \sum_{i=2}^{n-1} \varepsilon_{i,local}(\phi_i,\phi_{i+1}) + \sum_{j=1}^{n-4}\sum_{k=j+5}^{n+1} \Delta\varepsilon_{12}(r_{jk}) + \sum_{j=1}^{n+1} \Delta\varepsilon_s(d_j). \quad (10)$$

In equation (10), the sum over i is over skeletal bonds, which are numbered 1 to n, and the sums over j and k are over segments, which are numbered 1 to n + 1. Thus, bond p lies between segments p and p + 1. $\varepsilon_{i,local}$ is equal to 0 if $\phi_i = \phi_t$, ε_σ if $\phi_i = \phi_{g\pm}$ and $\phi_{i+1} \neq \phi_{g\mp}$, or $\varepsilon_\sigma + \varepsilon_\omega$ if $\phi_i = \phi_{g\pm}$ and $\phi_{i+1} = \phi_{g\mp}$. For the initial configuration,

$E_{non-local}$ and $E_{segment/surface}$ were finite, that is $r_{jk} \geq r_o$ for all j and k, and $d_j \geq d_o$ for all j. These conditions need not be met by subsequent configurations, which would then be excluded because of their infinite energy.

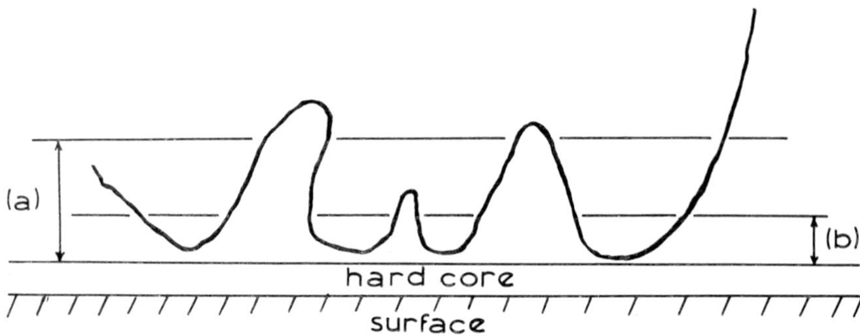

Fig. 2. Effects of criterion on train-loop-trail description of a configuration. Criterion(a): 3 trains, 2 loops, 1 tail. Criterion(b): 4 trains, 3 loops, 2 tails. $y_{(a)} < y_{(b)}$.

To generate a new configuration, the conformational angles of a randomly chosen sequence of v contiguous bonds were changed, with v equal to 3 or 4. These values were found to give the fastest convergence. A lower value gave a higher probability of a configuration penetrating the surface. For example, with v = 1 part of the molecule is rotated in a fixed internal configuration about a *single* skeletal bond. A higher value of v meant that too large a perturbation would be made between sequential configurations in the Metropolis sample(6). The difference in energy(ΔE) between the new and the existing configuration was then found, with

$$\Delta E = E_{config,\ new} - E_{config,old}. \tag{11}$$

If $\Delta E \leq 0$ then the new configuration was accepted and its characteristics added into the samples. If $\Delta E \geq 0$ then the new configuration was accepted and its characteristics added into the samples with probability $\exp(-\Delta E/RT)$. The procedure was continued until sufficiently constant answers for the various chain characteristics were obtained. The sample sizes required varied but were of at least 8 million configurations, with slower convergence for longer chain lengths, poorer solvent conditions and weaker adsorption.

Numerous characteristics of the adsorbed chains were evaluated, averaged over the samples of configurations generated. Those to be considered in the present paper are the distribution of distances of segments from the surface; the average lengths, $<L_{tr}>$, $<L_{\ell}>$, and $<L_e>$, in terms of numbers of segments, of trains, loops and tails; and the mean-square end-to-end distance ($<r^2>$) and the radius of gyration ($<s^2>$) of adsorbed chains in relation to their values in the absence of a surface, as found from separate calculations(7).

The samples generated were of adsorbed chains. Hence, different samples were obtained for the different adsorption criteria. For example, a configuration may have segments within the widest criterion considered (d_a = 0.983nm) but not within any of the others. That configuration would then only be considered for the sample for the widest criterion.

For weak adsorption (the smaller negative values of c_a) there was a tendency for the chain to drift away from the surface. This was avoided to some extent, without biasing the sample, by allowing the perturbation between configurations to be of 3 or 4 conformational angles, as already described, or by translation of the chain towards or away from the surface by a random amount less than some maximum value (0.1 to 0.5nm). The relative probability of the two types of perturbation (conformational and translational) in the configurational space of the whole system could be set *ab initio* by a parameter in the programme and was typically 0.5 : 0.5. Thus, some loss of computer runs was avoided since the net effect of allowing random translational shifts was to improve the efficiency of generating configurations near the surface, and hence possessing segments within the adsorption criteria.

RESULTS AND DISCUSSION

Segment Distribution Relative to Surface

Figure 3 shows the mean fractions of segments ($<f_a>$) within given distances from the surface for a chain of 21 segments in a poor solvent and subject to different strengths of adsorption. The distinct curves in figure 3 for different strengths of adsorption illustrate the range of segment distributions it is possible to describe with the present model. In general, $<f_a>$ increases more rapidly to unity the stronger the adsorption. In addition, other results show that, for a given strength of adsorption, $<f_a>$ increases to unity over larger distances as the chain length increases.

Figure 4 shows the mean fraction of segments adsorbed ($<f_a>$) versus strength of adsorption($-c_a$) for the adsorption

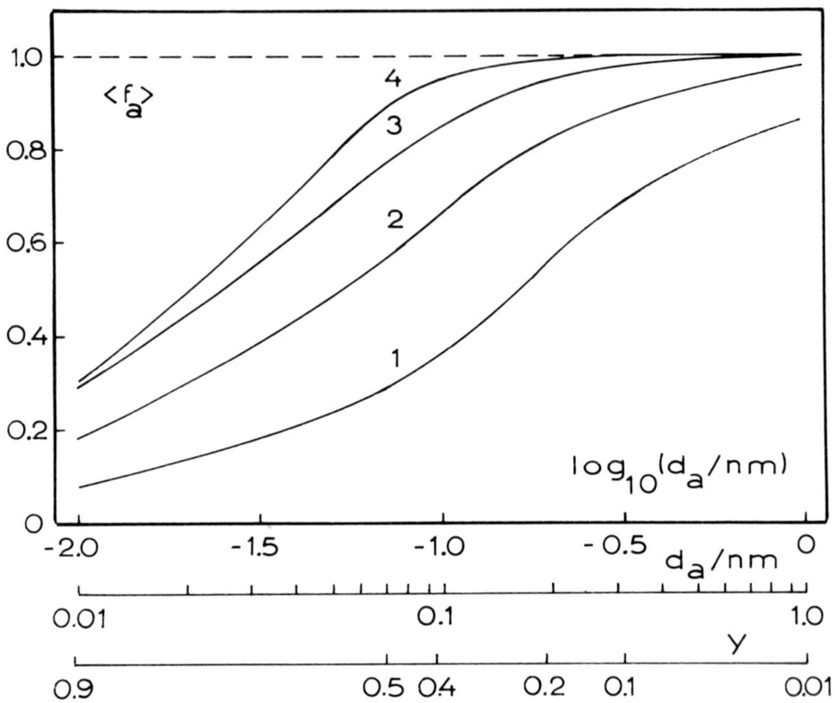

Fig. 3. Variation of mean fraction of segments ($<f_a>$) within distance d_a of 'hard-core' surface for a chain of 21 segments. Solution parameter $c/J\ nm^6\ mol^{-1} = -0.65$. Segment/surface parameter $c_a/J\ nm^3\ mol^{-1} = -50$ (curve 1), -60(curve 2), -75(curve 3) and - 100(curve 4).

criteria used. The differences between the curves illustrate the arbitrary nature of the criteria in their definition of boundaries between adsorbed and non-adsorbed segments. The curves have the same general shapes as those obtained from lattice calculations(4), when the average fraction adsorbed was plotted against $-\Delta\varepsilon_s/kT$, with $\Delta\varepsilon_s$ the effective energy of a segment on the surface. A distinction may be drawn with the lattice results, in that the present curve 1, which corresponds most closely to the lattice adsorption criterion, only increases to $<f_a> \cong 0.3$ under the strongest adsorption studied here. The lattice results, for 31- to 100 -segment chains were similar to curves 2-5. Obviously, the scale in $-c_a$ is different from that in $-\Delta\varepsilon_s/kT$ and larger values of $-c_a$ are required at y = 0.9 to cause

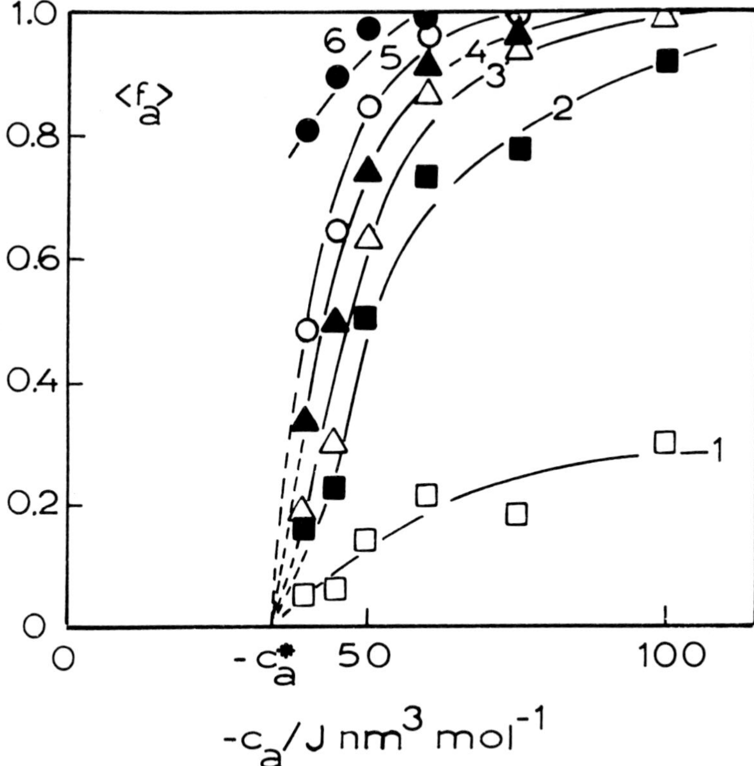

Fig. 4. Variation of the mean fraction of segments adsorbed ($\langle f_a \rangle$) with strength of adsorption ($-c_a$) for different adsorption criteria. Chain length 31 segements. Solution parameter c/J nm^6 mol^{-1} = -0.65. 1, y = 0.9(d_a = 0.010nm); 2, y = 0.5(d_a = 0.070nm); 3, y = 0.4(d_a = 0.096nm); 4, = y 0.2(d_a = 0.192nm); 5, y = 0.1(d_a = 0.312nm); 6, y = 0.01(d_a = 0.983nm). $-c_a^* \cong$ 35 J nm^3 mol^{-1} critical (minimum) parameter for adsorption.

$\langle f_a \rangle$ to increase to near unity. Of course, with the lattice calculations only one curve could result for a given chain length and solvent condition, as only one adsorption criterion was used.

An important use of the results for different criteria is the evaluation of $-c_a^*$, the critical adsorption-strength parameter which must be exceeded before adsorption can occur. The curves in figure 4 extrapolate to $-c_a \cong 35$ J nm^3 mol^{-1} at $\langle f_a \rangle = 0$. Thus, at smaller values of $-c_a$ the molecule will tend to move away from the surface because its loss of entropy (possible configurations) in being near the surface is not counteracted by the attraction of the surface. Dependence of $-c_a^*$ on chain length is expected.

Fig. 5 shows the mean thickness of adsorbed layer measured from the hard-core surface ($\langle d_a \rangle$) as a function of strength of adsorption ($-c_a$), for the various chain lengths, solvent conditions and criteria. The thickness decreases markedly as the strength increases. The curve is drawn through the points for n = 30 and poor solvent conditions (c/J nm^6 mol^{-1} = -0.65), with y = 0.5. From the data obtained there is no clear-cut dependence on chain length, although this needs further investigation, and the results are insensitive to solvent condition. Approximately the same curve results independent of the criterion used. Thus, the differences in samples for the different criteria, due to chains penetrating one criterion but not another, are not significant except perhaps under weak adsorption conditions.

Configurational Description in Terms of Train-Loop-Tail Model

Figure 6 shows the mean segment lengths of trains, loops and tails versus chain length for the relative case of strong adsorption (c_a = -100 J nm^3 mol^{-1}). The difference between the results for different criteria again illustrates the arbitrariness of using absorption criteria and also the arbitrariness of the train-loop-tail model. For example, the widest criterion shows train lengths approaching the full length of the chains, and the narrowest shows short trains which do not increase with chain length. Conversely, tails increase with chain length for the narrowest criterion and are shorter and insensitive to chain length for the widest criterion. Loops are longer than tails for the widest criterion, and the converse is true for the narrowest criterion. Results similar to those for the narrowest criterion were obtained from lattice calculations(4) under the weaker adsorption condition studied in that case ($\Delta\varepsilon_s/kT = -0.5$). Again, the present results are relatively insensitive to solvent conditions.

Size and Shape of the Adsorbed Molecule

Figure 7 shows the variation of $\langle r^2 \rangle/n$ and $\langle r^2 \rangle/\langle s^2 \rangle$ with chain length in comparison with the variation of these quantities in solution. Under both the stronger and the weaker adsorption the surface is seen to dominate the chain statistics. Differences due to solvent quality are within the errors of the computed points. The values of $\langle r^2 \rangle/n$ show that the molecules are more extended than in solution and

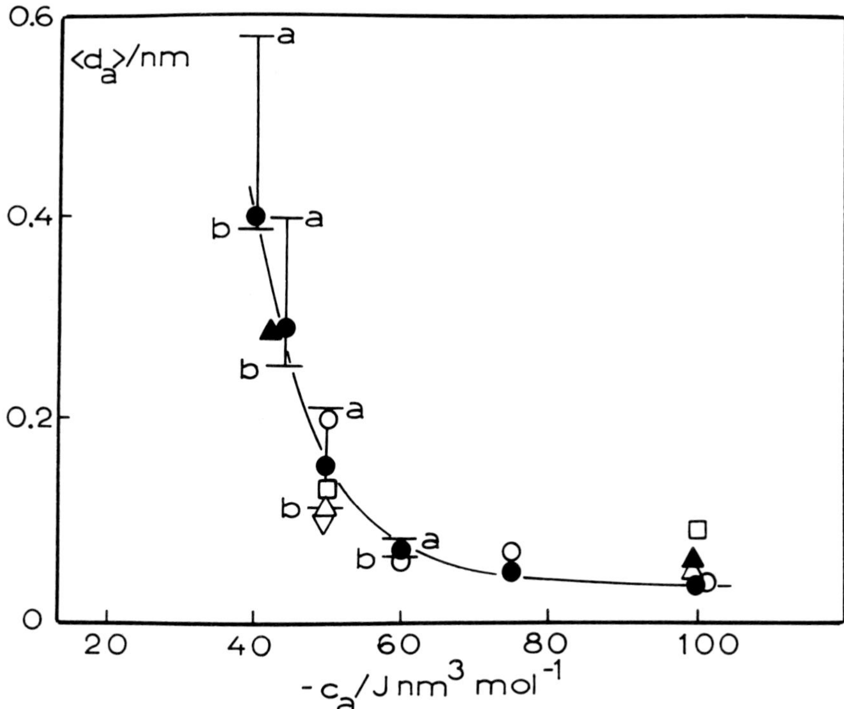

Fig. 5. Mean thickness of adsorbed layer ($\langle d_a \rangle$) versus strength of adsorption ($-c_a$). Open symbols, $c = 0$; closed symbols, $c = -0.65$J nm^6 mol^{-1}. △, ▲ ....$n = 10$; ○, ● .... $n = 30$; ▽ ....$n = 50$; □ ....$n = 80$. Criteria: a, $y = 0.01$; b, $y = 0.9$; points plotted for $y = 0.5$.

that the extension relative to that in solution increases with strength of adsorption and with chain length. This behaviour is confirmed by the values of $\langle r^2 \rangle / \langle s^2 \rangle$, remembering that this quantity is equal to 12 for a rigid rod, and for unperturbed, flexible chains of sufficient length it is equal to 6.

CONCLUSIONS

The present paper has explored the use of a realistic chain model and of continuous functions for segment/solvent and segment/surface interactions in describing the configurational behaviour of a chain in solution interacting with a

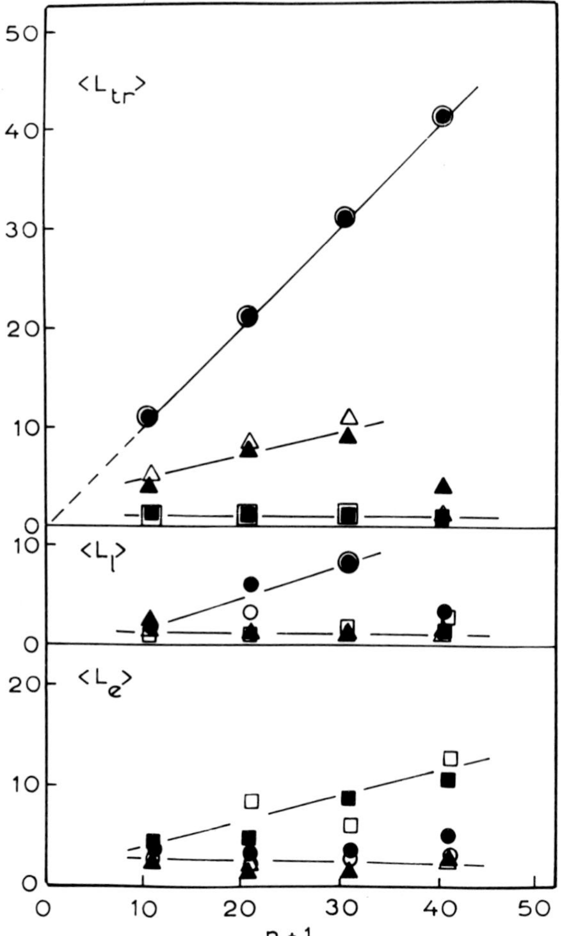

Fig. 6. Mean segment lengths of trains(<L_{tr}>), loops(<L_ℓ>) and tails(<L_e>) versus number of segments in chain(n+1) for strong adsorption (c_a/J nm^3 mol^{-1} = -100). Open symbols, c = 0; closed symbols, c = -0.65 J nm^6 mol^{-1}. ○,●... *y = 0.01;* △,▲... *y = 0.5;* □,■... *y = 0.9.*

surface. By incorporating several criteria for adsorption in the computations, it has been possible to show that the results obtained are generally consistent with previous ones from a lattice model(4) and to illustrate the arbitrary nature of such criteria, particularly with regard to the seemingly attractive train-loop-tail model of an adsorbed

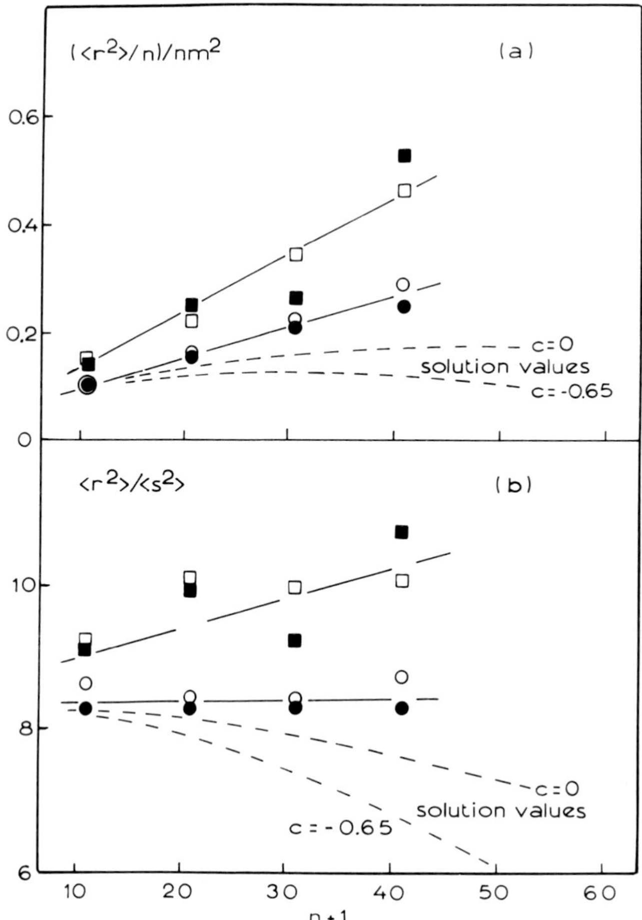

Fig. 7. (a) $\langle r^2\rangle/n$ *and (b)* $\langle r^2\rangle/\langle s^2\rangle$ *versus chain length. Open symbols, c = 0; closed symbols, c = −0.65 J nm*^{6} *mol*$^{-1}$. ○, ●... c_a/*J nm*^{3} *mol*$^{-1}$ = −50; □, ■... c_a/*J nm*^{3} *mol*$^{-1}$ = −100. --- *values in absence of surface(7).*

chain. Such criteria may have significance for systems with relatively strong and short-range interactions which yield relatively thin adsorbed layers, such as poly(ethylene oxide) and chromium surfaces(8), and may also be useful for correlating with data on adsorbed segments from different techniques(9).

The data presented are not comprehensive with regard to

range of chain lengths and strengths of adsorption. However, they indicate a domination of segment/surface interactions over segment/solvent interactions in determining the molecular configuration (fig.5 and 7). In addition, various average thicknesses of adsorbed layers can be calculated from the (cumulative) distributions of segments relative to the surface plane (fig. 3 and 5). Such distributions give the possibility of correlations with experimental data on adsorbed-layer thickness at low surface coverage (e.g. (8,10)), provided the appropriate chain model is used. Finally, the use of adsorption criteria in determining a critical segment/surface attraction for adsorption has been demonstrated (fig. 4).

ACKNOWLEDGEMENTS

A.H. gratefully acknowledges support under the Tokyo Institute of Technology/University of Manchester Institute of Science and Technology Exchange Scheme, and D.R. the award of an S.E.R.C. Research Assistantship.

REFERENCES

1. Dickinson, E. and Lal M. (1980). *Adv. in Molecular Relaxation and Interaction Processes* 17, 1-87.
2. Winnik, M.A., Rigby, D., Stepto, R.F.T. and Lemaire, B. (1980). *Macromolecules* 13, 699-704.
3. Flory, P.J. (1969). "Statistical Mechanics of Chain Molecules". Chap. III. Interscience Publishers, London.
4. Lal, M. and Stepto, R.F.T. (1977). *J.Polymer Sci., Polymer Symposia* 61, 401-412.
5. Flory, P.J. (1953). "Principles of Polymer Chemistry". Chap. XII. Cornell University Press, London.
6. Wood, W.W. (1968). *In* "Physics of Simple Liquids" (Eds H.N.V. Temperley, J.S. Rowlinson and G.S. Rushbrooke). pp.117-230. North Holland Publishing Co., Amsterdam.
7. Rigby, D. (1982). Ph.D. Thesis, University of Manchester, U.K.
8. Killmann, E. and von Kuzenko, M. (1974). *Angew. Makromol. Chem.* 35, 39-55.
9. Day, J.C. and Robb, I.D. (1980). *Polymer* 21, 408-412.
10. Stromberg, R.R. and Smith, L.E. (1967). *J.Physic.Chem.* 71, 2470-2474.

AN EXPERIMENTAL STUDY OF POLYMER CONFORMATIONS AT THE SOLID/SOLUTION INTERFACE

Terence Cosgrove, Trevor L. Crowley and Brian Vincent

Department of Physical Chemistry
University of Bristol
Bristol BS8 1TS, U.K.

INTRODUCTION

Knowledge of the conformation of various types of macromolecules at the solid/solution interface is useful in the interpretation of several important practical phenomena, for example, the flocculation and stabilisation of particulate dispersions (1). In this paper, we report studies on the conformational behaviour of a linear, homopolymer, poly (ethylene oxide)(PEO), in the one case physisorbed and in the other terminally-bonded, to polystyrene latex particles dispersed in water. The stability behaviour of polystyrene latices carrying PEO chains in these two types of basic configuration has recently been contrasted by Cowell and Vincent (2). The adsorption isotherms (25°C) have been obtained for the homopolymer samples used. The two main techniques that have been used to probe the conformation of the chains in each case are (i) pulsed ^{1}H n.m.r., and (ii) small angle neutron scattering (s.a.n.s.).

N.m.r. relaxation times (3,4) reflect both molecular structure and motion. With polymers the spin-spin relaxation time (T_2) may be related to the local segment mobility. In principle, for an adsorbed polymer different values of T_2 should be obtained for "bound" segments (in trains) and for "free" segments (in loops or tails). Barnett and Cosgrove (3) have developed a multipulse technique to enable segments in these two states to be differentiated, and which, at the same time, eliminates contributions from any free protons associated with the solvent (H_2O in D_2O). The ratio of the segments in each state and, hence, <p>, the fraction of bound segments, may be calculated using this method.

Neutrons are scattered by atomic nuclei and the intensity of scattering at a given angle depends on the distribution of scattering centres within the sample. Hydrogen (^1H) and deuterium (^2D) atoms scatter to different extents so that by isotopic substitution the scattering from different regions of the sample may be highlighted. For studying polymers adsorbed onto colloidal particles in aqueous dispersions, one may, in this way, study only the scattering from the adsorbed polymer molecules by "contrast matching" of the core particles with the solvent (5-8). This is achieved by adjusting the D_2O/H_2O ratio of the solvent until the scattering from the (deuterated) bare particles is reduced to zero.

Under contrast match conditions, the intensity of scattering, I(Q), from n independent scattering particles, is given by,

$$I(Q) = nF(Q)^2 \tag{1}$$

where F(Q) is the form factor for the adsorbed layer and $Q = (4\pi/\lambda)\sin(\theta/2)$, where θ is the scattering angle and λ is the neutron wavelength. For particles of radius a, carrying an adsorbed polymer layer of thickness δ, it may be shown (8) that,

$$F(Q)^2 = \frac{8\pi a^2}{Q^2}\,|\hat{\rho}(Q)|^2 \tag{2}$$

where
$$\hat{\rho}(Q) = \int_0^\delta \rho(z)e^{iQz}\,dz \tag{3}$$

Thus, in order to obtain $\rho(z)$, the normalised segment density distribution normal to the interface, we require $\hat{\rho}(Q)$; this is related to the experimentally determinable quantity $|\hat{\rho}(Q)|$ through an undetermined phase factor (7,8). Crowley has, however, recently discussed (8) how $\rho(z)$ may be generated from I(Q) data at contrast match using a dispersion integral technique.

If $Q\delta < 1$, then the exponential function in eqn.(3) may be expanded in a Taylor series. This leads to the expression,

$$I(Q) = \frac{8\pi^2 a^2 m^2}{Q^2}\exp - \sigma^2 Q^2 \tag{4}$$

where m is proportional to the total number of adsorbed segments per unit area, Γ, and σ^2 is the second moment of the segment density distribution, $\rho(z)$, i.e.

$$\sigma^2 = \int_0^\delta \rho(z)\, z^2\, dz - \left(\int_0^\delta \rho(z)\, z\, dz \right)^2 \tag{5}$$

$$= \langle z^2 \rangle - \langle z \rangle^2$$

We may also relate $\langle p \rangle$ to $\rho(z)$, if we assume that segments in <u>trains</u> extend a thickness t from the surface, i.e.

$$\langle p \rangle = \int_0^t \rho(z)\, dz \tag{6}$$

In an earlier paper (6) we also showed how $\rho(z)$ may be obtained by measuring I(Q) at "slightly-off" contrast match conditions, where (8),

$$I(Q) = I_0(Q) + I_1(Q)\Delta\rho + I_2(Q)\Delta\rho^2 \tag{7}$$

I_0 is now the adsorbed layer contribution, I_2 the contribution from the particle cores, and I_1 the core/adsorbed layer interference term. $\Delta\rho$ is the difference in neutron scattering density between the particle cores and the solvent. The form of $I_1(Q)$ is given by (8),

$$I_1(Q) = \frac{16\pi^2 am}{Q^4} \left[\int_0^\delta \rho(z)\cos(Qz)dz - Qa \int_0^\delta \rho(z)\sin(Qz)dz \right] \tag{8}$$

Thus, the sine transform of $Q^3 I_1(Q)$ gives $\rho(z)$ if $Qa \gg 1$. This earlier method is much less accurate since it involves a greater number of subtractions of I(Q) sets of data, but does, as will be seen , give similar shapes for $\rho(z)$ to the more direct transform of the $I_0(Q)$ data, working at contrast match conditions.

Using these two relatively new experimental techniques we are able to determine, therefore, $\langle p \rangle$, $\rho(z)$ and σ for the two configurations of adsorbed polymer mentioned earlier.

MATERIALS AND EXPERIMENTAL TECHNIQUES

The techniques used for the preparation of the two (fully protonated and 96% by wt. deuterated) polystyrene latices (H-PS and D-PS, respectively) used in this work have been described elsewhere (7), as has that for the preparation of the PEO grafted latex (D-PS-PEO)(7). M_w for the tails was 4,800 (M_w/M_n = 1.3). The following particle diameters were obtained from electron microscopy: H-PS latex 237 ±6 nm; D-PS latex 214 ±6 nm; D-PS-PEO latex, 100 ±5 nm.

The PEO samples used were: a 50,000 molecular weight sample (M_w/M_n = 1.23) prepared (9) by anionic polymerisation of ethylene oxide in a sealed ampoule, using sodium methoxide as initiator; (ii) a 40,000 molecular weight sample (M_w/M_n = 1.03) supplied by Polymer Laboratories (Shrewsbury, Salop.)

The n.m.r. experiments were performed using a 60 MHz computer-controlled pulsed n.m.r. spectrometer, the spectra being averaged using a PDP 11 minicomputer (3).

The s.a.n.s. experiments were carried out at the Institute Laue-Langevin (Grenoble) using both the D.11 and the D.17 camera facilities, set at a suitable range of detector distances and neutron wavelengths.

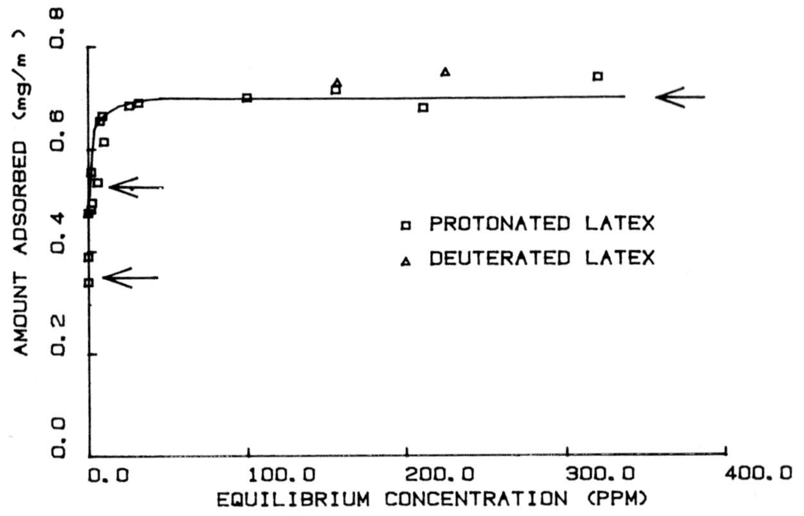

Fig. 1. Adsorption isotherm PEO/PS latex.

The adsorption isotherm for homopolymer PEO (M = 50,000) onto H-PS latex is shown in fig.1 (10). Also shown are several plateau points for the M = 40,000 used in most of the studies reported here. There is no apparent difference.

within experimental error. Also marked by arrows, are the three coverages studied in the n.m.r. and s.a.n.s. experiments.

TABLE 1

Adsorption Parameters For PEO (M = 40,000) Onto PS Latex

θ	<p>	σ (nm)	δ_h (nm)
0.5	0.2 ±0.1	1.4 ±0.1	-
0.75	0.12 ±0.05	1.5 ±0.1	-
1.0	0.12 ±0.05	1.7 ±0.1	19 ±3

The values of <p> obtained from the n.m.r. experiments for the adsorbed homopolymer PEO (M = 40,000) on D-PS latex are shown at the various coverages studied in table 1. The percentage errors involved become very large as θ decreases, and no meaningful value could be obtained for $\theta = 0.25$. The reasons for this are twofold. Firstly, the deuterated latex used contained ∿4% protonated styrene (in order to achieve contrast match conditions in the s.a.n.s. experiments) and this gave a residual ^1H signal similar in magnitude to that from the adsorbed PEO. Hence, <p> was obtained by the subtraction of two large numbers. Secondly, the maximum number of scans used was 3500 (ca. 13 hr) and the latex concentration was only 6.5%. Realistically, the maximum experimental time cannot exceed ∿48 hr; this would give an improvement of ∿2 in signal-to-noise. The use of a perdeuterated latex together with a higher solids content make it possible, however, to obtain spectra down to $\theta = 0.1$.

<p> values were obtained from the first solid echo and the last liquid echo by integration of the intensities between the radio frequency pulses as described in an earlier publication (3).

Also shown in table 1 are the σ values obtained from the s.a.n.s. experiments. A typical example of an experimental I(Q) plot is shown in fig.2 after background subtraction, ($\theta = 1$); σ values were obtained from the derived ln $Q^2I(Q)$ versus Q plots. The I(Q) plots for $\theta = 0.75$ and 0.5 were very similar to that shown in fig.2.

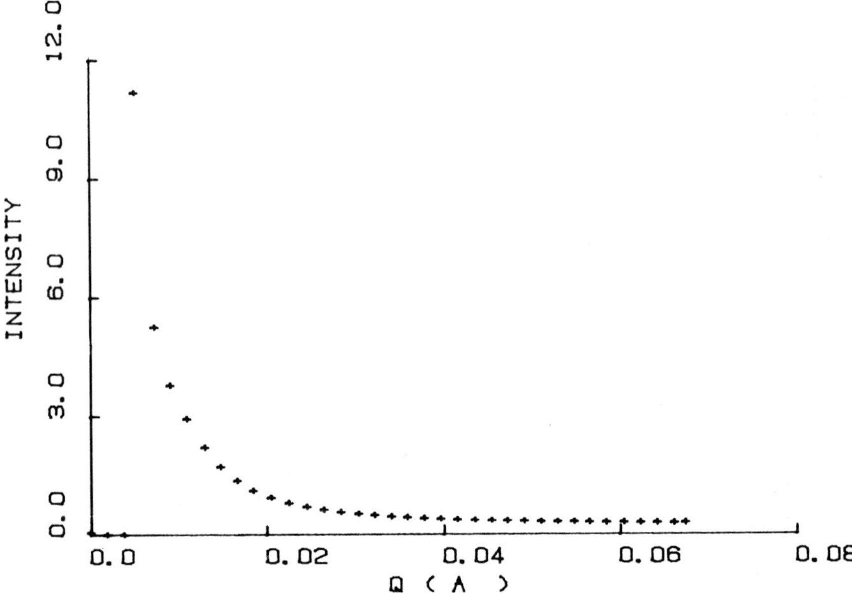

Fig. 2. I(Q) vs Q for PEO adsorbed on PS latex θ = *1*

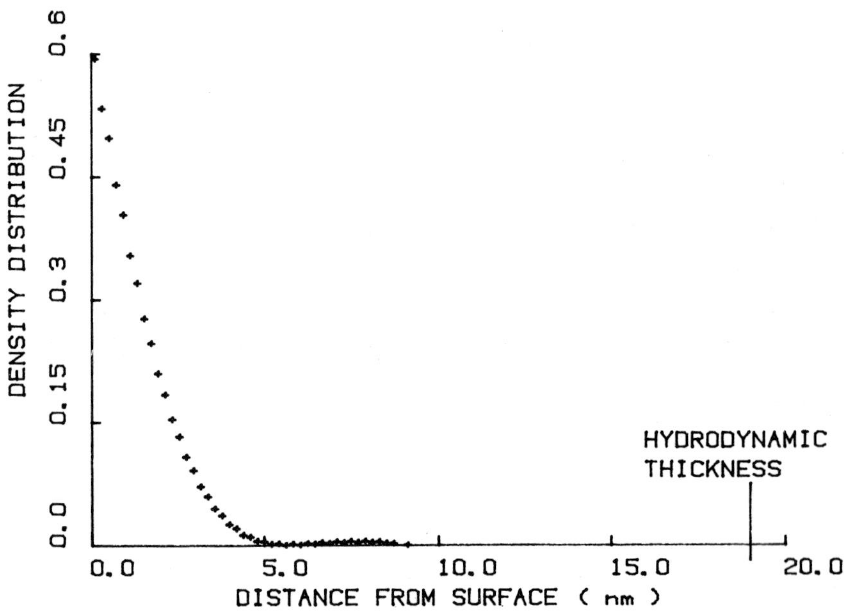

Fig. 3. ρ(z) *for PEO on PS latex* θ = *1*

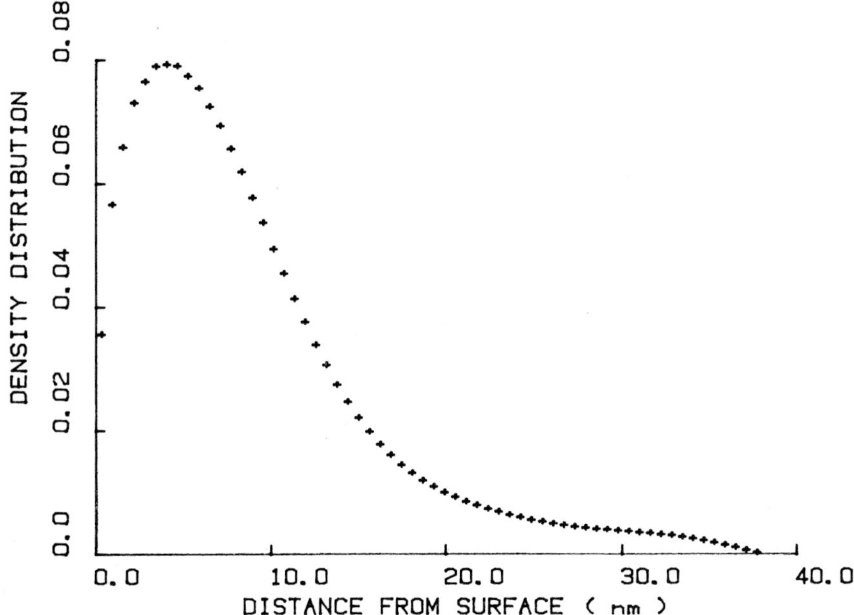

Fig. 4. ρ(z) for PEO tails on PS latex

The normalised ρ(z) versus z plot (θ = 1.0) derived from the I(Q) curve (fig.2)(contrast match condition) is shown in fig.3. Also shown in this figure is the value of the hydrodynamic thickness, δ_h, obtained from photon correlation light scattering; the value quoted is the average of values obtained by Turner (11) found for the same M = 50,000 sample used for the isotherm, and a more recent value obtained in this laboratory for the M = 40,000 sample. The value quoted by Kato *et al.* (12), δ_h = 20 ±2, for PEO (M = 46,000) adsorbed on a Dow polystyrene latex agrees well with this.

The normalised ρ(z) versus z plot for the PS-PEO latex (tails) is shown in fig.4. Unfortunately, a δ_h value could not be obtained in this case, since there is no equivalent of a "bare" latex in this case.

DISCUSSION

Tails

Cowell and Vincent (2) have shown, for a similar PS-PEO latex (but with M = 1,600 rather than M = 4,800), that the PEO tails are closely packed on the surface, with the chains extending out from the surface. This picture is confirmed by the form and extent of the $\rho(z)$ plot shown in fig.4. The form of the plot, with a maximum in $\rho(z)$, is similar to that predicted theoretically by Hesselink (13) and de Gennes (14) for tails. A fully-extended PEO chain having M = 4,800 would have a thickness of ∿30 nm; that a small number of segments are apparently observed at distances up to ∿38 nm in fig.4 probably reflects the somewhat polydisperse nature of this polymer, i.e. one is seeing the presence of chains of greater molecular weight.

Homopolymer at Full Coverage

From the plateau value of the adsorbed amount shown in fig.1 (0.7 mg m^{-2}), the area per adsorbed PEO molecule, A, is found to be ∿100 nm^2. The radius of gyration, $<s^2>^{\frac{1}{2}}$, of PEO 40,000 in water may be interpolated from the data of Vincent *et al.* (15) to be 8.3 nm, or from that of Kato *et al.* to be 9.0 nm. If the chains were adsorbed in their solution, random coil conformation, this would suggest a projected surface area per molecule (taken to be $\pi <s^2>$) of ∿200-250 nm^2, i.e. significantly larger than the A value derived from the adsorption isotherm. This implies that the chains are either "overlapping" somewhat on the surface or there is some extension (probably of the two end tails) normal to the surface. The latter explanation seems more likely. Lateral interpenetration would require significant osmotic work (12). It would appear from fig.3 that most of the segments are contained within a distance of 4-5 nm from the surface. Moreover, the form of this part of $\rho(z)$ is roughly exponential; this would agree with Hesselink's prediction (13) that $\rho(z)$ for a series of adsorbed loops/trains (but no tails) should be exponential in form. Clearly, however, the δ_h value (19 ±3 nm) implies that there must be some segments extending out much further than 4-5 nm. Indeed there is some evidence for this too from fig.3 itself, but here ρ is clearly very small (∿0) and the concentration of these segments must be very low. It is of interest in this context to examine the $\rho(z)$ plot previously reported (7) for PVA, i.e. poly (vinyl alcohol-co-acetate)(88% alcohol, 12% acetate), M = 37,000, adsorbed onto similar polystyrene latex particles. This is shown in fig.5.

It is significant to note that δ_h is of similar magnitude, but that, again, most of the segments lie within 4-5 nm of the surface. In this case, however, there is clear evidence from the s.a.n.s. data for segments in the region z = 5-20 nm. The <p> value obtained for PVA has not unfortunately been obtained for PVA, M = 37,000, at full coverage, but was found

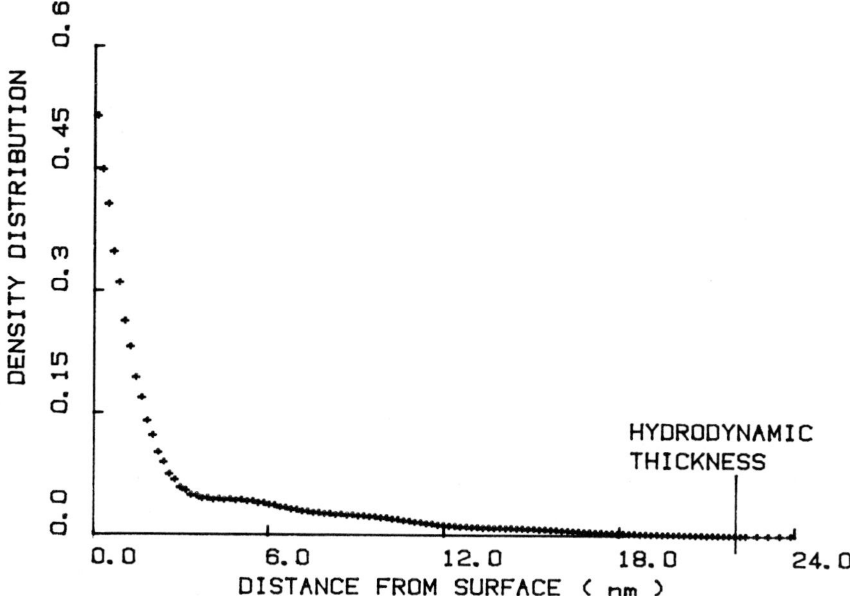

Fig. 5. ρ(z) for PVA on PS latex θ = 1.

to be 0.06 ±0.01 for PVA, M = 11,500 (6). This is somewhat less than that for PEO (table 1). However, the picture that emerges for both polymers (at θ = 1) is of a conformation of the type depicted schematically in fig.6, that is, most of the segments contained in loops or trains, with long tails extending into the solution. If it is assumed that it is the

Fig. 6. Schematic presentation of PVA/PEO on PS latex.

acetate groups (or more likely the short runs of acetate groups thought to be present in the PVA copolymer) which are the ones which are adsorbed in trains, then one might expect a $\langle p \rangle$ value closer to 0.12 if there were no tails. (It is of interest to note here that $\langle p \rangle$ values of this order of magnitude are obtained at low coverages, i.e. with isolated adsorbed chains (6).) The much smaller area per molecule value obtained for PVA (i.e. a $\sim$30 nm^2, based on the adsorption isotherm reported by Garvey *et al.* (17)), compared to A $\sim$100 nm^2 for PEO, may simply imply that the tails are longer in the case of PVA, i.e. there are fewer loops, but that these larger tails are somewhat "folded-back"; this would be consistent with the higher segment densities observed in the "outer" region of the adsorbed layer for PVA (fig.5) compared to PEO (fig.3).

The conformation suggested in fig.6 is consistent with the recent theoretical predictions of Scheutjens and Fleer (16), for $\theta = 1$.

Effect of Coverage

The effect of coverage on the $\langle p \rangle$ and σ values are shown in table 1. The decrease in $\langle p \rangle$ but increase in σ with increasing coverage are both consistent with theoretical predictions (16). Moreover, these trends are consistent with some results reported previously by us (10) for PEO adsorbed on PS latices. These earlier results were of a preliminary nature, however, and the s.a.n.s. data in particular were of a much inferior quality (much shorter data acquisition times being used to obtain the I(Q) plots).

The decrease in $\langle p \rangle$ with increasing coverage is also in agreement with the n.m.r. (4) and e.s.r. (18) data obtained for poly(vinyl pyrrolidone) adsorbed onto silica particles from aqueous solution. In this case, however, much higher $\langle p \rangle$ values (in the range 0.5 $\rightarrow$ 0.8) were obtained, reflecting, presumably, the much higher net segment adsorption energy in this case compared to PVA or PEO on polystyrene latex.

ACKNOWLEDGEMENTS

The authors wish to thank Dr. Th.F. Tadros of I.C.I. Plant Protection Division and Mr. K.G. Barnett of I.C.I. Corporate Laboratory for many helpful discussions during the course of this work. They are also grateful to the managements of I.C.I. Plant Protection Division and I.C.I. Corporate Laboratory, and to the Science and Engineering Research Council for financial support. The Institute Laue-

Langevin is thanked for providing s.a.n.s. facilities and Dr. D. Cebula for technical assistance at the I.L.L.

REFERENCES

1. Vincent, B. and Whittington, S.G. (1982). *In* "Surface and Colloid Science" (Ed. E. Matijević). Vol. 12, p. 1. Plenum Press, New York.
2. Cowell, C. and Vincent, B. (1982). *In* "The Effect of Polymers on Dispersion Properties" (Ed. Th.F. Tadros). p.263. Academic Press, London.
3. Cosgrove, T. and Barnett, K.G. (1981). *J. Magnetic Resonance* 43, 15.
4. Barnett, K.G., Cosgrove, T., Vincent, B., Sissons, D.S. and Cohen-Stuart, M. (1981). *Macromolecules* 14, 1018.
5. Cebula, D., Thomas, R.K., Harris, N.M., Tabony, J. and White, J.W. (1978). *Faraday Discuss. Chem. Soc.* 65, 76.
6. Barnett, K.G., Cosgrove, T., Crowley, T.L., Tadros, Th.F. and Vincent, B. (1982). *In* "The Effect of Polymers on Dispersion Properties" (Ed. Th.F. Tadros). p.183. Academic Press, London.
7. Cosgrove, T., Crowley, T.L., Vincent, B., Barnett, K.G. and Tadros, Th.F. (1982). *Faraday Symp. Chem. Soc.* 16, in press.
8. Crowley, T.L. (1982). D.Phil. thesis, Oxford, to be published.
9. Barnett, K.G. (1982). Ph.D. thesis, Bristol.
10. Barnett, K.G. *et al.* (1981). *Polymer Communications* 22, 283.
11. Turner, J.D. (1981). Ph.D. thesis, Manchester.
12. Kato, T., Nakamura, K., Kawaguchi, M. and Tabahashi, A. (1981). *Polymer I* 13, 1037.
13. Hesselink, F.Th. (1971). *J. Phys. Chem.* 75, 65.
14. de Gennes, P.G. (1980). *Macromolecules* 13, 1069.
15. Vincent, B., Luckham, P.F. and Waite, F.A. (1980). *J. Colloid Interface Sci.* 73, 508.
16. Scheutjens, J.M.H.M. and Fleer, G.J. (1982). *In* "The Effect of Polymers on Dispersion Properties" (Ed. Th.F. Tadros) p.145. Academic Press, London.
17. Garvey, M.J., Tadros, Th.F. and Vincent, B. (1974). *J. Colloid Interface Sci.* 49, 57.
18. Robb, I.D. and Smith R. (1974). *Eur. Polymer J.* 10, 1005.

THE ADSORPTION OF POLYMERS ONTO MICA: DIRECT MEASUREMENTS USING MICROBALANCE AND REFRACTIVE INDEX TECHNIQUES

H. Terashima,[a] J. Klein[b] and P.F. Luckham

Cavendish Laboratory, Madingley Road, Cambridge CB3 OHE, U.K.

1. INTRODUCTION

The forces acting between two solid surfaces in a liquid medium may be considerably modified by the presence of adsorbed polymer layers on the surfaces (1); this effect is widely used in controlling the stability of colloidal dispersions (2). Recently an optical technique was used to measure directly the force F(D) acting between layers of polystyrene adsorbed onto mica surfaces a distance D apart, immersed in cyclohexane, in the range $0 < D < 300$ nm (1). The technique is also capable of measuring the mean refractive index n(D) of the medium separating the mica surfaces; this yields information on the concentration of polymer in the gap separating the surfaces, and hence on the surface excess or *adsorbance* of polymer.

Muscovite mica is a particularly suitable surface for such studies, as it may be cleaved to give large, molecularly smooth sheets which permit approach of the surfaces down to molecular contact. It is clearly desirable, however, to have in addition an *independent* means of determining the adsorption of polymer onto plane mica. This would allow, i) a check that the adsorbance values obtained by the optical method were correct, ii) the possibility of pre-selecting polymer/solvent systems for the force

(a) Presently at Institute of Applied Physics, Tsukuba University, Japan.
(b) Also Polymer Department, Weizmann Institute of Science, Rehovot, Israel.

measurement experiments which had the desired adsorption characteristics, iii) independent monitoring of the effect of surface treatment (of the mica) on the adsorption of polymer from solution.

A number of techniques are available for measuring polymer adsorption from solution onto plane surfaces (for systems of low surface to volume ratio). Ellipsometry requires optically flat surfaces which are also highly reflecting (3) (mica is neither, though it is very smooth); attenuated total internal reflection and surface reflection techniques (4,5) also require optically flat surfaces. Radiolabelling (6) of adsorbing species is in principle suitable for mica but, in addition to the possibility of different adsorbing behaviour of the isotopically labelled adsorbants, there may be considerable difficulty in obtaining the required labelled polymers.

We have developed a novel approach suitable for measuring the adsorbance of polymer onto mica from solution, based on directly weighing the amount of adsorbed polymer, using a microbalance technique. To our knowledge this is the first study of polymer adsorption onto mica surfaces. This paper describes the determination of the kinetics of adsorption, and the adsorption isotherms of polystyrene adsorbing onto mica from cyclohexane at the Θ-temperature, using the microbalance techniques. The kinetics of *desorption* are also measured. The results are critically compared with adsorbance values for the same system evaluated from refractive index measurements using the optical approach.

2. MATERIALS AND EXPERIMENTAL METHODS.

a) *Materials:*

The solvent in the refractive index experiments was Fluka, Spectroscopic grade, used as received save for filtering through 0.22μm Fluoropore. In the microbalance experiments, BDH-Analar grade cyclohexane was dried over sodium and redistilled prior to use.

The characteristics of polymers used are summarised below.

Polystyrene:	M_w	M_w/M_n
PS1 (Pressure Chemicals)	5.99×10^5	1.10
PS3 (Toyo Soda)	1.07×10^5	1.05
PS4 (" ")	4.22×10^5	1.06

Polybutadiene:	M_w	M_w/M_n
LF4	9.60×10^4	1.06

The polybutadiene (PBD) sample was kindly synthesised, characterised and donated by L.J. Fetters . It had a microstructure of 92% 1,4 - and 8% 1,2-PBD.

The mica used throughout was Best Indian Ruby mica, FS/GS quality, grade 2, supplied by Mica and Micanite Supplies Ltd. (U.K).

b) *Microbalance Method.*

A fused quartz torsion microbalance was constructed for the purpose of the present study; its basic features are illustrated in fig. 1. A detailed description will appear elsewhere; a similar but more elaborate microbalance, for use in vacuum, has been described previously (7).

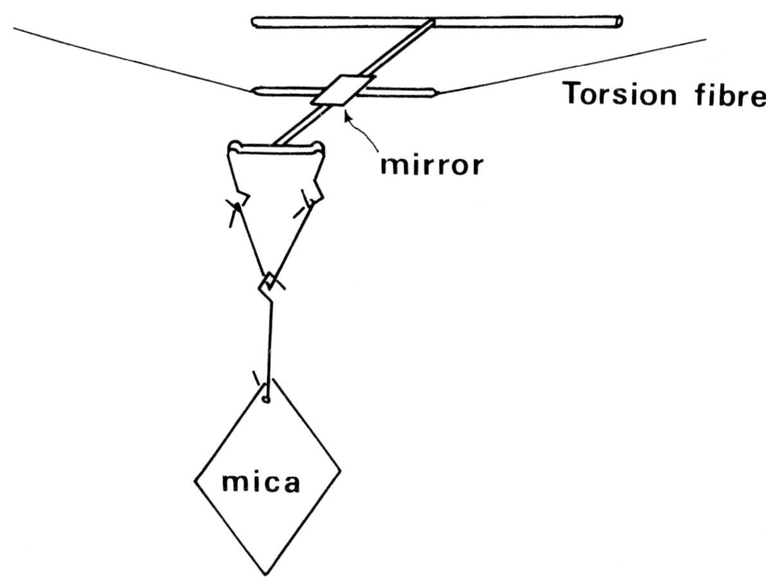

Fig. 1. Schematic illustration of the main features of the microbalance.

Torsional motion of the quartz fibre is monitored by observing the position of a hairline image, projected by a lamp and reflected onto a scale by a small mirror (fig. 1). The microbalance has a capacity of some 100mg, and its calibrated sensitivity is $(5.16 \pm 0.03)\times10^{-4}$mg per 1 mm deflection of the hairline. This is also approximately the resolution, since there is an uncertainty of about 1 mm in the hairline position on the scale. The mica sheets suspended on the balance (fig.1) are some 4.5x4.5 cm^2 in size, thus enabling a detection resolution in the amount of adsorbed polymer of some 0.2 $mg.m^{-2}$.

The experimental procedure is as follows: a thick sheet (∿ 200μm) of mica is cut to shape (∿ 4.5x4.5 cm^2) and thinner sheets (∿ 30μm) are cleaved from it. A freshly cleaved sheet is rinsed for a few seconds in pure cyclohexane, then withdrawn and dried. The mica is then hung on the microbalance (fig. 1) and the hairline position is noted: this is the reference position for zero adsorbance of polymer. The mica sheet is now immersed in a beaker (∿ 800 ml) of the polymer solution so that adsorption may take place. The incubating solution is maintained at constant temperature by a water bath, and is gently stirred by a vertically moving vane so that the mica sheets in solution are in constant motion throughout.

To determine *adsorption kinetics* the mica sheets were withdrawn after times t, and rinsed *once* by immersion (1-2 seconds) in a large excess of solvent. Sheets were then dried and weighed to give the adsorbance $\Gamma(t)$. Sheets once dried were not used again. The above procedure was followed for all adsorption kinetics experiments.

Desorption was monitored by immersing mica sheets, subsequent to prolonged incubation (∿ 17 hours) in solution, in a large excess of (stirred) pure solvent, and withdrawing and weighing at times t' to determine $\Gamma(t')$.

Adsorption *isotherms* were determined in two different ways: in both cases, the mica sheets were immersed for ∿17 hours in the solution at concentration c. They were then either rinsed *once*, as for the kinetics study, or immersed for a prolonged period (∿ 17 hours) in a large excess of stirred, pure solvent; $\Gamma(c)$ was then determined by weighing, in the two cases.

c) *Refractive Index Measurements.*

The optical method used for measuring the thickness D and refractive index n(D) of the medium separating two thin sheets of mica is described elsewhere (1,8). The experimental procedure is as follows: the mica surfaces are immersed in the pure solvent (cyclohexane in these experiments), and n(D) is determined in the range $0 < d \lesssim 300$ nm. Polymer is then introduced into the solvent to the required concentration, and the surfaces are left to incubate for 10 hours to allow adsorption from the solution to take place. After adsorption n(D) is again determined, following which the solution is almost entirely replaced by pure solvent; this results in an approximately hundred-fold dilution of the original concentration. n(D) is then measured again, over a period of up to 48 hours. The mica surfaces used in these studies are exposed to laboratory atmosphere for a total of about 1 hour between the original cleaving of the mica and initial immersion in the pure solvent.

3. RESULTS

a) *Microbalance Experiments.*

Several control measurements were carried out prior to the main study, in air and in polymer-free cyclohexane. The purpose of these was to ensure reproducibility of the zero position of the adsorption measurements - i.e., the scale-position of the reflected hairline image following the initial rinsing of the freshly cleaved mica sheets. These measurements showed that freshly cleaved mica would gain weight (up to $\sim$ 1 $mg.m^{-2}$) on prolonged standing (several hours) in air, but would always return to a fixed weight following immersion in cyclohexane, withdrawal and drying. This weight was reproducible to ± 2mm on the scale - corresponding to approximately ± 0.3 $mg.m^{-2}$ of mica surface - even after several cycles of rinsing, drying and standing, over a total period of a day or so.

The kinetics of adsorption for polystyrene sample PS3 (M_w = 1.07×10^5) adsorbing onto mica from cyclohexane at the Θ-temperature (34.5±0.5°C), for various solution concentrations, are shown in fig. 2. The adsorbance values for the case of incubating solution concentration c = 0.02 $mg.cm^{-3}$ rise at once ($t \lesssim 5$ minutes) to a reasonably steady plateau value of 2-2.5 $mg.m^{-2}$; on the other hand, for c=0.13 and 0.5 $mg.cm^{-3}$, Γ(t) is higher and also more scattered. This effect probably arises because in the adsorption kinetics experiments only a *single rinse* of the incubated mica sheets in pure solvent is carried out prior to drying and weighing. As a result not all of the fluid adhering to the mica sheets upon withdrawal from the solution is removed; some solution, containing free polymer, is left on the once-rinsed mica sheets and, on drying, this free polymer collapses onto the mica surface and contributes to the weight increase. Thus (for the case of the higher concentrations c = 0.13 and 0.5 $mg.cm^{-3}$ of polymer) part of the weight increase is due to surface-adsorbed polymer, and part to the free polymer (in solution) remaining on the surface following the single

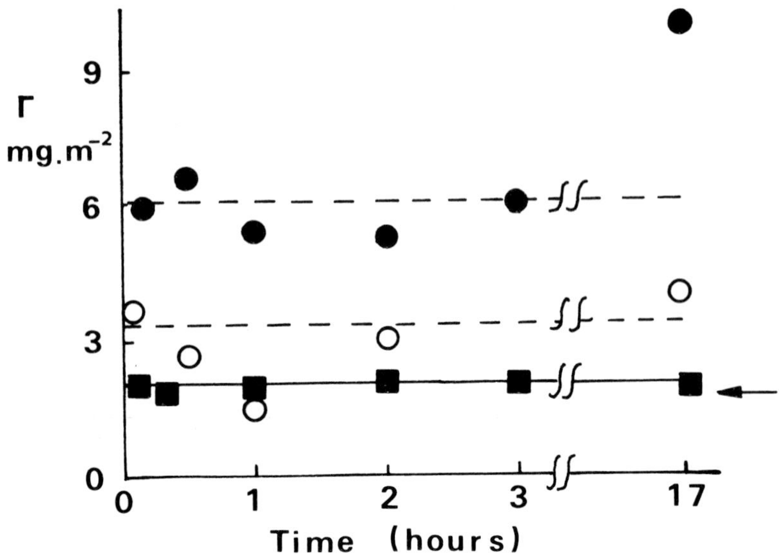

Fig. 2. Kinetics of adsorption of PS3 ($M=1.07x10^5$) onto mica at various solutions concentrations. ■ = *0.02;* ○ = *0.13;* ● = *0.5 $mg.cm^{-3}$. T = 34.5°C. The weight excess following 17 hours incubation in 0.13 and 0.5 $mg.cm^{-3}$ solutions is reduced to value indicated by arrow after prolonged immersion in cyclohexane.*

rinse in pure solvent. To illustrate this, also shown on fig. 2 (arrow) are the weight increases of the mica sheets (which had been incubated in 0.13 and 0.5 $mg.cm^{-3}$ solutions) following a longer immersion (about ½ hour) in the pure solvent: the net remaining excess weight of polymer now corresponds closely to the amount adsorbed from the low c solution of 0.02 $mg.cm^{-3}$. For mica sheets incubated at the lower concentration, prolonged immersion in the pure solvent did not lead to a further reduction in the weight excess relative to once-rinsed sheets. This point is considered further in the following section.

The general trend of the adsorption measurements at the higher incubating concentrations was that, once the mica sheets had been thoroughly rinsed of solution following incubation (whether by multiple-rinsing or by prolonged immersion), the amount of firmly adsorbed polymer remaining on the sheets was more or less constant, and equal to the adsorbance values at the lowest concentration (0.02 $mg.cm^{-3}$). This is more clearly shown on the *desorption* curve shown in fig. 3.

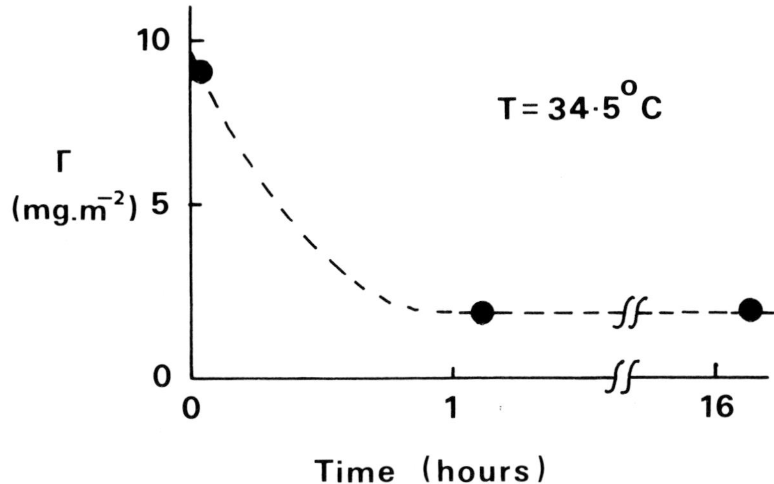

Fig. 3. Kinetics of desorption of PS3 from mica. t = 0 corresponds to a weight excess after 17 hours incubation in 0.5 mg.cm^{-3} PS3 solution followed by a single rinse.

The figure shows the excess weight of polymer on mica sheets, subsequent to 17 hours immersion in a 0.5 mg.cm^{-3} solution of PS3 in cyclohexane, following which they were immersed in pure solvent; the 'loosely-adhering' polymer (i.e. that in the adherent film of solution following one rinse) is rapidly removed ($\sim$ ½ hour), leaving the firmly adsorbed molecules which undergo no further desorption on up to 18 hours immersion in pure solvent. This behaviour is representative of all mica sheets which had been incubated at c $\gtrsim$ 0.15 mg.cm^{-3}, for both PS3 and PS4 solutions.

Fig. 4 shows the kinetics of adsorption of polystyrene sample PS4 (M_w = 4.22x10^5) from a 0.06 mg.cm^{-3} solution in cyclohexane at 34.5 ± 0.5°C onto freshly cleaved mica. The plateau adsorbance value $\Gamma \gtrsim$ 4 mg.m^{-2} is reached very rapidly (t $\lesssim$ 5 minutes). As in the case of PS3 adsorbing from a 0.02 mg.cm^{-3} solution, the weight excess of polymer shown in fig. 4 is the 'firmly adsorbed' value. i.e. further prolonged immersion did not result in a decrease in Γ.

Figs. 5 and 6 show the adsorption isotherms for samples PS3 and PS4 adsorbing onto mica from cyclohexane at the Θ-temperature. Shown is the polymer weight excess measured following a single rinse, and also following prolonged ($\sim$ 17 hours) immersion in pure solvent. The plateau

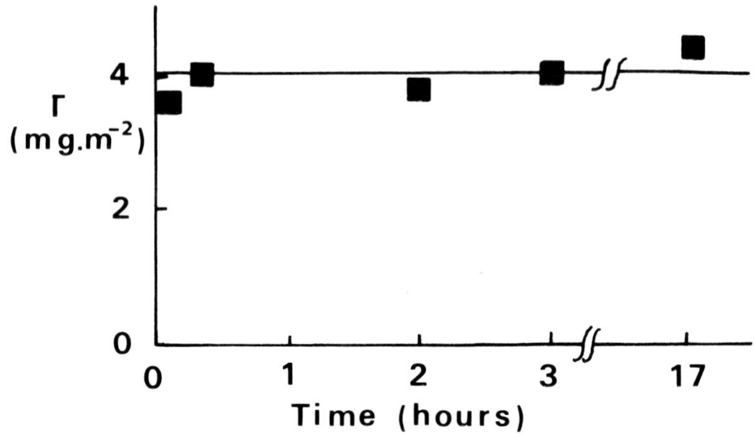

Fig. 4. Kinetics of adsorption of PS4 ($M = 4.22 \times 10^5$) onto mica at polymer concentration 0.06 mg.cm^{-3}. $T = 34.5$°C.

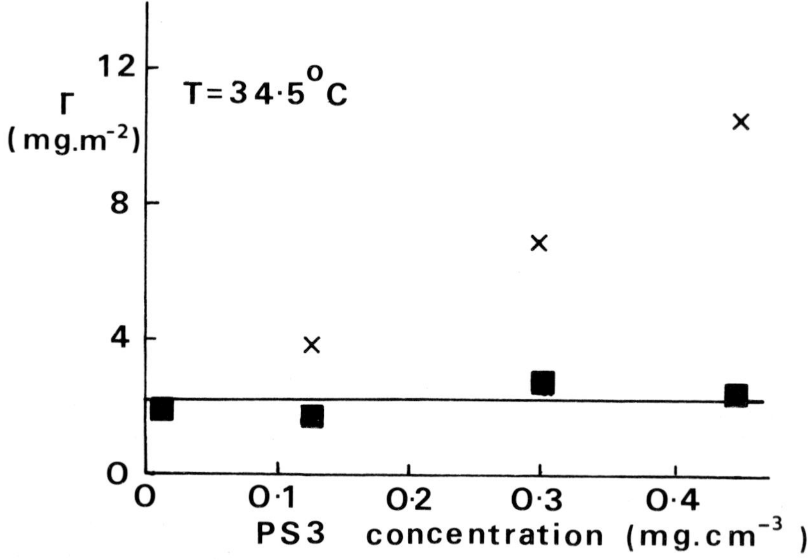

Fig. 5. Adsorption isotherm for PS3: ✕ = weight excess after one rinse; ■ = adsorbance following prolonged immersion in solvent.

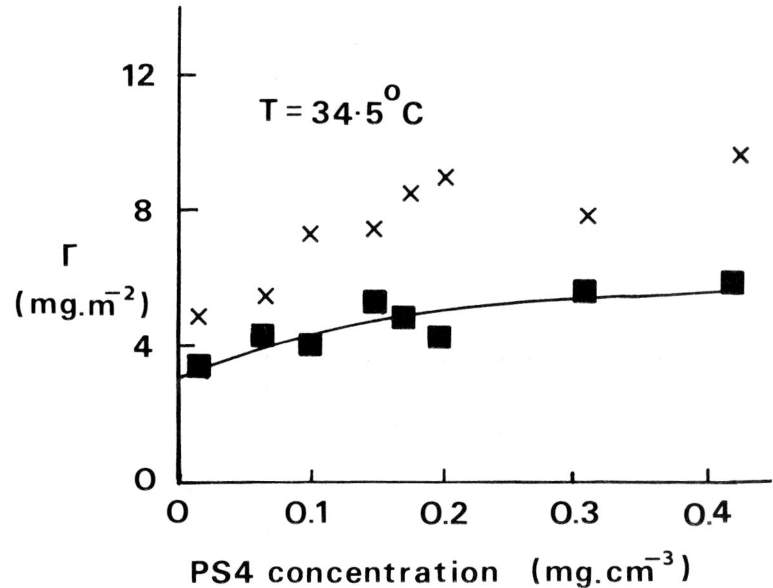

Fig. 6. Adsorption isotherm for PS4: × *= weight excess after one rinse;* ■ *= adsorbance following prolonged immersion in solvent.*

adsorption values indicated are $\Gamma_{PS3} = 2.5\pm0.5$ mg.m^{-2} and $\Gamma_{PS4} = 5\pm0.8$ mg.m^{-2} for the PS3 and PS4 samples respectively; there is some indication that in the latter case the isotherm has not reached its plateau value at the highest concentration, though the scatter in the data obscures this to a certain extent.

Fig. 7 shows the adsorption kinetics for the PBD sample LF4 ($M_w = 9.6\text{x}10^4$) adsorbing from a 0.1 mg.cm^{-3} solution in cyclohexane at 25±0.5°C onto freshly cleaved mica. At this temperature cyclohexane is a *good solvent* for PBD. As before, $\Gamma(t)$ is seen to reach a quasi-plateau value, $\Gamma_0 \simeq 2$ mg.m^{-2}, very rapidly ($t \lesssim 5$ minutes); prolonged incubation appears to increase this slightly (t - 15.5 hours, fig. 4), but on longer immersion in the pure solvent (as opposed to a single rinse) the 'firmly-adsorbed' component is again seen to be close to Γ_0.

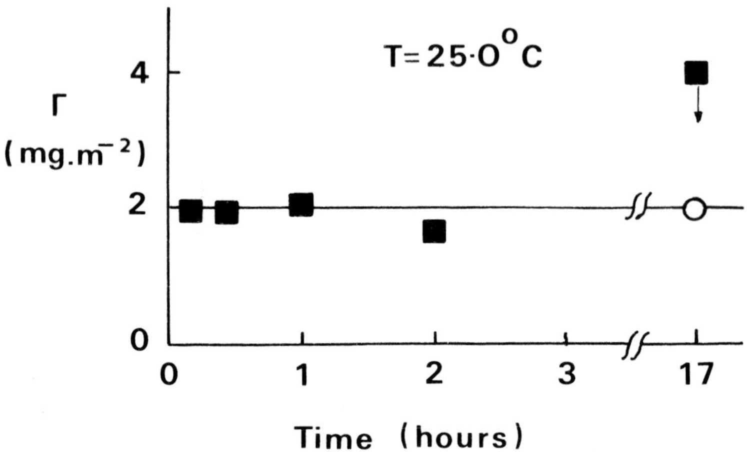

Fig. 7. Kinetics of adsorption of LF4 (polybutadiene) onto mica at concentration 0.1 mg.cm^{-3}. ■ - *after one rinse;* ○ - *17 hours incubation followed by overnight immersion in pure solvent.*

b) *Refractive Index Measurements.*

These are reported in detail elsewhere (1).For completeness we note here the main features: the values of n(D) were determined in the range of mica surface separations 120 nm < D < 300 nm, with a layer of polystyrene sample PS1 (M_W = 5.99x10^5) adsorbed onto the mica from cyclohexane (c = 7x10^{-3} mg.cm^{-3}) at 24±0.5°C (poor solvent conditions). From the refractive index n(D) at D = D_o = 14 nm the mean volume fraction of polymer in the gap, $\phi(D_o)$, was estimated and the adsorbance was evaluated as

$$\Gamma = \rho_{PS} \cdot D_o\phi(D_o)$$

where ρ_{PS} is the density of bulk polystyrene. A value for Γ of some 6 mg.m^{-2} of mica surface was determined. The values of n(D) were determined with the mica surfaces immersed in the incubating solution, and remained unchanged when the polymer solution was diluted 100-fold, as described, up to periods of 48 hours.

4. DISCUSSION

Before discussing the results in detail, it is appropriate to consider the nature of the 'loosely-adhering' and 'firmly-adsorbed' components of the excess weight measured following incubation of the mica sheets in the high concentration ($c \gtrsim 0.15$ mg.m^{-2}) solutions. As noted in the previous section, a single rinse of the mica sheets following incubation is not sufficient to remove all the free solution adhering as a film to the surface. (This may be demonstrated qualitatively by rinsing, in clear water, a mica sheet which had previously been immersed in ink solution: the colour is finally removed only after several rinses or, alternatively, a prolonged immersion in the water). This suggests that, in the adsorption experiments, the excess weight measured (after a single rinse) following immersion in the $c \gtrsim 0.13$ mg.cm^{-3} solution (for PS3) is partly due to polymer which had adsorbed onto the surface, and partly due to free polymer in the solution film adhering to the surface. This latter component was invariably and rapidly removed following a longer immersion in the pure cyclohexane, as noted (also fig. 3). For the incubating solutions for which $c \leq 0.06$ mg.ml^{-1}, the amount of excess free polymer in the adherent solution film is probably only a small fraction of the adsorbed amount, so that further prolonged immersion in pure solvent has little additional effect. The variation of the amount of this 'loosely-adhering' component as a function of concentration, as indicated in figs. 5 and 6, also supports this view. When referring to the adsorbance measured using the microbalance technique, we shall therefore mean the *firmly adsorbed* component of the total weight excess.

The central features of the present investigation, and their relation to the adsorbance measured optically in the force-measurement study, are as follows: the adsorption of both PS3 and PS4 from cyclohexane at the Θ-temperature onto freshly cleaved mica is rapid, and the plateau value of Γ at all concentrations is reached already at the first measured point on the adsorption v. time curve (5 minutes, figs. 2,4). This is in general accord with the results of other studies of polymer adsorption kinetics onto plane surfaces (9). It also suggests that the adsorbance of PS1 onto the mica surfaces, as determined from refractive index measurements, following 10 hours of incubation is the equilibrium value for the conditions of that study ($c=7\times10^{3}$ mg.cm^{-3} at 24°C, i.e. worse-than - Θ temperature).

The adsorption isotherm plateau values for PS3 and PS4, figs. 5 and 6, are comparable with those determined in ellipsometric (3) and radiometric (11) studies of polystyrene adsorption onto various plane surfaces from cyclohexane under Θ-conditions. The plateau adsorbance values of 2.5±0.5 and 5.0±0.8 $mg.m^{-2}$ for PS3 and PS4 respectively are also proportional to the square-root of the polymer molecular weights M, as previously observed for other surfaces (10). If this proportionality holds also for PS1, of M = $6x10^5$, one would expect an adsorbance for this polymer of some $(6/4.22)^{\frac{1}{2}}$ x 5 ≃ 6 $mg.m^{-2}$, as is indeed observed. While adsorbance values determined in the two studies are thus consistent, we should bear in mind a) the PS1 adsorbance was determined at somewhat worse-than-Θ conditions, and b) the polystyrene samples, while reasonably monodispersed, are from different sources.

The extent of desorption of PS3 and PS4 (as typified in fig. 3) is negligible over periods of up to 17 hours. This is in accord with the observation that the PS1 adsorbance, as determined from n(D), did not change over a day or two when the polymer bearing surfaces were immersed in essentially pure solvent. Again, this quasi-irreversible behaviour is in accord with many previous studies of polymer adsorption (6,11).

The adsorption of PBD onto freshly cleaved mica, in *good solvent* conditions, is also rapid, and appears to be quasi-irreversible. Experiments to measure this adsorbance using the optical technique are currently in preparation .

Finally, we note that very recently the adsorbance of polyethylene oxide from aqueous solution onto mica surfaces has been measured using the refractive index approach. The results indicate a considerably higher adsorbance than previously measured for PEO adsorbing onto colloidal silica particles. The microbalance approach is presently being adapted to check this result by measuring such adsorption directly.

ACKNOWLEDGEMENT

We thank Professor D. Tabor for useful discussions.

REFERENCES

1. Klein, J. (1980). *Nature* 288, 248-250; *JCS Faraday I* (in press).
2. Vincent, B. (1974) *Adv.Colloid Interface Sci.*, 4, 193-277.
3. Stromberg, R.R. Tutas, D.J. and Passaglia, E. (1965). *J.Phys.Chem.*, 69, 3955-3964.
4. Peyser, P. and Stromberg, R.R. (1967). *J.Phys.Chem.*, 71, 2066-2074.
5. Yang, R.T., Low, J.D.,Haller, G.L. and Fenn, J. (1973). *J.Colloid Interface Science* 44, 249-258.
6. Peterson, C. and Kwei, T.K. (1961). *J.Phys.Chem.*, 65, 1330-1341.
7. Fujiwara, S. and Terashima, H. (1970). *J.Phys.E.*, 3, 695-698.
8. Israelachvili, J.N. (1973). *J.Colloid Interface Science* 44, 259-269.
9. Stromberg, R.R. and Smith, L.E. (1967). *J.Phys.Chem.*, 71, 2470-2474.
10. Takahashi, A., Kawaguchi, M., Hirota, H. and Kato, T. (1980). *Macromolecules* 13, 884-889.
11. Grant, W.H., Smith, L.E. and Stromberg, R.R. (1976). *Disc.Faraday Soc.*, 59, 209-217.

SUBJECT INDEX